Global Correlation of Tectonic Movements

INTERNATIONAL GEOLOGICAL CORRELATION PROGRAMME

PROJECT 107

Global Correlation of Tectonic Movements

Edited by

YU. G. LEONOV
Geological Institute of USSR Academy of Sciences, Moscow, USSR

and

V. E. KHAIN
Geological Faculty of Moscow State University, Moscow, USSR

A Wiley–Interscience Publication

JOHN WILEY & SONS
Chichester · New York · Brisbane · Toronto · Singapore

Library of Congress Cataloguing-in-Publication Data:

Global correlation of tectonic movements.
'International Geological Correlation Programme. Project 107'—
Includes index.
1. Geology, Stratigraphic. 2. Stratigraphic correlation. 3. Orogeny I. Leonov, Yuri Georgievich. II. Khain, Viktor Efimovich.
III. International Geological Correlation Programme. Project 107 'Global Correlation of Phases and Epochs of Tectogenesis.'
QE501.G557 1986 551.1′36 86-13361

ISBN 0 471 90547 X

British Library Cataloguing in Publication Data:

Global correlation of tectonic movements.—
(IGCP project; 107)
1. Geology, Structural
1. Leonov, Yuri G. II. Khain, V. E.
III. Series
551.1′36 QE601

ISBN 0 471 90547 X

Printed and bound in Great Britain

Contents

List of Contributors

K. BREZSNYÁNSZKY	*Central Office of Geology, Budapest, Hungary*
G. CSÁSZÁR	*Magyar Állami Földtani Intézet, H 1442 Budapest, Népstadion út 14, Postbox 106, Hungary*
YU. G. GATINSKIY	*VNII Zarubezhgeologia, USSR Ministry of Geology, Moscow, USSR*
G. A. GRINBERG	*Institute of Geology, Siberian Department of USSR Academy of Sciences, Iakutsk, USSR*
G. S. GUSEV	*Institute of Geology, Siberian Department of USSR Academy of Sciences, Iakutsk, USSR*
J. HAAS	*Magyar Állami Földtani Intézet, H 1442 Budapest, Népstadion út 14, Postbox 106, Hungary*
J. HALMAI	*Magyar Állami Földtani Intézet, H 1442 Budapest, Népstadion út 14, Postbox 106, Hungary*
G. HÁMOR	*Magyar Állami Földtani Intézet, H 1442 Budapest, Népstadion út 14, Postbox 106, Hungary*
M. A. ITURRALDE-VINENT	*Geological Institute of Cuban Academy of Sciences, La Habana, Cuba*
V. E. KHAIN	*Geological Faculty of Moscow State University, Moscow, USSR*
N. V. KORONOVSKY	*Geological Faculty of Moscow State University, Moscow, USSR*
L. KORPÁS	*Magyar Allami Földtani Intézet, H 1442 Budapest, Népstadion út 14, Postbox 106, Hungary*
V. I. KOVALENKO	*IGEM, USSR Academy of Sciences, Moscow, USSR*
I. KRYSTEK	*Prírodoved. fakulta Unoveristy J E Purkyne, 600 00 Brno, Kotlářska 2, Czechoslovakia*
M. G. LEONOV	*Geological Institute of USSR Academy of Sciences, Moscow, USSR*
YU. G. LEONOV	*Geological Institute of USSR Academy of Sciences, Moscow, USSR*

A. V. LUKYANOV — *Geological Institute of USSR Academy of Sciences, Moscow, USSR*

R. MALARODA — *Istituto di Geologia, Torino University and CNR, Centro di Studio sui Problemi dell'Orogeno delle Alpi Occidentali, Torino, Italy*

M. S. NAGIBINA — *Geological Institute of USSR Academy of Sciences, Moscow, USSR*

N. OSZCZYPKO — *Geological Institute, Carpathian Branch, 31-560 Krakow, ul.Skrzatów 1, Poland*

U. C. PATI — *Geological Survey of India*

B. N. RAINA — *Geological Survey of India*

M. SĂNDULESCU — *Institutul de Geologie si Geofizica, Str Caransebes nr 1 sector 8 of. 32, Bucuresti 7000, Jud. Ilfov, Romania*

W. SCHWAN — *Institut für Geologie und Mineralogie der Universität Erlangen-Nürnberg, Schlossgarten 5, 8520 Erlangen, Federal Republic of Germany*

I. G. SCHERBA — *Geological Institute of USSR Academy of Sciences, Moscow, USSR*

N. SRIMAL — *Geological Survey of India*

Z. STRÁNIK — *Ústřední ústav geologický, 602 00 Brno, Leitnerova 22, Czechoslovakia*

D. VASS — *Geologický ústav D. Štúra 809 40 Bratislava, Mlynská dolina 1, Czechoslovakia*

K. ŻYTKO — *Geological Institute, Carpathian Branch, 31-560 Krakow, ul.Skrzatów 1, Poland*

Preface

The work on the project, 'Global correlation of epochs of tectogenesis', lasted from 1975 until 1980. In 1975 the programme of the project was worked out and an organizational *ad hoc* meeting was called (Grenoble, France, August 1975). The investigations proper began in 1976. The work was generally organized in a similar manner to other International Geological Correlation Programme (IGCP) projects: national studies were carried out by individuals or working groups: representatives of participant countries formed the international working group that co-ordinated the investigations and controlled the results. Several conferences were held and joint field trips were arranged.

The following countries took part in the project (national representatives and persons participating in the meetings and excursions at different stages of the works are in parentheses): Bulgaria (Yu. Karagjuleva, I. Haidoutov); Canada (N. Rast); Czechoslovakia (D. Vass); West Germany (W. Schwan, R. Rossner); Hungary (G. Császár, K. Brezsnyánszky); India (B. Raina, D. K. Ray); Italy (R. Malaroda); Japan (N. Kitamura); Romania (M. Săndulescu); Turkey (E. Bingöl, E. Demirtaşli); United Kingdom (F. W. Dunning, A. M. Spencer); USSR (Yu. G. Leonov, V. E. Khain). In addition, various other people occasionally took part in the work by submitting materials, making reports, and so on: these included J. Goguel (France); E. Niggli, R. Trümpy (Switzerland); F. Doutch, E. Scheibner (Australia); H. Zwart (the Netherlands); R. Oberhauser, P. Plöchinger, A. Matura, W. Schnabel (Austria); R. Whyte (Singapore); V. D. Nalivkin, Yu. G. Gatinsky, I. G. Scherba, E. E. Milanovsky, V. G. Varnavsky, B. V. Grigoriants, E. Sh. Shikhalibeili, G. S. Gusev (USSR) and others.

The scientific direction and organization of the project was undertaken by V. E. Khain (Project Leader, USSR) and Yu. G. Leonov (General Secretary, USSR).

Work was financed by the participant countries, covering national investigations, complete or partial reimbursement of the travel expenses of national representatives to the meetings and excurions, and so on, and the meetings of the International Working Group were annually subsidized by the IGCP Secretariat of UNESCO.

Sincere acknowledgement for financial and organizational support in the arrangement of the meetings and excursions is due to the Geological Survey of Austria, the Institute of Mineral Resources (MTA) of Turkey, the USSR Academy of Sciences and the Azerbaidjan Academy of Sciences, the Organizing Committee of the VI Colloquium on the Geology of the Aegean Region, and the organizing committees of the 25th and 26th sessions of the International Congress.

The main scientific objective of the project was to solve a major geotectonic problem: that of the existence and role of global tectonic movements, or rather the global pulses of tectonic (and magmatic) activity in the process of tectogenesis. The notion of such pulses is the basis of a widely known concept of orogenic phases (phases of folding) or phases and epochs of tectogenesis. Investigations concerning the main objective were carried out along several interrelated directions, mainly: (a) the development of the methodology of correlation of tectonic and magmatic events and distinguishing phases and epochs of tectogenesis; (b) study of successions of tectonic and magmatic events in certain regions in order to determine periods of intense tectonic and magmatic activity; (c) interregional correlations to ascertain whether phases and epochs of tectogenesis are global or regional; (d) studies of the duration of these epochs and their types of manifestation in different regions of the earth and in different tectonic environments (mobile belts, platforms, and so on); (d) the compilation of maps of tectonomagmatic activity for some epochs of tectogenesis.

The general problem of the existence of global tectonic pulses (orogenic phases of similar events) is closely connected with the related problem of lateral continuity of diverse, although simultaneous, tectonic and magmatic events. This leads to a still more fundamental problem, that of the existence of a global mechanism or tectogenesis. Such relationships compelled workers in the project to consider this problem as well.

The availability of regional data largely depended on what countries took part in the project and on the personal interests of the participants. The composition of the working group being what it was, it was impossible to carry out systematic regional investigations of the earth's crust without leaving large gaps. The investigations were therefore selective. Nevertheless, the analysis of material on different regions and time spans and the study of different features of tectonic movements and magmatism compensated for the fragmentary investigations and permitted the study of a fairly wide range of problems.

The papers included in this volume reflect the results of part of these investigations. In the main, the papers need no introduction. However, those dealing with the so-called maps of tectono-magmatic activity must be singled out. These papers have a unique common objective and therefore constitute a separate series of contributions.

The compilation of the maps of the types and intensity of tectonic and magmatic activity during epochs of tectogenesis (maps of epochs of tectogenesis) was originally one of the main objectives of the project. It was intended that such maps would impart a clear picture of the distribution of different tectonic and magmatic events, associated with selective epochs of tectogenesis.

As far as we know, such maps were never really compiled. Therefore the explanation the method of approach and the wording of the captions are most important. As for the methodology, there are two major concepts: (a) epochs of tectogenesis are manifested by different phenomena (deformation, magmatism, metamorphism, and so on); (b) these epochs are not instantaneous, but continue for a fairly long period and movements in different regions may not coincide in time. Consequently a group of workers under the general editorship of Yu. G. Leonov prepared a legend which was repeatedly discussed before final ratification for the compilation of regional drafts.

Maps of the Pyrenean and Savic phases were compiled for Cuba (K. Brezsnyánzky, M. A. Iturralde-Vinent—Hungary, Cuba); for the British Isles and the adjacent North Sea area, the English Channel and a part of the Atlantic shelf (F. W. Dunning, A. M. Spencer—Great Britain); West Germany (R. Rossner—FRG); the Polish Carpathians (N. Oszczypko, K. Żytko—Poland); the Czechoslovakian Carpathians (D. Vass, Z. Stránik—Czechoslovakia); Hungary (G. Császár, J. Haas, J. Halmai, G. Hamor, L. Korpás—Hungary); Romania (M. Săndulescu—Rumania); the Soviet Union (the shelf and Arctic islands: Yu. Ya. Livshits, V. A. Vinogradov; the Soviet Northeast: G. S. Gusev; the Soviet Far East: V. G. Varnavsky); the Indonesian region and Indochina (Yu. G. Gatinsky, A. V. Mishina—USSR).

In addition, some countries, maps or series of maps were prepared for other time intervals. This work was done for the territory of the Federal Republic of Germany,

Hungary, the Indian Himalayas (B. Raina *et al.*). Maps of these territories were compiled for periods of intense tectonic activity lasting from the Cretaceous to the Neogene and therefore these map series, together with the maps for the Pyrenean and Savic epochs, give a complete idea of the sequence of epochs, phases and episodes of tectogenesis and magmatism in the areas under review. Some of these maps and papers that analyse the data on which the maps are based are included in this book.

V. E. KHAIN

YU. G. LEONOV

Global Correlation of Tectonic Movements
Edited by Yu. G. Leonov and V. E. Khain

1

Problems of a Systematic Approach to the Study and Correlation of Tectonic Phases

YU. G. LEONOV AND V. E. KHAIN*

GLOBAL AND REGIONAL CONCEPTS

Ever since Stille proposed the concept of orogenic phases, two controversial geotectonic concepts have been competing for ascendancy. Known as the global (planetary) and the regional concepts, they not only reflect different approaches to the identification of orogenic phases, but also to the explanation of the general tectonic development of the earth.

The global concept is based on the idea that global processes play a leading role in tectogenesis, simultaneously involving the whole earth and generating analogous rhythms of tectonic movements on its surface. The idea of the global character of movements is closely connected with that of the episodic manifestation of orogenic movements. Both considerations (global and episodic) are most strikingly presented in the concept of orogenic phases by H. Stille. Later, these concepts were supported by many geologists, including Kunin and Sardonnikov (1976), Pronin (1969a,b) and Rubinstein (1967) among Soviet researchers. Orogenic phases represent the main element of this concept although, as mentioned earlier, this has only a general significance. The concept is also shared by the present authors, who regard the development of the earth as an integrated process, arising not only from orogenic manifestations, but also from other geological events such as, for instance, major palaeogeographic changes, transgressions and regressions (Kazarinov, 1976), which point, directly or indirectly to tectonic movements.

Stille's ideas are known to have been widely recognized. However, some data accumulated since then are either not quite in accord with his concepts or are even at variance with them (Leonov, 1976a,b; Khain, 1971, 1973; Leonov and Khain, 1980).

Firstly, it has been shown (e.g. Donbass, Southeast Caucasus, Fergana, etc., in the USSR) that a vital role in the development of a folded structure belongs to long synsedimentary (Schatskiy, 1937, 1951) and syndenudational fold deformation (Schultz, 1970). Hence many geologists have come to the conclusion that the history of the development of separate areas and, especially, that of the earth as a whole, cannot be divided into short orogenic and long non-orogenic periods. An opposite conclusion was drawn, that tectonic movements are continuous in space and time (Schatskiy, 1939, 1951; Gilluly, 1949; Goguel, 1965; Rutten, 1949). It was pointed out that breaks marking the phases are

* Geological Institute of USSR Academy of Sciences, Moscow; Geological Faculty of Moscow State University, Moscow.

produced not by folding, but by episodic epeirogenic uplifts superposed on what are effectively continuously growing folds.

Secondly, numerous regional departures from Stille's orogenic phases law have been observed. These include unconformities disregarded by him (pointing to deformation occurring in periods of time formerly considered 'anorogenic') and the frequent absence of deformation features at the times where they might be expected according to Stille's law.

Thirdly, durations of the 'phases' were revised. The duration of epochs of intense tectonic activity that can be distinguished on a global scale proved to be, as a rule, millions of years (Rubinstein, 1967), probably up to 10–15 million years (m.y.) (Pronin, 1969a,b). In other words, they are not short or almost instantaneous (from the geological point of view) pulses, but long periods of intense activity with complicated histories. Does it follow, therefore, that the concept of phases is erroneous, or do the above-mentioned considerations merely suggest certain limits to the theory?

As for the continuity of the movements, Stille himself explained that it must not be identified with the uniformity of their style and intensity and that he considered as 'orogenic' not only those periods that were devoid of any tectonic movements, but also those devoid of certain intensity and types of movements (Stille, 1950a, pp. 1–2).

The existence of slow synsedimentary deformations in itself is not at variance with the existence of episodic climaxes of tectonic activity, since synsedimentary deformations are irregular and their velocities differ (Khain, 1973). Such deformations display periods of both relative (or absolute) slowing down of structural growth ('anorogenic'), and of particularly intense development ('orogenic'). Therefore the concept of episodic intense tectonic movements (synsedimentary and orogenic) cannot be rejected for these reasons, though the movements must be specified. According to the Soviet geologist Mazarovich (1940), who correctly grasped the essence of the question, 'the idea of regular-uniform development of the earth crust may be an anachronism nowadays. . . . Not to be led by H. Stille and not carrying N. S. Schatskiy's ideas through to their logical end, we have to accept that in the history of the earth there are epochs when orogenic forces are manifested most intensely and spectacularly and when the major turning points in the development of the vast areas of the earth crust are created. . . . Orogenesis is not a catastrophe, but only a fairly long epoch of extremely intense tectonic activity' (p. 21). Later, Khain (1971, 1973) formulated, in accord with Mazarovich, the concept of continuous-intermittent tectonic processes. According to this theory, the process of tectogenesis is both continuous and intermittent with periods of relatively weak tectonic activity alternating with periods (phases, epochs) of intense tectonic activity.

The essence of the idea of global movements is not disproved by deviations from the law of orogenic phases, be it Stille's initial canon or any of its modernized versions. The existence of deformations not taken into account by the canon does not discredit the concept but, as a rule, points to the existence of regional or local pulses concurrent with the global ones. In any event, the conclusion of worldwide (at least for the continents) distribution of tectonic movements in certain periods of time is supported by Leonov (1976b) and Pronin (1969a,b).

Nor is the revision of the duration of orogenic phases and epochs a decisive factor. It indicates that corrections must be introduced into the original concepts, but does not disprove them. In general, it has to be admitted that the opponents of the global concept have detected some of its weaknesses, but have not disproved its two basic tenets. The first of these is that processes of tectogenesis develop intermittently with the alternation of the more and less active periods (whatever they be called). The second is that some periods of intense tectonic activity (disregarding the terminology), even if not of the character and duration assumed by Stille, may be of global distribution.

It should be stressed, however (and will be discussed later) that the tectonic life of the earth should not be accepted as simple or characterized only by global events. The regional concept was formulated, to a certain extent, in response to the ideas of Stille and his followers. According to this concept, the time and character of tectonic movements vary from place to place so much that we should not speak of a general worldwide rhythm of movements, including global tectonic phases and epochs. Elaborating his concept of orogenic phases, Stille used mainly European and North American data, since those on other regions were more scarce. Subsequently, fresh information often contradicted the harmonious picture of a universal sequence of tectonic movements. The proponents of the new trend lay stress on regional events and show the diversity of the tectogenetic processes and the impossibility of summarizing them in simple schemes. In this respect a series of investigations has been carried out by Soviet geologists (Tectonics of Eurasia, 1964; Yanshin, 1965). Most advocates of this concept, by emphasizing regional features of tectogenesis, reject the existence of global tectonic processes. This extreme viewpoint, expanded into a system of ideas, has the same methodological defect as Stille's concept. It is striving too uncompromisingly to put the tectonic life of the earth in the strict framework of regional processes. Thus, on the basis of the data restricting the global concept, the latter is unreasonably being entirely rejected.

A paradoxical situation results, indicating that both the global and the regional concepts are well-founded. The paradox, however, does not involve an irreconcilable contradiction. It mainly arises from methodically and logically unjustified postulates according to which the two concepts automatically and utterly exclude one another. The interpretation of some terms such as 'global character', 'synchroneity', and so on, also plays a certain role. Hence the problem of synchronous global movements is only vaguely set out and different authors may draw opposite conclusions from the same data.

The resolution of such a situation seems to depend on discarding the above-mentioned postulates and attempting a synthesis of the global and regional concepts. All that we know today indicates that the concepts are mutually complementary rather than exclusive. They reflect different aspects of the process of tectogenesis, the overall effect of which is the interaction of the worldwide (global concept) and regional (regional concept) components (Leonov, 1976a). Therefore, in studying global processes, one may not, regardless of the underlying causes, reject the existence of either the globally episodic worldwide simultaneous movements, or those that occurred on a regional scale. Owing to the former movements, the main tectonic discontinuities occur in strict order throughout the world and synchronous tectonic and magmatic processes are laterally related. The regional movements produce a diversity of tectonic regimes changing more or less from region to region. Thus the global movements represent a general basis on which an intricate and changeable design is traced, both complicating and concealing the general regular pattern.

SOME ASPECTS OF DISTINGUISHING GLOBAL OROGENIC EVENTS

By orogenic events we mean the variable forms of tectonic activity with which uplift, formation of positive structures and folding are associated. They comprise all such notions as orogenic phases and epochs, phases and epochs of folding, tectogenesis, tectono-magmatic phases and epochs, and so on. In the geological literature these terms are commonly used arbitrarily. Nevertheless, it is the study of events like orogenic phases and epochs that is of particular importance in the investigation of global processes. This is accounted for by the fact that they are comparatively abrupt pulses breaking the course of the geological development, initially sedimentation, as is attested by different features of rock sequences.

In distinguishing orogenic events, particularly if investigated globally, some questions need special discussion: (a) criteria of global character; (b) features by which they are distinguished; (c) their duration; (d) terminology.

Criteria of Global Character

The standpoint of a geologist on the existence of global events often depends on how he understands the expression 'global distribution'. An identical pattern of distribution of movements on the surface of the earth may be taken as evidence both for and against their global character. This is true of both orogenic and other geological events, as shown by the following examples.

Many continental areas are at the end of Silurian–Early Devoniantinal are characterized by uplift and regression. It was one of the most geocratic Palaeozoic epochs. Geocratic tendencies are particularly pronounced in some areas of the Caledonian orogeny and on most of the cratons (Leonov, 1976b). The South American craton, however, stands apart; the Early Devonian epoch here involved a transgression. Moreover, the environments within the territory of the mobile belts, such as the Late Palaeozoic, Mesozoic and Cenozoic orogenic areas of today, were very diverse. In other words, these events were manifest over a large area and in different regions, but not everywhere, even within the continents. The history of the oceans of that time is unknown. A question arises: can these events be considered global? The answer obviously depends on the meaning of the term 'global'.

Another example concerns Palaeozoic unconformities. Pronin (1969a,b) estimated the percentage of sections with stratigraphic breaks and unconformities among all the sections that he reviewed. Six unconformities were recognized in 76 per cent (pre-Middle Devonian unconformity) and 88 per cent (Sudetic unconformity) among the total of the studied sections. Other unconformities are rarer and usually detected in no more than 50 per cent of the sections. Of course marine areas, regions devoid of sediments, or poorly known regions were not taken into account. Interpretations of these data provoke the same question, as in the previous example. Pronin considers the distribution of unconformities as evidence in favour of global movements and the concept of orogenic phases, although not all researchers are convinced.

At present it is apparently accepted that there are no movements that are registered absolutely everywhere. There is always a certain number of regions and tectonic units which show no traces of movement in their sediments, apart from oceans and vast land areas for which neither positive nor negative data exist.

It seems, however, that such a literal interpretation of 'global character' reflects only the appearance, but is not enough to explain the essence of the process. Therefore we suggest that an orogenic event (phase, epoch) should only be considered global if the whole complex of its manifestations (vast area, occurrence in widely separated regions) can be considered as a result of a general process involving the whole of the earth simultaneously, but not necessarily leaving distinct or analogous traces everywhere.

From this standpoint we may consider as global both the Late Silurian–Early Devonian epoch of uplift and regression and the movements that produced the unconformities mentioned above. Indeed, a similar behaviour of vast and widely separated areas at the same periods of time can be attributed only to the general rather than local processes. It should be borne in mind that tensional and compressional deformations may proceed simultaneously in different areas, thus compensating each other. The same phase of tectonic activity may therefore be manifested at one place as folding and uplift and at another as rifting and increase of subsidence (e.g. spreading and subduction in plate tectonics).

All this conforms to the basic interrelation of global and regional concepts. It is therefore quite natural that global processes, including events of orogenic type, are complicated or even suppressed by regional effects.

Characteristics of Orogenic Events

From the initial stage of the concept of orogenic phases (Stille, 1924) such phases were recognized by angular unconformities indicating tectonic deformations (orogenesis in the broad sense). Even today, this approach often prevails in publications on this subject, the phases being identified with unconformities. At the same time, it is generally understood that the assemblage of phenomena included among orogenic processes is larger. Such processes are therefore manifest not just as breaks and unconformities and consequently the terms 'tectonic phase', 'tectono-magmatic phase', 'tectogenesis', and so on have been introduced (details in Leonov and Khain, 1980).

A more fruitful approach to the study of orogenic events is to consider them as a complex assemblage of time-interconnected more 'elementary' phenomena. Orogenic phases and epochs or 'phases and epochs of tectogenesis' are treated not only as periods of intense deformation or orogeny *sensu stricto*, but also as periods of intense tectonic and magmatic activity in general. Their particular manifestations, depending on many circumstances, may vary from place to place, for instance in terms of lateral heterogeneities of the core and mantle, orientation of the global stress field, and so on. Such interpretation of phases and epochs is a consequence of observations and is in line with the already propounded thesis of combination of global and regional tendencies in tectogenesis.

In studying and correlating orogenic events in small areas, it is often enough to trace a single phenomenon. As a rule, this may be folding and uplift reflected in angular unconformities and stratigraphic breaks. However, even in homogeneous tectonic zones this is often insufficient for clarifying the whole picture of the orogenic process. This is true of both more intense movements, when orogenic deformations are combined with other manifestations of endogenous activity such as magmatism and metamorphism, and of less intense movements such as synsedimentary deformation, lithological changes, etc. Thus it is not permissible to confine oneself to unconformities in interregional and global studies. The investigation of unconformities alone diminishes final conclusions; firstly, it gives no integrated idea of the event, its manifestations in different regions, their lateral and vertical ratios, the total timespan of the event, and secondly, it produces a false impression of its more limited distribution and role compared with what can be obtained by studying all the components.

Features to be taken into account in the study of orogenic events can be divided into several groups.

(1) Tectonic deformation and uplift determined from the mode of occurrence and breaks. These include, on the one hand, breaks without angular unconformity indicating uplift without deformation (the synorogenic and epeirogenic movements of Stille) and, on the other, angular unconformities pointing to deformation that either took place during uplift or finished in an absolute uplift (orogenic movements, folding). A group of movements marked by angular unconformities comprises different types of deformation varying from weak folding and block (germanotype, Saxonian) deformations to the most intense deformations with general flowage of rocks, isoclinal folding, nappes, and so on.

(2) Synsedimentary deformations without absolute uplift. In many regions (e.g. the Southeast Caucasus, the Dnyepr–Donyets depression and the Volga–Urals area in

the USSR) it has been shown that regional periods of increased synsedimentary deformation often coincide with epochs of reactivized tectonic movements distinguished by breaks and unconformities. Synsedimentary deformations are, certainly, an integral part of the orogenic process, as are the movements producing uplifts and unconformities. The main limitation in using data on synsedimentary deformation is imposed by difficulties in their detection, particularly in the interior zones of geosynclinal areas. Radiometric age data are sometimes useful. One of the examples is the continuous Devonian and Carboniferous geosynclinal sequence in Cornwall, England, where Middle and Upper Devonian slates and phyllites yielded a K/Ar date of 340–360 m.y. (Dodson and Rex, 1970). These figures are interpreted as the age of dynamic metamorphism, with some argon having escaped. The figures and stratigraphic position of the rocks permit the correlation of the metamorphism with the Bretonic deformation at the Devonian/Carboniferous boundary. To judge from these data, the deformation here was intense, but developed during continuous sedimentation.

(3) Some sedimentary formations indicate certain types of movements. These are, in the first place, the serpentinite melange and olistostrome complexes. In many cases both mark the time of formation of nappes (see Chapter 3 of this book). Sometimes, olistostromes may be due to gravity sliding of rocks in areas of rugged relief. Evidence for orogenic movements (folding, mountain building, uplift) may be provided by synorogenic clastics in the section. In interregional and global analysis, all these formations should be appraised and those phenomena playing an essential role in the development of major tectonic units and local phenomena related to specific features in the development of individual structures should be differentiated.

(4) Essential changes in the tectonic regime result in changes of palaeogeographic environments and corresponding changes in the composition and thickness of rocks. Of interest are, firstly, major changes, such as transition from flysch into molasse in the change over from a geosynclincal to an orogenic regime; secondly, transitions of platformal formations into continental molasse in areas of epi-platformal mountain building, and so on.

(5) Magmatism and metamorphism may well correlate with other forms of tectonic activity, granitoid rocks being particularly revealing. Times of granitoid intrusions often coincide, though usually rather broadly, with that of tectonic movements, including uplift, folding and reactivation of epi-platformal movements. Such correlations are recognizable on the global scale too, as indicated by the correlation of at least some epochs of granitization with epochs of intense orogenic movements (Rubinstein, 1967). Distribution of volcanism in time may also be instructive, for instance volcanism in rift zones, its evolution being correlatable with stages of rift development (Kazmin, 1980).

The examples above are not exhaustive. However, they seem to present the main phenomena that reflect orogenic activity and may be components of global events such as phases and epochs.

There is one more source of information on orogenic events: the radiometric ages of igneous and metamorphic rocks, the dates reflecting not only the time of origin of such rocks, but also their transformational history. A particularly precise indicator of tectonic processes accompanied by changes in thermal regime is K/Ar data.

Radiometric age data and orogenic events can be correlated both in separate areas and on a larger scale. Rubinstein (1967) analysed time distribution of radiometric age data in

Phanerozoic granites and metamorphic rocks and concluded that intervals where dates concentrate are discrete and of global character. In subsequent years new data were obtained to supplement and correct Rubenstein's somewhat ideal scheme; however, they do not give any reason for a fundamental revision of the scheme.

Duration of Orogenic Events

In distinguishing orogenic phases, Stille considered them as short events lasting no longer than tens or hundreds of thousands of years. Further investigations enlarged the time intervals corresponding to phases, or rather, according to many investigators, the events of minimal duration, that can be distinguished on a global scale. Thus, citing only Soviet researchers, Rubinstein estimates their duration as about 3–5 m.y. and Pronin, who called them epochs, 15–20 m.y.

It was mentioned above that a global orogenic event consists of a complex assemblage of phenomena. Therefore, its total duration is a sum of durations of its components—individual phenomena. In interregional correlations it is often observed that ages of movements do not entirely coincide, the movements being of different duration in different regions. However, this shift of unconformities and other forms of activity in the chronological scale are not at variance with the idea of the existence of general, planetary orogenic events. Shift in time in accordance with regional conditions is as natural as the different intensity and type of manifestations mentioned above.

The principal scheme of summarizing regional intervals is given in Fig. 1. Indices T_1, T_2, . . . show regional duration of movements, index T being the total duration of the event ($T = T_1 + T_2 + \ldots$). It is noteworthy that the view held by many of a regular repetition or even similar duration of different age phases or epochs (Kunin and Sardonnikov, 1976; Rubinstein, 1967) has not as yet been proved. Therefore the duration of phases can in each case be estimated only empirically on the basis of rather voluminous data from different regions. The result of such an analysis is shown in the diagram (Fig. 2) illustrating the results of the study and the correlation of tectonic movements and magmatic activity during the latest Silurian to Devonian time.

The estimation of duration in absolute figures encounters great difficulties. Firstly, radiometric data for small subdivisions of the geochronological scale (boundaries of stages and substages are scarce). As a rule, the duration of these subdivisions has to be estimated conventionally, usually by dividing the age of the division or system into equal parts according to the number of the stages, thus allowing a possibly great error from the very beginning. Limitations are involved in imprecise stratigraphic (biostratigraphic) correlation in order to determine the duration of movements or magmatic manifestations from

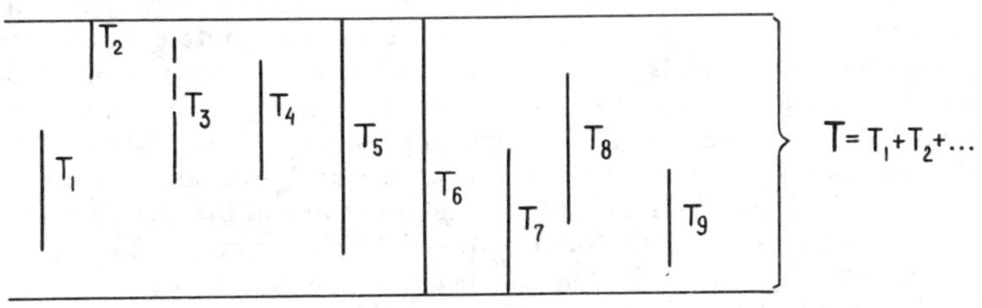

FIG. 1 Scheme of relation of total and regional duration of an orogenic event.

stratigraphic amplitudes of breaks. In the latter case the possibility of error always lies in the danger of overestimating age ranges of movements. Taking into account all possible errors, the duration of most of the global orogenic events may hardly be estimated with an accuracy of more than ±5 m.y. Higher accuracy seems to be possible only for the Neogene-Quaternary times.

Orogenic events may occasionally be reflected not as separate unconformities, but as a series positioned one above the other (e.g. T_3 in Fig. 1). Gilluly, in his criticism of Stille, gave an example of the Neogene in California where a batch of about 40 local unconformities corresponds to the five or six global phases of Stille. Gilluly thought that such multiple unconformities were at variance with the idea of episodic orogenic movements and precluded the distinction of global phases. However, these unconformities are situated regularly. Most of them occur in the section at the levels corresponding exactly to Stille's phases. Von Bubnoff (1954) in pointing to this fact stated: 'a phase of H. Stille is not a

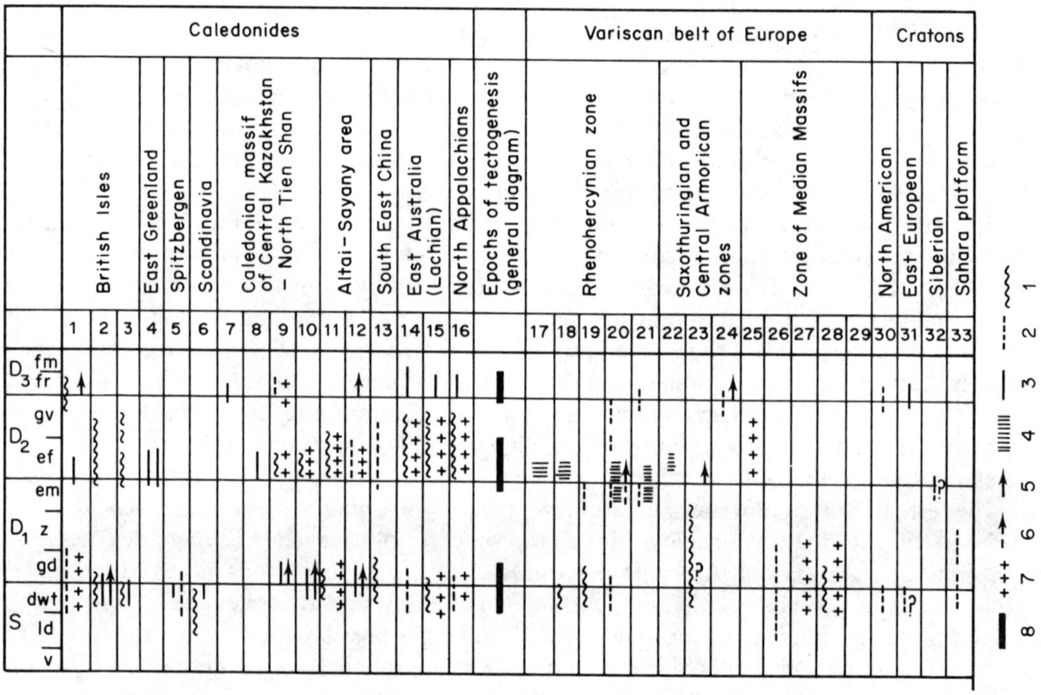

FIG. 2 General diagram of sequence of Mid-Palaeozoic events in Caledonides (orogenic areas) compared with movements in areas of other tectonic regimes (compiled by Yu. G. Leonov).

1 – folding of different intensity and type; 2 – uplift, sometimes gently folded; 3 – orogenic impulse; 4 – essential change in tectonic regime; 5 – beginning (locally, short outbursts of orogenic volcanism; 6 – beginning of geosynclinal (usually spilite-diabase) volcanism; 7 – epochs of granitoid intrusions, granitization and metamorphism; 8 – duration of epochs of tectogenesis. *Note*: Length of symbol shows not the real duration of the event, but maximal possible determination by stratigraphic data.

Figures in the diagram: 1 – region of the Orcadian basin; 2 – region of the Caledonian basin; 3 – Wales; 7 – zone of the Karatau-Talass fault; 8 – interior parts of the massif; 9 – volcanic belt and perivolcanic depressions; 10 – Dzhalair-Naiman zone; 11 – W. Sayany; 12 – region of Minussinsk basins; 14 – W. Lachlan belt; 15 – E. Lachlan belt; 17 – S.W. England; 18 – High Ardennes; 19 – Condroz anticline; 20 – Rhein State Mts; 21 – Harz, Harzgerode zone; 22 – Central-Armorican zone; 23 – N. Vosges; 24 – Saxonia and Thuringia; 25 – Cornouaille Anticline (Armorican massif); 26 – Ligerian zone (Armorican massif); 27 – Central Massif; 28 – Central Vosges; 29 – Barrandov synclinorium (Bohemian massif).

single pulse, but a sum of subsequent pulses. This sum is an active episode in the history of the earth.' Such cases may not be the rule, but neither are they very rare. In German geological literature closely spaced unconformities distinguished in some regions and in total corresponding to a more general phase or epoch of movements are called 'Teilphase'. For example, the Bretonic phase in the horizons transitional from the Devonian to the Carboniferous is marked in the Harz Mountains by several such 'Teilphase'. Therefore such phenomena must be taken into account in estimating the total duration of global events.

Thus, to estimate the duration of tectonic phases and epochs, one should take into account that such a duration is: (a) a sum of regional manifestations; (b) often of a fairly large timespan (actual or apparent due to inaccurate age definition and correlation); (c) often rather different for events of different ages. It is particularly desirable to emphasize the necessity of a realistic approach to the problem of global events. Their study should involve the consideration of all possibilities. It seems feasible to pose and solve the problem of short movements by referring to the data on the Late Tertiary–Quaternary time (mainly Pliocene–Quarternary) that permit the application of their detailed age subdivisions and correlation. It appears that for the time being older epochs cannot be studied in such detail since present-day methods of global investigation are inadequate for the older events. The existence of short global movements cannot be proved from the available material. Therefore the task of the investigation should be formulated separately for each short timespan, not as a proof or disproof of the existence of global events of a given duration, but as the investigation of the minimal duration of the events recognizable on a global scale.

Terminology

The terminology of orogenic phenomena is very imprecise. This is partly due to the fact that the term 'phase' is preserved in the literature in spite of the fact that its meaning (e.g. the idea of phase duration) has changed in the course of time and in the works of different authors. Nevertheless, the same terms ('phase' and, later, 'epoch') are often applied to events of different scope as regards their duration, area of manifestation and features by which they are distinguished. In this respect certain limiting definitions may be helpful, for example 'global', 'regional', 'orogenic', 'uplift', 'deformation', and so on. But they are also used rather unsystematically.

The terms describing orogenic events should apparently reflect firstly, a 'complexity' of the event and secondly its areal distribution. From the viewpoint of the complexity it is reasonable to deal with simple events represented by one group of single events (e.g. uplift deformation) and complex events, involving several simple events (e.g. deformation and magmatism). As regards the areal importance, it is reasonable to distinguish local (characteristic of small areas, of parts of tectonic units), regional (characteristic of major tectonic units, e.g. the Greater Caucasus, Urals, Rhenohercynian zone, and so on) and global (superregional, inclusive) events. Accordingly, the following system of terms is proposed:

	Orogenic events	
	Elementary and relatively elementary	Complex
Global (and superregional)	?	Epoch of tectogenesis
Regional	Phase, regional phase	Phase of tectogenesis
Local	Episode (of uplift, deformation, etc.)	

Elementary orogenic events can be distinguished on a local and, apparently, regional scale. The question remains open whether elementary events of a global scale exist and can be distinguished.

It is proposed that local events which are recognized by individual structures, parts of tectonic units or relatively small units (e.g. some platformal structures, individual zones of fold structures) be called 'episodes'. Depending on their definite manifestations, they may be called episodes of uplift, deformation episodes, episodes of synsedimentary movements, volcanism, and so on. As a rule, they are not only local, but also short-lived manifestations of tectonic activity. In addition, they do not point to intensive tectogenesis that may have initiated such processes as large-scale magmatism, metamorphism, intensive alpinotype deformation. The presence of such processes points not to a local episode, but to movements of higher rank when the process of tectogenesis develops over a large territory and within a great volume of the crust and mantle.

It is reasonable to reserve 'phase' for elementary and relatively simple regional orogenic events involving major tectonic units: fold structures (such as the Greater Caucasus, Urals, and so on), platforms and other structures of similar scale and importance. In tectonically inactive areas, first of all on platforms, the phases manifest uniformly over a large area (e.g. regional unconformity). In this case they differ from episodes only in the extent of manifestation, their division being largely conventional and subjective. However, apparently regional events of the scale appropriate to the suggested term 'phase' are often somewhat more complex than episodes. Complex orogenic events are distinguished on a regional and global (superregional inclusive) scale. Local events seem as a rule not to be complex. The term 'phase of tectogenesis' is proposed for complex regional events and 'epoch of tectogenesis', for global and superregional events. Epoch of tectogenesis is a complex event whose duration is a sum total of durations of its regional phases and regional phases of tectogenesis (see the previous section). It is noteworthy that a global (and superregional) epoch of tectogenesis is a synthetic notion. Such epochs are distinguished indirectly, that is they are determined by interregional correlations. Accordingly, they have no *sensu stricto* tectonotypes.

In compliance with the growing tendency, we have chosen the term 'phase (epoch) of tectogenesis' that lays stress on the complexity and diversity of an event. Other definitions could equally well be used, namely the often used 'orogenic', 'of folding', 'tectonic', 'tectono-magmatic', 'of intense activity or mobility', and so on. Their choice is not a matter of principle, but the evolution of the definition reflects, to a certain extent, a change in the understanding of the process described by such terms.

In some cases it may be useful to distinguish orthotectonic phases, manifested as intense compressional deformations (with the formation of linear folds, overthrusts, nappes), and paratectonic phases displayed as other manifestations of tectonic activity, including those of tension (rifting, etc.). It is clear that orthotectogenesis in one region may correspond to paratectogenesis in another, as Stille meant in dealing with synorogenic movements. These definitions apply to phases of tectogenesis (i.e. for regional units), whereas according to the definition, global epochs of tectogenesis comprise different manifestations of tectonomagmatic activity of both ortho- and paratectonic types.

REFERENCES

Bubnoff, S. von (1954). *Grundprobleme der Geologie*, 3rd edn, Berlin: Akademie-Verlag.

Dodson, M. H. and Rex, D. Ch. (1970). Potassium-argon ages of slates and phyllites from south-west England, *Quart. J. Geol. Soc.* (London), **126**, part 4, no. 504, 465–500.

Gilluly, J. (1949). The distribution of mountain building in the geological time, *Bull. Geol. Soc. Amer.*, **60**, no. 4, 561–590.

Goguel, J. (1965). *Traité de Tectonique*, 2nd edn, Paris: Masson et Cie.

Kazarinov, V. P. (1976). Upon the path of finding palaeogeographic regularities in the development of the earth, *Bull. Mosk. obshch. isp. prirody*, Geol. section, **51**, issue 2, 5–32 (Russian).

Kazmin, V. (1980). Geodynamic control of rift volcanism, *Geol. Rundsch.*, **69**, no. 3.

Khain, V. E. (1971). On general regularities in the development of tectonic processes in time—the problem of discontinuity-continuity-cyclicity-direction, *Vestnik Moscow State University*, geol. section., no. 4, 3–22 (Russian).

Khain, V. E. (1973). *General Geotectonics*. Moscow: Nedra (Russian).

Kunin, N. Ya. and Sardonnikov, N. M. (1976). Global cyclicity of tectonic movements, *Bull Mosk. obshch. isp. prirody*, geol. section, **51**, issue 3, 5–27 (Russian).

Leonov, Yu. G. (1976a). On some methodological premises in the correlation of tectonic movements, *Geotektonika*, no. 6, 3–14 (Russian).

Leonov, Yu. G. (1976b). *Tectonic Nature of Devonian Orogenesis*, Moscow; Nedra (Russian).

Leonov, Yu. G. and Khain, V. E. (1980). General problems of the study of global tectonic processes, in *Problems of Global Correlation of Geological Events*, Moscow; Nauka, Trudy GIN AN SSSR, pp. 6–21 (Russian).

Mazarovich, A. N. (1940). Rhythm in the history of the earth, *Bull. Mosk. obshch. isp. prirody*, geol. section, **18**, issue 5/6, 4–23 (Russian).

Pronin, A. A. (1969a). *Hercynian Cycle of the Tectonic History of the Earth*, Leningrad: Nauka (Russian).

Pronin, A. A. (1968b). *Caledonian Cycle of the Tectonic History of the Earth*, Leningrad: Nauka (Russian).

Rubinstein, M. M. (1967). Orogenic phases and periods in folding in the light of data on absolute geochronology, *Geotektonika*, no. 2, 35–49 (Russian).

Rutten, L. M. P. (1949). Frequency and periodicity of orogenic movements, *Bull. Geol. Soc. Amer.*, **9**, 1755–1770.

Schatskiy, N. S. (1937). Orogenic phases and foldings, in *Proceedings of the 17th Session of the Intern. Geol. Cong., USSR*, Moscow: GONTI, vol. 2, pp. 323–30 (Russian).

Schatskiy, N. S. (1951). Duration of folding and phases of folding, *Izv. AN SSSR*, ser. geol., no. 1, 15–53 (Russian).

Schultz, S. S. (1970a). Epeirogenic folding and its role in the development of structure and relief of the earth, *Geotektonika*, no. 4, 3–18 (Russian).

Stille, H. (1950a). Nochmals die Frage der Episodizität and Gleichzeitigkeit der orogenen Vorgänge, *Struktur und Zeit. Geol. Rundschau*, **38**, part 2.

Stille, H. (1950b). Bemerkungen zu James Gilluly's 'Distribution of Mountain Building in Geologic Time', *Struktur und Zeit. Geol. Rundschau.*, **38**, part 1.

Global Correlation of Tectonic Movements
Edited by Yu. G. Leonov and V. E. Khain

2

Palaeogene Volcanism in the Near and Middle East and its Tectonic Setting

N. V. KORONOVSKY*

Palaeogene volcanogenic rock associations, particularly Eocene, are widespread in the Mediterranean belt, occurring from the meridian of Erzincan, Turkey, in the west, to the Kwetian ophiolite suture of Pakistan, in the east. The volcanites exceed 200,000 km^3 in volume, which greatly exceeds the bulk of all volcanic rocks produced during the preceding Phanorozoic and following Neogene–Quaternary history of the region. Volcanic eruptions occurred in different structural zones and were short-lived. Palaeogene volcanism was widespread in the adjacent areas of the Mediterranean belt—in the Hindustan, Arabian, African and Turanian platforms and in the north-eastern part of the Indian Ocean. It therefore appears to be a general phenomenon and requires further investigation to explore the cause of such strong volcanic activity, and an attempt should be made to reconstruct the palaeotectonic environments that existed during the period of volcanic activity.

GEOLOGICAL SETTING OF VOLCANIC ROCK ASSOCIATIONS

Palaeogene volcanism is manifested in the Mediterranean belt, within the study area, in different structural zones and tectonic environments (Fig. 1). Space does not allow for detailed descriptions here of the volcanogenic rock series of all areas, and therefore only the most essential features of its geological position and structure are briefly outlined below. On the one hand, volcanic activity was restricted to relatively narrow superposed Middle and Late Alpine geosynclinal troughs, that underwent closing at the end of the Palaeogene. This refers to the Adzhara-Trialeti, Talysh, northwestern Iran, Elburz, eastern Iranian and its eastern extension (Chagai), and also the Katawaz fold zones. All these zones were generated over a consolidated basement that underwent disruption either at the end of the Early or during the Late Cretaceous time. Volcanic action was manifested chiefly in shallow sea or, occasionally, in subcontinental environments. Manifestations of strong Eocene and Oligocene volcanism are reflected in the degraded geosynclinal troughs of the Sevan-Akera zone, particularly in its northwestern part—the Sevan-Shirak synclinorium. Incidents of less powerful volcanic action are reflected in the Somkhet-Karabagh and Aiotsdzor-Ordubad zones. On the other hand, strong volcanic eruptions occurred during the Eocene within the rigid stabilized structures, as, for instance, in the southwestern part of the Central Iranian

* Geological Faculty of Moscow State University, Moscow.

zone (Urmia-Dokhtar volcanogenic belt), in the Central Iranian zone, within the vast Lut Massif, and even transgressed on to the epi-Palaeozoic Turanian Platform (Badhiz).

In the east of the Mediterranean segment under discussion, manifestations of Eocene volcanism are reflected in the Farahrud and Argandab zones, which experienced folding during the Cretaceous. Thick sheets of continental volcanites comprise the orogenic complexes in these zones. Middle Eocene volcanic associations are known in the Gerirud trough, that was first formed during Danian–Palaeocene times on a pre-existing Hercynian and Cimmerian basement. As yet undefined is the position of the Palaeogene volcanites in the Sabzevar zone, south of the Sabzevar fault, that delimits to the south the Aladagh-Binalud Mountains, where no manifestations of Palaeogene volcanism have been found. Lower and Middle Eocene tuffs, basalts and andesite-basalts of the calc-alkaline series are known in this zone, where they rest with a sharp discordant unconformity on Cretaceous volcanogenic-sedimentary series of strong deformational pattern, containing lenses of the 'coloured melange' (Förster, 1968).

Eocene volcanic rocks exhibit only weak block faulting, volcanic action having occurred subaerial or shallow marine environments.

PALAEOGENE VOLCANISM AND OPHIOLITE BELTS

No definite spatial relationship is apparent between the ophiolites and Palaeogene volcanic rocks (see Fig. 1). Ophiolite belts are of wide extent over vast areas, where Palaeogene

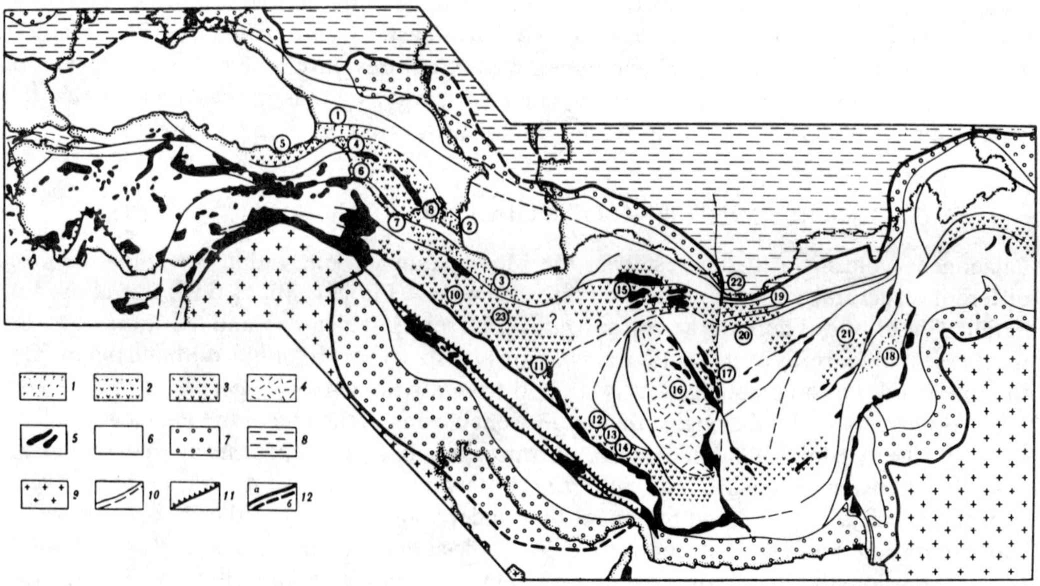

FIG. 1 Geological setting of Palaeogene volcanic areas in the Near and Middle Eastern sectors of the Mediterranean belt.

1–2 – volcanites in geosynclinal troughs: 1 – subalkaline, weakly differentiated basalt formation; calc-alkaline andesite formation; 3–4 – subaerial volcanites: 3 – sequentially differentiated basalt-andesite-dacite-liparite formation, 4 – contrastic liparite-basalt formation; 5 – localities of 'coloured melange'; 6 – Mediterranean fold belt; 7 – Alpine marginal troughs; 8 – epi-Palaeozoic platforms; 9 – Precambrian platforms; 10 – major fractures: (a) – definite, (b) – probable; 11 – nappes and overthrusts; 12 – confining boundaries of the Mediterranean belt. Figures show areas of volcanic action (see Fig. 2).

volcanites are completely absent, for instance in Anatolia, west of the Erzindzhan fold convergence, in the Quetta ophiolite suture, and so on. Conversely, powerful volcanic action is reflected in other areas where ophiolites are absent, for example in Adzhara-Trialeti, Talysh, Badhiz, the Gerirud trough, Argandab and Central Iranian zones. In some places Palaeogene volcanites are in close proximity to the ophiolite belts, but in various relationships to them. The volcanites almost invariably rest with sharp unconformity on ophiolite assemblages, seemingly 'sealing' them, and extend in different directions from the ophiolite belts, as, for instance, in the Sabzevar, Sevan-Akera and eastern Iranian zones. At the same time, the Urmia-Dokhtar volcanogenic belt parallels the trend of the ophiolite belt of the Zagros Interior Zone. In the southeastern part of the belt another ophiolite belt can be traced further north along the eastern edge of the Hamadan zone, but near Nain it trends northward, whereas the volcanogenic belt extends to the northwest and intersects the ophiolites. The Veda ophiolites delimit at the south the Palaeogene volcanites of the Aiotsdzor-Ordubad zone. Hence no regular spatial relationship can be established between the ophiolites and Palaeogene volcanism.

In most of the ophiolite belts the major tectonic movements that caused the formation of the 'coloured melange' and nappe-overthrust structures occurred prior to the Palaeocene or Eocene. In the east, in the Quetta ophiolite belt, compressive events occurred at the end of the Late Cretaceous (during the Maastrichtian), and during the Palaeocene and Eocene flysch troughs were formed behind the belt, where signs of poorly studied basic volcanism are evident. The Maastrichtian was characterized by the most powerful compressive events and the creation of the Oman ophiolite suture. The Zagros ophiolite suture originated in the Pre-Campanian. At the end of the Late Cretaceous, folding and nappe formation occurred in the East Taurus, Amanus and Cyprus ophiolite belts. Consequently, the marginal ophiolite suture, which extends over vast areas of the Near and Middle East, originated mainly during the Maastrichtian (Lomize, 1972). In the Iranian ophiolite belts, after the major compression phase at the end of the Late Cretaceous, another short-lived phase of relatively weak movements with the same direction can be recognized in the Early Paleocene—the period of the first orogenic block movements in northern Iran (Stöcklin, 1974). In the Sevan ophiolite belt the major compressional phase occurred at some time during the Late Cenomanian or Turonian, and in the Veda zone, further south, during the middle of the Coniacian, providing that the Veda ophiolites are of autochthonous nature.

Thus during the Maastrichtian, and even at an earlier period, tectonic movements were powerful, causing a strong compression in the zones of the ophiolite assemblages, the creation of the 'coloured melange', its overthrust and partial imbrication. It should be noted that incidents of volcanic action occurring during the period of tectonic compressive events are not known in any of the ophiolite belts. Certain areas reveal evidence which distinctly indicates that compression in the ophiolite belts coincided with tension in the neighbouring areas. For instance, the formation of the Sevan ophiolite suture was practically contemporaneous with the disintegration and subsidence of the southern edge of the Transcaucasian Median Massif, accompanied by strong eruptions of a differentiated series (Adamia *et al.*, 1974). Fold movements in the Kerman-Tebes zones of the Naiband trough, south of the Lut Massif, coincided with the formation of the eastern Iranian trough at the east of the Massif (Stöcklin *et al.*, 1972). Folding and overthrusting in the Zagros internal zone were probably partially compensated by tension in the Hamadan zone. In all the above-described areas the Eocene volcanogenic and volcanogenic-sedimentary series rest with sharp unconformity on the ophiolite assemblages, thus giving an unambiguous indication to the occurrence of strong tectonic movements and a significant time break between the compressive events in the ophiolite belts and the manifestations of Eocene volcanic action. This break spans no less than 20–25 m.y., but is more often longer—30–40 m.y.

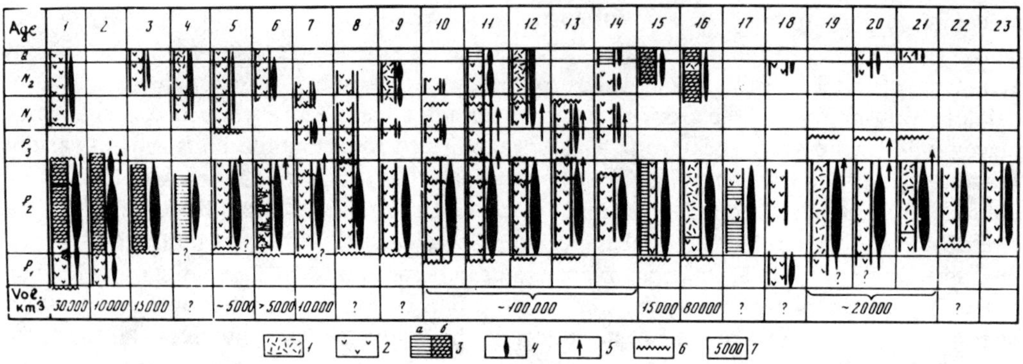

FIG. 2 Cenozoic volcanism of the Mediterranean belt, in the Caucasus, Near and Middle East.
1 – liparites and dacites; 2 – calk-alkaline sequentially differentiated basalt-andesite-dacite-liparite formation, with a prevalence of andesites; 3 – undifferentiated basalt formation; (a) – of normal alkalinity, (b) – subalkaline; 4 – provisory volcanic intensity index; 5 – injection of intrusives; 6 – folding movements; 7 – volume of Palaeogene products, in km^3.
Areas of volcanic action: 1 – Adzhara-Trialeti; 2 – Talysh; 3 – Elburz; 4 – Arthvin-Bolnisi Block; 5 – Eastern Pontides; 6 – Sevan-Akera zone; 7 – Aiotsdzor-Ordubad zone; 8 – Iranian Azerbaidzhan; 9 – Balgash; 10–14 – Urmia-Dokhtar belt (10 – Cum, 11 – Yezd; 12 – Kerman-Sirdjan; 13 – Kahar-Ganbad; 14 – Taht-Bidha); 15 – Sabzevar; 16 – Lut Massif; 17 – Eastern Iranian zone; 18 – Katawaz zone; 19 – Gerirud trough; 20 – Farahrud zone; 21 – Argandab zone; 22 – Badhiz (Turanian Plate); 23 – Central Iranian zone; 24 – age; 25 – volume, in km^3.

Consequently, volcanic action was of powerful manifestation during the Palaeogene in the Near and Middle East segments of the Mediterranean belt, being delimited by the Erzindzhan fold convergence, to the west, and the Penjab fold convergence to the east. The maximum volcanic outbursts occurred during the Eocene, and in some areas, allowing for more precise stratigraphic correlation of the sequence, this timespan is reduced to the Middle Eocene (Fig. 2). Thereby, a definite synchronism in the volcanic activities occurring over vast areas of the belt can be established. The volcanogenic and volcanogenic-sedimentary rock series almost everywhere rest with disconformity, angular and, at times, discordant unconformity on older deposits, including the ophiolite assemblages. In many places the Eocene volcanogenic series lies directly on the metamorphosed Precambrian (?). Even in those areas, where Palaeogene volcanism was preceded by Late Cretaceous volcanism, evidence is found of pre-Palaeocene and pre-Eocene movements of varying intensity. All of this serves explicitly to prove the superposed nature of the volcanic activities.

Palaeogene volcanism was restricted to different structural zones and manifested in various types of environments: submarine, shallow marine and continental. In some instances it was confined to troughs of geosynclinal type, in others to rigid structures, and in still others to areas that underwent folding and began their orogenic development. A specific feature of Palaeogene volcanism is the great predominance of explosive, as compared to effusive proper, processes. In most cases the pyroclastic materials comprise 70–80 per cent and more of the total volume of the ejecta, which implies that the magma was rich in volatile matter. Most prevalent were eruptions of the central volcanian type, and more seldom of the fissure type. Volcanic activity in many instances was contemporanous with the emplacement of hypabyssal and subvolcanic intrusions of medium and acid composition. The formation of these intrusions was either concurrent with volcanic activity or was somewhat later.

THE PETROCHEMISTRY OF EOCENE VOLCANIC ROCKS

Analytical treatment of petrochemical data available on the Eocene volcanogenic rock series of the Alpine belt in the study area served to establish certain general regular patterns. For petrochemical studies over 800 chemical analyses of the volcanic rocks were selected from various sources (Aslanyan, 1958; Azizbekov, 1961; Dzherbashyan, 1966; *Geology* . . ., 1970; Kochetkov *et al*., 1972; Adamia *et al*., 1974; Lordkipanidze and Zakariadze, 1974; Dovgal (unpublished data); Nadareishvili, 1974; Stalder, 1977; Förster *et al*., 1972; Darvishzadeh and Ardzhangravesh, 1975; Darvich-Zad, 1976; Dimitrijevič (unpublished data)). In the northern zone of Eocene geosynclinal troughs—Adzhara-Trialeti, Talysh, Elburz—of predominant occurrence are subalkaline series, either weakly differentiated or completely undifferentiated and consisting chiefly of basic rocks that are characterized by low titanium content (particularly in the Adzhara-Trialeti trough), slight increase in time of the K/Na ratio, and low rate of oxidation. Although common features are apparent in these volcanic rocks, they usually reveal certain important differences, consequential upon the inherent properties of the parent magmas. For instance, the Middle Eocene effusive rocks of Adzhara-Trialeti are characterized by a tholeiite differentiation trend (Fenner trend), with a rapid increase of iron index (for basalt constituents), high magnesia and low alumina content (Fig. 3,a). The Elburz volcanites characteristically are of a similar, but poorly expressed toleiite differentiation trend and, as a whole, of higher iron content (Fig. 3,b). Apart from this, they are typically of low magnesia and high alumina content. The Talysh basalts and andesite basalts are highly leucocratic and have a very low iron content. All this taken together serves to suggest that the subalkaline basaltoids of Adzhara-Trialeti are, most likely, derivatives of the least differentiated, hypabyssal, relatively 'dry' magma. Support for this is found in their high magnesia content, since ordinarily a low magnesia content is atypical of the first magmatic melts. The Elburz subalkaline basaltoids, which characteristically are of high potash and alumina content and low magnesia content, were formed under the profound influence of volatile components, seemingly involving relatively impeded and intermittent magma flowage towards the surface. The Talysh basaltoids, being of low magnesia content and highly leucocratic, in these features closely resemble the Elburz effusive rocks, thus implying similar conditions of formation.

Owing to the progressive evolution of the magmatic melts, the Late Eocene volcanic associations, occurring in all these troughs, reveal a calc-alkaline differentiation trend that characteristically demonstrates a slow rate of Fe accumulation, increased K_2O and Al_2O_3 content and high oxidation coefficient. The effusive rocks of Elburz, Talysh and, to a lesser degree, Adzhara-Trialeti, are distinguished by the wide occurrence of hornblendic types. Calculations of the discriminant functions, according to V. A. Kutolin, serve to indicate that the basaltoids, occurring in these troughs, can be assigned to continental olivine-basalts. It can be presumed that the parent magma of the Adzhara-Trialeti volcanic associations was low in titanium, relatively 'dry', and of subalkaline olivine-basalt type. The Elburz and Talysh magmas were generated characteristically under the influence of volatile components. The occurrence of such magmatic melts at the surface is attributed to their rapid flowage from the source chambers through zones of large fractures, apparently without the formation of intermediate cells.

Eocene volcanism in all other areas of the Near and Middle East differs markedly from that described above, and is characterized by sequential differentiation series of calc-alkaline, more seldom subalkaline rock associations, which have a high total iron content that is considerably greater than the total iron content in the calc-alkaline series of the

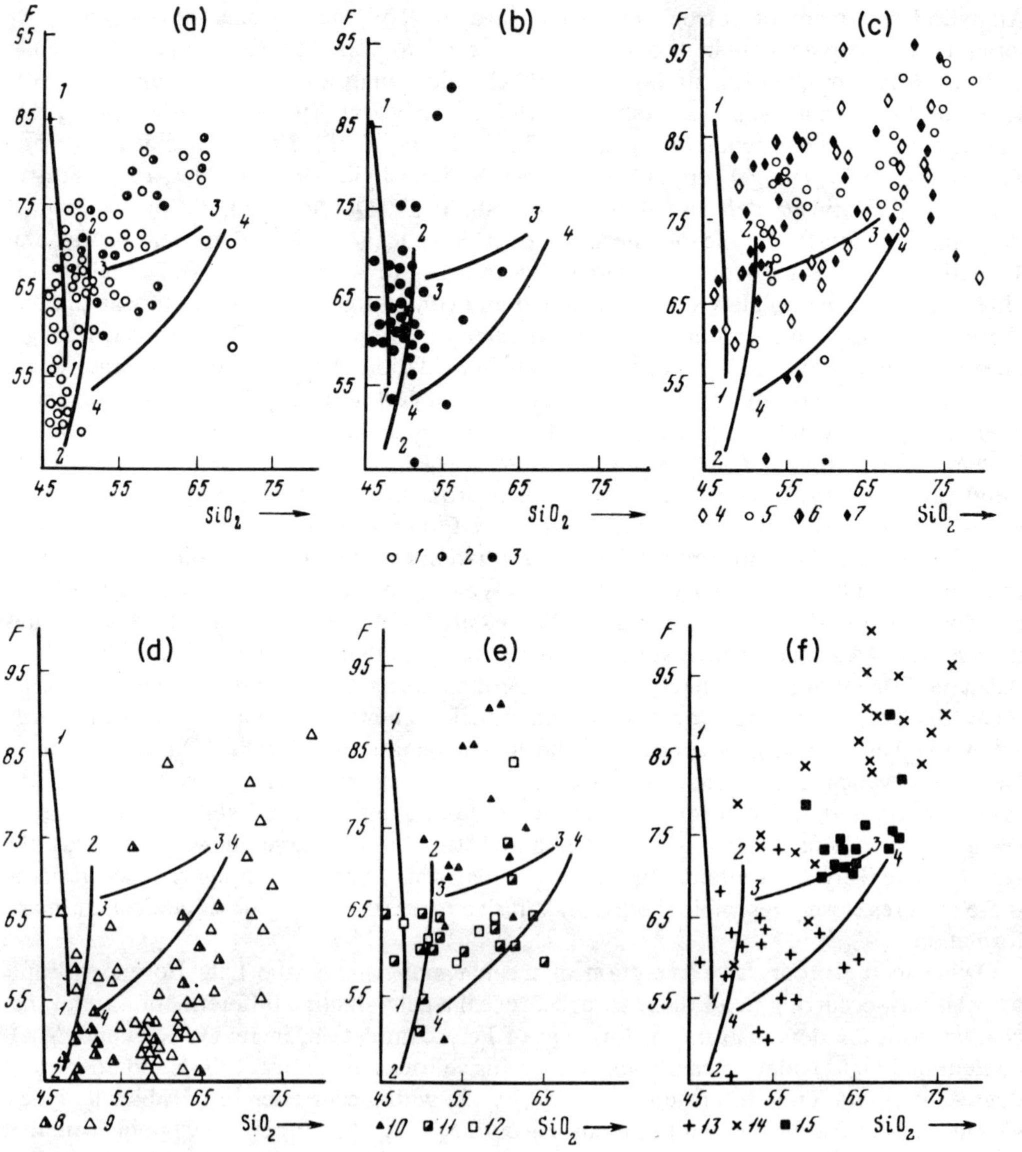

FIG. 3 Diagram: Total iron (Fe_2O + FeO)/(Fe_2O_3 + FeO + MgO)–silicate content of the Cenozoic volcanic rocks in the Caucasus, Near and Middle East.

Composition trends: 1–1 – Skaergaard intrusion, 2–2 – tholeiite series, 3–3 – andesite-dacite-liparite, 4–4 – calc-alkaline series of the Cascade Mountains. a – Adzhara-Trialeti: 1 – Middle and 2 – Upper Eocene; b – Elburz: 3 – Eocene; c – Sevan-Shirak zone: 4 – Lower Eocene, Agarcenian formation; 5–6 – Middle Eocene: 5 – Shirak and 6 – Kirovokan formation, 7 – Upper Eocene, Pambak formation; d – 8 – Aiotsdzor, Upper Eocene; 9 – Vardenis, Oligocene; e – Urmia-Dokhtar Belt: 10 – Eocene, 11 – Miocene, 12 – Pliocene–Anthropogene; f – 13 – Lut Massif, Eocene–Anthropogene, 14 – Argandab zone, Eocene, 15 – Desht-I-Navar, Afghanistan, Pliocene–Anthropogene.

Cascade Mountains—the classical example of such rock series (Fig. 3,c–f). This regular pattern is interrupted only by the Aiotsdzor Upper Eocene and Vardenis Oligocene volcanites that, being of very low iron and extremely high magnesia content, reveal a typical calc-alkaline differentiation trend with a slow rate of Fe increase (Fig. 3,d). Taking into account the fact that on the AFM diagram the Vardenis effusive rocks plot almost entirely along the hypersthene series, coupled with the fact that they have the highest titanium ratio, as compared with all other Eocene volcanites, and are also highly leucocratic, the inference can be drawn that the crustal substance exerted a profound influence on the parent magma.

The evolution of volcanism and nature of magmatic differentiation, notwithstanding their general similarity, differ essentially from area to area. In the Sevan and Aiotsdzor-Ordubad zones the sequential differentiation series of basalts-andesites-dacites-liparites occurs, which is distinguished by highly leucocratic basic and intermediate members of relatively low titanium and magnesia content and rate of oxidation. The Eocene-Oligocene volcanites of the Lesser Caucasus show a variable distribution of K_2O. On the one hand, the lowest potash content is typical of the volcanic associations in the southern zones, particularly of the Aiotsdzor-Ordubad trough. On the other hand, the K_2O values show such a large spread, even within a single volcanic formation (in rocks of equal silica content), that the inferences concerning the regular lateral distribution in the K_2O content are quite vague. These differences can possibly be explained by the effects that the consolidated crust and volatile components exerted on the evolution of the magmatic melts. Calculations of the discriminant functions, according to Kutolin, serve to indicate that these volcanic associations are examples of the typical andesite formation, where the increase in the K_2O content is not a function of the source depth.

The Urmia-Dokhtar volcanogenic belt is comprised characteristically of volcanic rocks that, notwithstanding certain differences in their inherent properties, are most closely similar to the typical representatives of the calc-alkaline series. These volcanites are commonly of low magnesia content and potash-soda ratio, with a wide occurrence of hornblendic rock types. From this it can be inferred that the parent magma was closely similar to that of a leucobasalt. Within this belt the communication between the magma chamber and the surface was impeded, due to the pressure that the rigid crust and abundant volatile components exerted on the source. Conducive to this were different trends in the evolution of individual isolated igneous sources. In some instances the evolution was single course, in others multi course, but all the volcanoes were of powerful explosive type. The igneous sources, associated with the Eocene volcanoes, lasted a long time and despite gradual extinction, continued to feed the volcanoes throughout to the end of the Quaternary.

The volcanism in the Lut Massif is, as a whole, characterized by divergent differentiation of the magma and, during the Eocene, by the formation of huge and thick sheets of acid ignimbrites; although in some places (Shotori range) andesite basalts and andesites are present at the base of the sequence. However, the possible occurrence of a contrasting rock series in this area cannot be discarded, considering that Neogene–Quaternary volcanic associations, recently discovered in extensive areas, are represented by basalts and andesite basalts. Seemingly, a long interval of time occurred between the formation of the acid and the basic volcanogenic rocks of the Lut Massif. The huge volumes of dacite and liparite-dacite volcanites, judging by their chemical properties, are products of a palingenetic magma.

In eastern Afghanistan (Argandab zone), the Palaeogene volcanic associations are represented by a calc-alkaline sequential differentiation series of basalt-andesite-dacite-liparite rocks, which characteristically are of very high iron content (up to 85–105) in the acid differentiates (see Fig. 3,f); high K_2O content; high titania ratio; weakly leucocratic. According to the data advanced by Dovgali, sodium differentiates predominate in the lower sections and potash differentiates in the upper sections. The acid effusive rocks are closely similar in their inherent chemical properties to the liparites of the Northwestern Hindustan trap formation. The evolution of the magma melt was of a single-course type, involving distinct emplacement of intermediate igneous sources and impeded magma flowage toward the surface.

DISCUSSION OF THE RESULTS

The evidence available on the relation between volcanism and tectonic movements gives positive proof of the fact that volcanic action was invariably manifested under tensile stress conditions. Not a single region where volcanic action was associated with firmly established compressional structures is known in the Mediterranean Belt. As to whether compressive stresses at the surface correspond on all occasions to similar stresses at depth is a question for which no convincing answer can at present be offered. However, one important point emerges: tension occurs at depths where the magma melt can originate, and these depths can differ greatly. If this supposition is true, then an attempt should be made to analyse the conditions of tectonic movements that existed during the period of Eocene volcanic activities. Reconstruction of the palaeotectonic stress patterns is a difficult task, but can be reliably accomplished by means of pertinent geological data.

As was previously noted, the Eocene volcanogenic and volcanogenic-sedimentary series rest with disconformity and often pronounced angular unconformity on the underlying rocks, including the ancient metamorphosed basement. The Palaeogene volcanic associations also unconformably overlay ophiolite assemblages. In other words, prior to the volcanic activity, in different areas strong tectonic movements occurred that varied in intensity and were characterized by compressional stresses.

The formation of the strongly deformed ophiolite belts with the 'coloured melange' is commonly regarded as a result of the closing of basins with oceanic crust, caused by plates moving from opposite directions. The consequence of this is the formation of subduction zones that are related to the Benioff zones and volcanic activity associated with the latter. The literature contains several publications that emphasize this concept. Practically all types of volcanic action (excepting early-geosynclinal volcanism), which manifest in various structural settings, including median massifs, are explained in terms of the Benioff zone activity, the zone dipping gently and 'penetrating' beneath the rigid lithosphere over several hundreds of kilometres (Adamia *et al.*, 1977; Jung *et al.*; 1976, Gerod and Conrad, 1975). From this standpoint, volcanism is a function of the horizontal movements of isolated 'micro' and 'macro' plates. In the light of this, volcanic activity should, obviously, be of the 'island-arc' type and occur either during tectonic compressive events or directly following them. Thus was the case considered here, at some time near the end of the Early Cretaceous, at the beginning and end of the Late Cretaceous. However, the major eruptions of differentiated calc-alkaline andesite volcanics started during the Eocene, which is significantly removed in time (over 7–20 m.y. later) from even the latest compressional fold phases in the ophiolite zones. Worthy of note is the fact that andesitic volcanism is completely absent from Anatolia, where the ophiolites are considerably more widespread and compressive events again occurred during the Late Cretaceous. A view is current, that the Palaeogene and even more recent volcanic activities in the Caucasus are related to the residual activities of the Cretaceous Benioff zones (Adamia *et al.*, 1977; Khain, 1975). However, Late Cretaceous volcanic activity was much weaker than in the Eocene, and of only local occurrence. Hence the reason for the 'residual' activity being much more powerful than the initial activity remains unexplained.

In connection with the reconstruction of the palaeotectonic stress patterns, emphasis should be placed on the fact that periods of strong ubiquitous compression, which could have been generated by many incompletely defined causes, were followed by tectonic 'pauses', characterized either by the relaxation of compressive strain or by an irregular increase in the intensity of opposing tensile stresses. Such incidents were accompanied by powerful volcanic eruptions. The manifestations of tension, concomitant with volcanic activity, are of different expression in each particular area of the Mediterranean Belt. However, the existence of a period of relaxation or decrease in the intensity of tectonic activities, occurring between the periods of strong compressive events, can be definitely established. This tectonic 'pause' spans approximately 30 m.y. in the Early-Middle Eocene. Compressive events correspond in time to intervals of 80–70 m.y. (end of Late Cretaceous) and 37–26 m.y. (Late Oligocene–Early Miocene), respectively (Kasmin, 1976). Thus, in the Mediterranean belt, during the Early-Middle Eocene, olistostromes did not form; as is well known, they are associated with the formation of the frontal nappe structures, that is, with compressional structures (Leonov, 1976). The concept of plate movements and conjugated rift zones likewise gives supporting evidence to the occurrence of an Eocene epoch of tension. The relatively rapid relaxation of the compressive strain, immediately following the strongest Late Cretaceous compressive events, generated tension and, to a certain degree, the break up of consolidated structures of various ages. This break up and the resulting subsidence are well reflected in different tectonic units of the Mediterranean belt, in the pronounced unconformity with which the Eocene volcanics overlie the heterogeneous, including the rigid crystalline basement in the Arthvin-Bolnisi block, the Adzhara-Trialeti, Northwestern Iran, Elburz, Farahrud and Argandab zones, the Lut Massif, Turanian platform (Badhiz), and so on. The superposition of volcanic processes is a feature that gives explicit support to the ubiquitous occurrence of tension. Evidently, the formation of deep faults, the drop in the total lithostatic pressure, and the outpouring of the magmatic melts could have been most complete in those places that were subjected to the greatest tension. In this respect probably not of the least significance were the pre-existing zones of crustal weakness, which favoured the development of inherited forms of volcanism, as, for instance, in Adzhara-Trialeti.

PALAEOGENE VOLCANISM AND ISLAND ARCS

Attention is now focused on the palaeotectonic conditions that existed in the Caucasus, the Near and Middle East during the Palaeogene. As a necessary prerequisite, comparative analysis should be made of the Palaeogene tectonic environments and those existing in the present day island arcs, considering that the available up to date reconstructions give direct indication of the analogy between the Palaeogene volcanic zones of the Lesser Caucasus, the Urmia-Dokhtar belt, other areas, and the present island arcs (Adamia *et al.*, 1977; Jung *et al.*, 1976). The present island arcs are characterized by a combination of structural elements including vast oceanic domains, lack of a granitic metamorphic layer; the presence of an abyssal trough and island arc and, as an almost indispensable feature, the presence of a marginal sea with an oceanic or suboceanic crust. There is a strong belief that only the combination of all these specified elements makes an actualistic comparison possible between the conditions existing at present with those of ancient island arcs. One or two features, taken separately, do not permit such a comparison; as, for instance, the occurrence of only folding, or basic rocks, or the existence of only a thick rock series, does not provide adequate evidence for a former geosyncline in the area. Therefore, the question follows: to what extent are the conditions of the present day island arcs reflected in the Palaeogene of the study area?

In the Lesser Caucasus oceanic domains with non-granitic crusts (if they ever existed) should have disappeared either at the end of the Early Cretaceous (Albian) (Belov *et al.*, 1976), or during the Cenomanian times. Indeed, during this period, the Erzindzhan-Sevan ophiolite suture originated, where the rocks although of strong deformational pattern are of steep inclination, which is stressed by all investigators. The Veda ophiolite suture (provided it is not of allochthonous nature) was formed during the Coniacian. All the Eocene volcanogenic and volcanogenic-sedimentary series lie with a sharp discordant unconformity over the ophiolite sutures. It is immediately apparent that during the Eocene abyssal troughs, as well as marginal seas with non-granitic crust, did not exist, although marine conditions undoubtedly existed. Thus not one of the necessary conditions (in the sense of indispensable elements) for the reconstruction of the palaeotectonic environments in the Palaeogene of the Lesser Caucasus is in any sense similar to those of present day island arcs. What, then, remains? An indisputable fact is the presence of a discontinuous belt of calc-alkaline volcanic rock associations, which as a whole are represented by a sequential differentiation series of high alkalinity in some places and particularly at the end of its formation. This volcanic belt is located both north of the Erzindzhan-Sevan ophiolite zone, and south of the latter. Within the zone proper, the Eocene volcanogenic series is of low grade and rests with a pronounced unconformity on the Upper Senonian and Lower Palaeocene carbonate deposits. Westward of the Lake Sevan, the volcanites of the Shirak trough completely cover the ophiolite zone. North of the Veda ophiolite suture volcanic rocks are relatively rare; the Late Eocene–Oligocene effusive series of the Daralagez trough are located much further east. Hence, during the Palaeogene the conditions in the study area were in no way similar to those in the present island arcs. Even allowing for the Eocene volcanic activity, which occurred north of the Erzindzhan-Sevan ophiolite suture and was related to a Benioff palaeozone, then what explanation can be offered for the occurrence of the southern volcanic associations and the absence of volcanic rocks of the same age? For instance, in Anatolia, where ophiolite zones are more widespread and numerous, andesite volcanic belts of the 'island-arc type' should be found.

The interpretation of the lateral zonation of the K_2O content lacks cogency. The outward increase of the K_2O content of an island arc, and also in certain continental areas, is a long-known fact, but is subject to controversy. According to the data obtained in recent years, the universality of this pattern is doubtful. For instance, the Lesser and Great Kurile Island arcs show reverse relations: high alkilinity is typical of the effusive rocks in the Lesser Kurile Arc (Frolova *et al.*, 1977). The transverse alkalinity zonation of the Kamchatka volcanic rocks is now refuted, although a zone of low potash basalts can be traced along the Pacific coast. A more intricate distribution of alkalis has been recognized in the Cenozoic of western North America. Considering that andesitic magmas originated, most probably, as a result of magmatic replacement processes in the earth's crust, then, besides the influence that the crust exerted on the alkali content of the magma, of essential importance is the alkalinity of the parent magma, which in a broad sense depends upon the depth of the source. Of strong effect on the K_2O content is fractional crystallization, during which the variation in the K_2O content depends more upon the silica content of the rocks, than the potassium content of the initial melt. Besides, as is well known, during volcanic activity the potassium content shows wide variations. This is well illustrated by the Eocene volcanic associations of the Sevan-Shirak synclinorium. If the K_2O content depends only upon the depth of molten magma in the Benioff zone, then these differences in its content cannot be adequately explained (Cawthorn, 1977). Apart from this, the explanation for the lateral zonation in the K_2O content, based on the increase in depth of the igneous source, is cogent only for the trend: pigeonite basalts–alkali basalts–trachytes. Otherwise, if the trend leads towards hypersthene basalts and andesites, the mechanism responsible for the increase of the K_2O content is of a fundamentally different nature (crustal trend). It should be noted that the relation between the igneous sources and seismic focal zones is not simple. All of this indicates that the lateral increase of the K_2O content in the calc-alkaline series (given it is firmly established) occurs in isolated volcanic sheets and cannot be explained merely by the effects of the Benioff zone and subduction.

The generation of the Urmia-Dokhtar volcanogenic belt and the Elburz Eocene volcanites is regarded by some authors as a result of subduction in the Zagros zone and the gentle northward dip of the Benioff zone; as a consequence of this the Elburz magmas originated at deeper levels and are of higher alkilinity (Jung *et al.*, 1976). It has already been shown that in the internal Zagros zone, where the ophiolite assemblage is present in some places, and also within another narrow ophiolite suture located along the northern margin of the Hamadan zone, the major compressive events occurred either prior to the Maastrichtian, or during the Middle Maastrichtian. However, volcanic activity started much later during the Middle Eocene time. The Elburz volcanics are unconformably superposed on the northern rigid margin of the Central Iranian median zone, which during this epoch underwent faulting that caused the formation of graben-type troughs. Compressive events occurred here during the Palaeocene and Oligocene. The Urmia-Dokhtar belt is likewise restricted to a linear graben-type structure, that is superposed on the southern edge of the Central Iranian zone (Dimitrijevič, 1973). Extension and disruption of the marginal parts of this consolidated zone are apparent, specifically in the Eocene period. The igneous activity, parent magma types and their evolution were all essentially different in the Urmia-Dokhtar belt and Elburz. Apart from this, the conditions were not even remotely similar to those existing in the present island arcs. Thereby, the association of the volcanic activities in these areas with one and the same conjectural Benioff zone, or even with two different zones, is not warranted. This is also the case since the volcanic rocks show essential differences in their inherent petrochemical natures. What is unconfirmed is the assertion that the Eocene volcanic activity of the Lut Massif is associated with the movement of the Hindustan plate under the Massif (Darvich-Zad, 1976). No other more or less cogent evidence has been offered for the explanation of the volcanic activity by subduction in the Badhiz, Gerirud, Farahrud, Arganda and other zones cited above.

Attention is called to the fact that comparison of volcanic action of any type with that of the 'island-arc' type in practice lacks cogency, owing to the diversity of the island arcs. Virtually any volcanism, with the exception of the undifferentiated basalt type, can be assigned to the 'island-arc' type. For example, the Neogene-Quaternary volcanics in the Carpathians, the Caucasus, Iran and Afghanistan are distinctly unrelated to island arcs but exhibit close similarity in their petrographical and petrochemical properties to the 'island-arc' types, and in this case ignorance of the geological conditions that existed during the volcanic activity can lead to the wrong conclusions.

CONCLUSIONS

During the Middle Palaeogene practically the entire territory of the Mediterranean belt in the study area experienced either the effects of tectonic tensile stresses or pronounced relaxation of the compressive stresses. Tension was of a maximum during the Middle Eocene, and therefore this interval spans the time of the most powerful volcanic eruptions, which differed in behaviour and composition of erupted products relevant to the specific structural conditions. At the end of the Late Cretaceous and during Oligocene times, when compressive events occurred, volcanism was either absent or of only weak local expression. A new phase of compression began at the end of the Eocene. The periodicity in the occurrence of compression and tension cannot, as yet, be explained. Probably, global processes or even the movements of individual plates played an appreciable role. Nevertheless, it is noteworthy that during the Palaeogene volcanic activity was manifested over vast territories of the Hindustan platform, South Arabia (Yemen), North Ethiopia and the adjacent northern part of the Indian Ocean. Incidents of Palaeogene volcanic activity are reflected even in the Tien Shan (Dobretsov and Zagruzina, 1977). Probably a causal relationship exists between these phenomena. At any rate, the highly active hot state of the upper mantle in large areas is an indisputable fact. Possibly all of this can be explained in terms of the general tectonic expansion, the origin of the northern part of the Indian Ocean, 'cracking' of the platform and the northward propagation of this process into the Mediterranean belt, where it underwent gradual extinction.

REFERENCES

Adamia, Sh. A., Gamkrelidze, I. P., Zakariadze, G. S. and Lordkipanidze, M. B. (1974). The Adzhara-Trialeti trough and the problem of the formation of the Black Sea abyssal basin, *Geotektonika*, no. 1. (Russian).

Adamia, Sh. A., Zakariadze, G. S. and Lordkipanidze, M. B. (1977). Evolution of the ancient active continental margin, with reference to the Alpine history of the Caucasus, *Geotektonika*, no. 4. (Russian).

Aslanyan, A. T. (1958). *Regional Geology of Armenia*, Erevan: Aipetrat (Russian).

Azizbekov, Sh. A. (1961). *Geology of Nakhichevan ASSR*, Moscow: Gosgeoltekhizdat (Russian).

Belov, A. A., Knipper, A. L. and Ruzhentsev, S. V. (1976). Historical-tectonic structural interpretation of the granite-gneiss massifs of the Alpine fold region, in *Tectonics of Median Massifs*, Moscow: Nauka (Russian).

Cawthorn, R. G. (1977). Petrological aspects of the correlation between potash content of orogenic magmas and earthquake depth, *Miner. Mag*., **41**, no. 318.

Darvishzadeh, A. and Ardzhangravesh, B. (1975). Geochemical study of volcanic rocks of Sahl-Abad region (South of Birjand, Iran), *Bull. Fac. Sci. Tehran. Univ*., **6**, no. 3/4.

Darvich-Zad, A. (1976). Les caractères du volcanisme tertiaire de l'est Iranian-bloc du Lut, *Cah. geol*., no. 92.

Dimitrijevič, M. D. (1973). *Geology of Kerman Region*, Beograd-Yugoslavia: Inst. Geol. and Mining, rept. N.Yu-52.

Dobretsov, G. L. and Zagruzina, I. A. (1977). Specific features of young basaltoid magmatism of the eastern Tien Shan, *Dokl. AN SSSR*, **235**, no. 3. (Russian).

Dzherbashyan, R. G. (1966). Petrochemical particularities and evolution of Eocene volcanism of the Bazum Ridge, in *Problems on Mineralogy and Petrography of Armenian SSR*, Erevan: Izd-vo AN ArmSSR (Russian).

Förster, H. (1968). Association of volcanic rocks in the mountains south of Sabzevar (NE Iran), Report of 23 Sess. Intern. Geol. Congr. Proc., Sect. 2: Volcanism and Tectogenesis, Prague.

Förster, H., Fesefeldt, F. and Kursten, M. (1972). Magmatic and orogenic evolution of the central Iranian volcanic belt, 24th Intern. Geol. Congr., Sect. 2: Petrology, Montreal.

Frolova, T. I., Burikova, I. A., Frolov, V. T. and Guschin, A. V. (1977). Specific features of basalt volcanism of the Lesser Kurile Arc, *Bull. Mosk. obshch. isp. prirody. geol*. otd., no. 4. (Russian).

Jung, A., Kürsten, M. and Tarkian, M. (1976). Post-Mesozoic volcanism in Iran and its relation to the subduction of the Afro-Arabian under Eurasian, Intern. Union. Commis. Sci. Stuttgart.

Geology of the Armenian SSR (1970). Erevan: Izd-vo AN SSSR, vd, IV: *Petrography, Volcanic Rocks*. (Russian).

Gerod, M. and Conrad, G. (1975). Les formations volcaniques récentes due sud de l'Iran (Kouh-e-Schahsavaran) données pétrologiques préliminaires; implications structurales, *Bull. Volcanol*. **39**, no. 4.

Kasmin, V. G. (1976). Geodynamics of the African-Arabian rift system, *Geotektonika*, no. 6.
Khain, V. E. (1975). The major tectono-magmatic developmental stages of the Caucasus: a geodynamic interpretation, *Geotektonika*, no. 1 (Russian).
Kochetkov, A. Ya., Demin, A. N., Stazhilo-Alexeev, K. F. and Chalyan, M. A. (1972). Igneous rock formations of Central Afghanistan, *Izv. AN SSSR*., ser. geol., no. 9 (Russian).
Leonov, M. G. (1976). Tectonic regime of the Epochs of formation of olistostromes, *Geotektonika*, no. 3. (Russian).
Lomize, M. G. (1972). Marginal ophiolite sutures of the Near and Middle East Alpides, *Izv. vuzov. geologiya i razvedka*, no. 2. (Russian).
Lordkipanidze, M. B. and Zakariadze, G. S. (1974). Palaeogene volcanism of Adjaria, in *Problems Bearing on the Geology of Adzhara-Trialeti*, Tbilisi: Meschniereba (Russian).
Nadareishvili, G. Sh. (1974). Eocene volcanism of southwestern Guria, in *Problems Bearing on the Geology of Adzhara-Trialeti*, Tbilisi: Meschniereba (Russian).
Stalder, P. (1977). Magmatismes tertiare et subrécent entre Taleghan et Alamout, Elbourz Central (Iran), *Schweiz. miner. und petrogr. Mitt.*, **51**.
Stöcklin, J., Eftekhar-Nezhad, G. and Hushman-Zadeh, A. (1972). Cental Lut reconnaissance, *Geol. Surv. Iran*, rept. no. 22.
Stöcklin, J. (1974). Northern Iran: Alborz Mountains, In *Mesozoic-Cenozoic Orogenic Belts*, Edinburgh: Data of Orogenic Studies.

Global Correlation of Tectonic Movements
Edited by Yu. G. Leonov and V. E. Khain

3

Tectonogravitational Mixtites and Types of Horizontal Displacements of the Earth's Crust

M. G. LEONOV*

Space-time correlation of various geological events such as tectonic displacements, magmatism, sedimentation, and so on, has in recent years become a most trenchant geotectonic approach. The correlation envisages studies in two directions: first, a study of interregional time correlation, or the identification of narrow time intervals marked by specific geological processes; and second, the identification of genetic correlations among various geological processes, as for instance, between tectonic displacements and sedimentation.

No special research has been attempted which includes both aspects of correlation. However, in recent years, geologists have stated studies dealing with a peculiar, although widely-occurring, group of chaotic formations known as olistostromes. As a result, their genetic links with some geological processes have been identified so as to establish correlations of these processes in space and time. This paper will discuss the genetic relationships between horizontal displacements of blocks in the earth's crust and the formation of olistostrome-type coarse fragmental complexes, as well as their geotectonic implications. However, prior to getting down to a study of such coarse fragmental complexes one has to find their place both among a whole series of similar formations and among sedimentary formations in general.

CHAOTIC ROCK COMPLEXES AND THEIR CLASSIFICATION

Many regions in the world are geologically composed of specific rock complexes, represented by chaotically arranged unsorted coarse fragmental material making up interbeds, lenses and thick layers, extending over dozens and hundreds of kilometres. Those geologists who became interested in these complexes as far back as the late nineteenth century are still studying them. Many authors described rock complexes with a chaotic structure and gave them different names—melange, tillite, tilloid conglomerate, wildflysch, megabreccia, olistostrome and others—and views on their origin differed. Special studies of chaotic complexes that are currently being carried out have resulted in their classification, identification of their types and the outlining of features inherent in the whole class of these complexes and their particular types. A comprehensive analysis of these problems has been undertaken by the author of this paper in a special monograph (Leonov, 1981) and therefore only basic information will be given here.

*Geological Institute of USSR Academy of Sciences, Moscow.

The very name of the rock association known as 'chaotic complexes' points to the characteristically chaotic nature of their inner structure. Their most essential structural features are as follows:

(a) heterogeneity—association with other types of deposits, the presence of two heterogeneous components, matrix and inclusion;
(b) chaotic inner structure—absence of sedimentary structures and textures, of lamination and stratification, as well as irregular distribution of inclusions and variable ratios of matrix and inclusions;
(c) the presence of considerable amounts of coarse fragmented materials of all sizes and shapes;
(d) a mixture of rocks of various genetic types including sedimentary, magmatic and metamorphic.

All these features lead to the identification of these chaotic complexes among other geological formations. These features also justify their treatment as a specific category and their name of 'mixtite' (a mixture), a term that has already been used by many geologists with respect to similar rock complexes (Schermerhorn, 1966; Ruzhentsev and Khovorova, 1973; Sokolov, 1977; Chumakov, 1978). This term may be defined, with only a slight deviation from the definition adopted by Schermerhorn and Chumakov, as: coarse fragmental rocks of any lithological-petrographic composition and genesis that consist of unsorted and unstratified lumps of any size and form irregularly scattered in finer-grained matrix.

Mixtites can be divided into various types (Table 1), each type possessing a set of inherent features. The basic principles involved in mixtite classification and identification of their characteristic features are comprehensively studied in Leonov (1981). This paper will deal mainly with problems related to tectonogravitational studies of mixtites.

It follows from Table 1 that two varieties are now distinguished in what was formerly considered a single type of olistostrome (Leonov, 1978, 1979, 1981). These are gravitational mixtites whose formation is attributed to a complex of exogeneous factors, and tectonogravitational mixtites, formed mainly as a result of tectonic processes. The formation of the latter is due to the destruction and disintegration of moving nappes and charriages. The generation of fragmental material takes place in different ways (Fig. 1):

(a) collapse of frontal escarpments of overthrusts or nappes as a result of tectonic disintegration of allochthonous rocks that is concurrent with submarine or subaerial weathering;
(b) tectonic crushing, imbrication, brecciation of the bottom parts of nappes while they move in a sedimentary basin, and at the interfaces of various nappes;
(c) tectonic crushing, imbrication, brecciation and formation of tectonic breccias in the parent allochthons with subsequent emergence or tectonic extrusions of breccias and sheets onto the ground surface or the sea-bed.

Subsequently, the fragmental material either remains directly at the base of a disintegrating nappe, retaining some features of its tectonic origin, or moves under gravitational forces into a basin where deposition of other types of sediments takes place. In the course of the sliding and burying of the coarse fragmental materials, rock masses acquire new features characteristic of sliding bodies, but still retain in their inner texture the original tectonic features. Finally, in many instances, tectonogravitational mixtures become overthrust by those same nappes whose destruction contributed to their formation and undergo a second tectonic transformation, turning again into a new tectonic mixtite.

Table 1 Classification of Chaotic Rock Complexes (Mixtites)

Fragmental formations	Fragmental material transport								
		Gravitational (β)							
		collapses talus		landslides		mudflows			
	tectonic (α)	sub-marine	over-ground	sub-marine	over-ground	sub-marine	over-ground	glacial (γ)	magmatic (δ)
tectonic (α)	α-mixtites (tectonic melange, tectonic breccia)		$\alpha\beta$-mixtites (tectonogravitational)						
exogenous (β)			β-mixtites (gravitational olistostromes)					$\beta\gamma$-mixtites (tillites, lateral moraines)	
glacial (γ)			$\alpha\beta$-mixtites (glacial, gravitational, allotillites)					γ-mixtites (glacial, tillites)	
magmatic (δ)			$\delta\beta$-mixtites (lahars)						δ-mixtites (magmatic, eruption breccias, etc.)

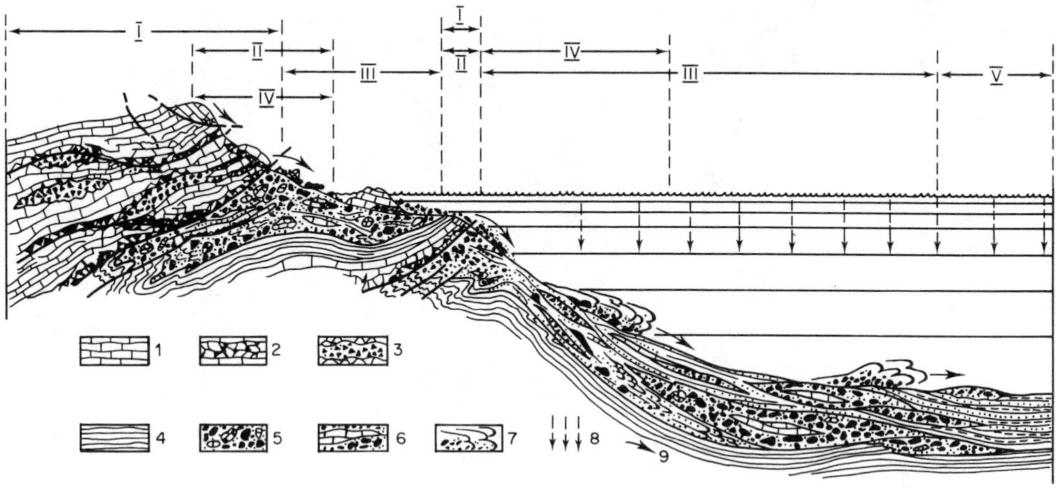

FIG. 1 Principal scheme of tectonogravitational mixtite formation.
1 – allochthonous rocks; 2 – zone of tectonic crushing; 3 – tectonic breccias; 4 – deposits underlying chaotic complexes; 5 – tectonogravitational and attendant gravitational mixtites; 6 – large sheets and consedimentation covers; 7 – slides and mudflows; 8 – pelagic sedimentation; 9 – direction of fragmental material dislocation.
I – zone of tectonic disintegration; II – zone of tectonic material outcrop on the earth's surface or the seabed; III – zone of deposition and burial of fragmental material; IV – zone of collapse-sliding processes development.

All the above leads to the conclusion that tectonogravitational mixtites are complex indicators of horizontal rock displacements.

Formation of gravitational mixtites (olistostromes proper) is related to collapse and sliding processes in conditions of gravitational instability of rock masses which can be attributed to various causes: accumulation of a considerable amount of sediments, an increase in submarine slope, growth of anticlines, and so on. Yet the main prerequisite here is the existence of a morphologically pronounced escarpment or a continuous slope, namely a difference in height between the area where a collapse or slide originates and the area where the disrupted materials are laid down.

The basic genetic difference between the gravitational and tectonogravitational mixtites, however, consists in that coarse fragmental materials of the latter are of tectonic origin (tectonic crushing to produce brecciation, imbrication, melange formation), while fragmental materials of the former are of collapse-sliding origin and tectonic processes only indirectly contribute to their formation. It is because of this that the gravitational mixtites (olistostromes proper), unlike tectonogravitational mixtites, reflect mainly vertical displacement of crustal blocks. Consequently, gravitational mixtites can be referred to a category of complex indicators of vertical displacements.

TECTONOGRAVITATIONAL AND GRAVITATIONAL MIXTITES OF THE ALPINE FOLD BELT

Tectonogravitational and gravitational mixtites are widespread in many regions. No exception in this respect is the Alpine fold belt with dozens of major deposits of mixtite masses

(Fig. 2). Chaotic complexes of coarse fragmental materials, earlier referred to as olistostromes, are now classified as tectonogravitational and gravitational mixtutes, but their place in the structure of individual regions for long remained obscure, while their areal distributions and their profiles seemed devoid of any regular pattern. The 2β- and β-mixtites of the Jurassic, Cretaceous, Palaeogene and Quaternary periods are developed in the Alps, the Caucasus, the Balkans, the Dinaric Alps, the Himalayas and in Turkey, Iran and Oman. They extend in the rocks of ophiolitic association, among flysch and molass, of geosynclinal troughs, in median massifs and platformal margins, but their extent is very variable. Thus, in the deposits of one age they may occur in great amounts and continue for many dozens and hundreds of kilometres, while in the deposits of another age their occurrence may be sporadic and they are represented by thin bodies of a limited extent. Systematic studies of 2β- and β-mixtites indicate that their position in space and time is governed by specific factors that are much less chaotic than their structure.

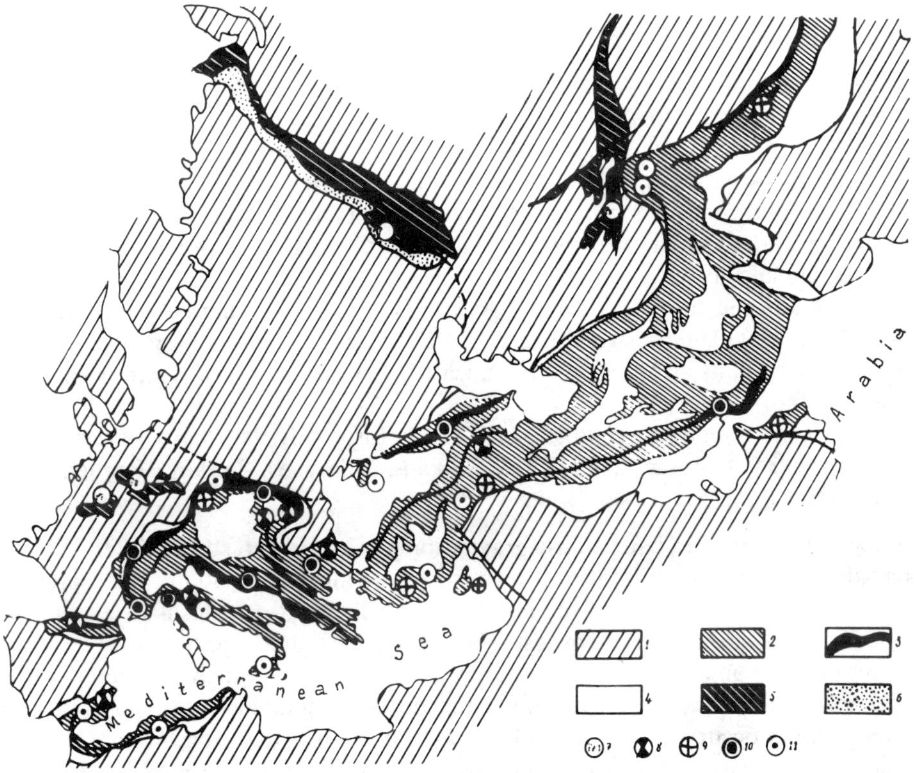

FIG. 2 Distribution of basic complexes of $\alpha\beta$- and β-mixtites in the Alpine-Himalayan fold belt and Palaeozoic fold areas (Ardennes, Harz, Urals, Tien Shan).

1 – epi-Baikal and pei-Palaeozoic platforms; 2 – Alpine-Himalayan fold belt; 3 – Alpine flysch depressions; 4 – Alpine foredeeps and inner depressions; 5 – Palaeozoic fold areas; 6 – foredeeps in the Ural fold area ($\alpha\beta$- and β-mixtites); 7 – Palaeozoic, 8 – Middle Cretaceous, 9 – Upper Cretaceous (Campanian–Maastrichtian); 10 – Upper Eocene–Early Oligocene; 11 – Miocene–Quaternary.

Therefore, the genesis of tectonogravitational mixtites from their origin to their final textural shaping is as follows:

Tectonic process	*Landslide process*	*Tectonic process*
Imbrication of parent rocks, tectonic crushing and brecciation, appearance of coarse fragmental materials	Displacement down the slope of a sedimentation basin, additional mixing, crushing and trapping of basinal sediments	Overthrusting of tectonogravitational mixtites and their secondary transformation into a tectonic mixture

The objective of this section is to identify the areal and vertical distribution of $\alpha\beta$- and β-mixtites and to unravel the regular time-space factors affecting their formation. The mixtite formations in the Alpine fold belt are described at length in Leonov (1981), which contains a comprehensive bibliography on the subject. Only general patterns governing their distribution in cross-sections and over regions will be given here.

Patterns of Distribution of Tectonogravitational and Gravitational Mixtites in Time

The following age levels in the development of tectonogravitational ($\alpha\beta$) and gravitational (β) mixtites identified are (Fig. 3):

(a) the Late Jurassic through Early Cretaceous (Appenines, Alps, Outer Carpathians, Caucasus, Dinaric Alps);
(b) the Middle Cretaceous, Albian-Cenomanian (Pyrenees, Apennines, Balkans, Romanian Carpathians, Provence, Lesser Caucasus);
(c) the Late Cretaceous, Campanian-Maastrichtian (Carpathians, Pyrenees, the southern flank of the Greater Caucasus, Daghestan, Cyprus, western and eastern Turkey, Oman, Himalayas?);
(d) the Late Eocene through Early Oligocene (Pyrenees, northern Apennines, French and Swiss Alps, Dinaric Alps, External Carpathians, Rhodope Massif, Kraischtides, the southern flank of the Greater Caucasus, Lesser Caucasus, western and eastern Turkey, Iran?);
(e) the Neogene–Quaternary substages (Cordilleras, Sicily, Northern Africa, southern and northern Apennines, Calabria, Balkans, Caucasus, Turkey, Pamirs, Crimea, Carpathians).

Tectonogravitational and particularly gravitational mixtites also accumulated in between the stages; but such chaotic formations were volumetrically small and not widespread. The duration of the epochs of an intensive generation of $\alpha\beta$- and β-mixtites is determined from the moment of appearance of the earliest mixtites of a certain age to the latest of the same age and this involves one or two and, more rarely, three or four stratigraphic stages. In other words, the formation of tectonogravitational mixtites was not an instantaneous, but rather a prolonged process, occurring in stages of about 5 to 20 m.y. duration. Within every such stage a horizon was established when tectonogravitational mixtites formed in all regions of their development. For the Late Eocene–Early Oligocene stage such a horizon lies at the boundary between the Late Eocene and Early Oligocene, and for the Middle Cretaceous stage at the boundary between the Albian and Cenomanian stages.

The accumulation of $\alpha\beta$- and β-mixtites may in time shift with respect to identified time-intervals and sometimes does not coincide with them. In some regions mixtites can be traced continuously from one time-interval to another. Examples can be found in the Lesser

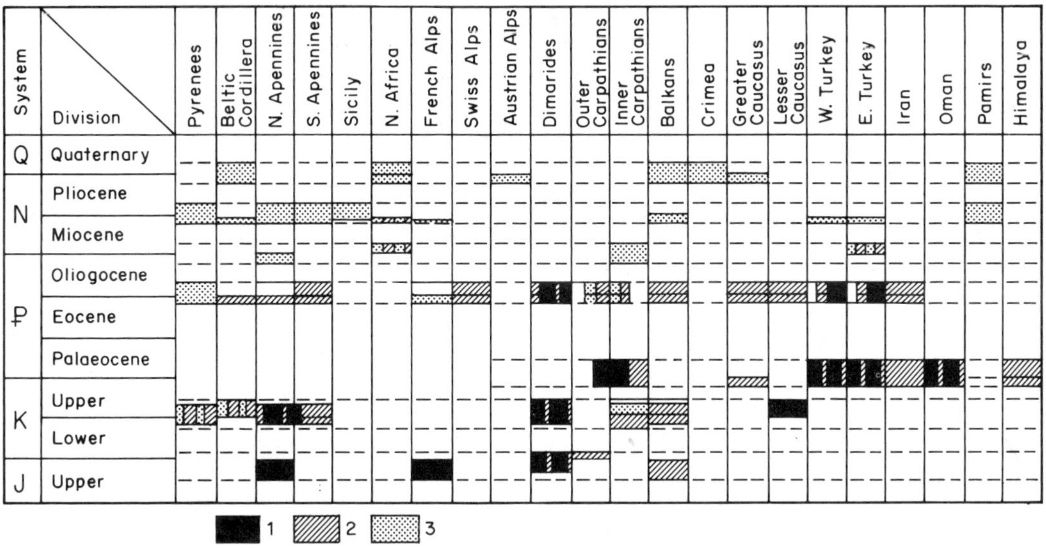

FIG. 3 Time-space distribution of tectonogravitational and attending gravitational mixtites in the Alpine fold area.
1 – ophiolitic and ophiolitic-fragmental; $\alpha\beta$- and β-mixtites; 2 – flysch-related $\alpha\beta$- and β-mixtites; 3 – molasse-related $\alpha\beta$- and β-mixtites.

Caucasus, where two stages were involved; the Pyrenees, where three stages were involved; and Daghestan, where the gravitational mixtite formation went on continuously from the Campanian–Maastrichtian to the Oligocene and Miocene. But the continuous occurrence of mixtites does not contradict the established stages, since in most cases gravitational mixtites, which reflect local features, form in between stages and the tectonogravitational mixtites are of limited extent.

Both tectonogravitational ($\alpha\beta$) and gravitational (β) mixtites are usually developed as unified complexes within one stage, while β-mixtites form individual bodies.

Patterns of Distribution of Tectonogravitational ($\alpha\beta$) and Gravitational (β) Mixtites in Space

Mixtites occur widely in various parts of the Alpine region. Tectonogravitational mixtites form belts some hundreds and, with intervals, even thousands of kilometres long. Gravitational mixtites either form individual bodies of insignificant length, or enter complexes with $\alpha\beta$-mixtites. Mixtites at various stages are usually confined to various zones of the Alpine region. The Late Jurassic-Early Cretaceous mixtites mainly occur in the inner parts like the Apennines and the Alps, while gravitational mixtites of this stage also occur in other regions such as the Caucasus and the Carpathians. Middle Cretaceous mixtites are developed still more widely, but are also associated with the inner regions of the Alpine belt (the Apennines, Internal Carpathians, Lesser Caucasus, Balkans). At this stage tectonogravitational mixtites prevail. Late Cretaceous $\alpha\beta$-mixtites are confined to more external parts of the belt (the Pennine klippe belt, the southern flank of the Greater Caucasus), but most of them form at the southern margins of the Tethys (western and eastern Turkey, Cyprus, Oman, possibly the Himalayas). Mixtites of $\alpha\beta$-type are also found in the northern branch of the Alpine area (Pyrenees, Cordillera, Caucasus). Of greatest extent are Late Eocene–Early Oligocene $\alpha\beta$-mixtites; they are linked with the outer, most often flysch, zones in the northern branch of the Alpine belt (French and Swiss Alps, Dinaric Alps, Apennines, Rhodope Massif, Kraischtides, the southern flank of the greater Caucasus). But they can also be found in more internal zones (the Bitlis, the Lesser Caucasus). Neogene–Quaternary tectonogravitational mixtites are spacially related to the most external molasse and flysch zones, especially at the southern margins of the Alpine belt (Calabrian-Sicilian arc, North Africa, Apennines, western and eastern Turkey), while in central and northern parts gravitational mixtites are mainly found. Therefore the principal pattern of their distribution in space is the shift of their zones of origin from the internal to the external parts of the Alpine fold belt. This is also characteristic of individual geosynclinal depressions. Thus, for example, in the Apennines, the Lycian Taurus, and some other regions the position where mixtites are found gradually shifted to the external parts of the geosyncline.

The Importance of Tectonogravitational and Gravitational Mixtites and their Relationships

$\alpha\beta$- and β-mixtites occur both in conjunction and individually. While gravitational mixtites often form individual bodies, tectonogravitational mixtites are more often associated with the gravitational since tectonic activity in an area where tectonogravitational mixtites are formed results in gravitational instability of sediments. To trace the patterns of distribution of identified varieties of mixtites is impossible in quantitative terms (the number of particular bodies), but the volume of the tectonically generated materials is much larger than that of slumping, although the volume of a particular body does not depend on its origin.

The relationships between gravitational and tectonogravitational mixtites change with time. Being poorly developed at the early stages of the geosynclinal development, the tectonogravitational mixtites reach their acme by the Middle Cretaceous–Oligocene, tending to lessen in extent in the Neogene–Quaternary period. The gravitational mixtites are of little importance in the early stages, then they gain strength in the Cretaceous–Eocene, but they are most widespread in the Neogene–Quaternary time. It is significant that in the period of molasse development $\alpha\beta$-mixtites are less developed than gravitational ones. This again indicates that the $\alpha\beta$-mixtite formation should not be associated only with sliding processes which are most pronounced at the molasse stage of mountain morphogenesis.

The formation of gravitational mixtites due to changes in local conditions (accelerated sedimentation, changes in bottom slope, greater number of individual uplifts and anticlinal structures, and seismic shocks), their small volume and length and the absence of strict time relation to stratigraphic periods prove that gravitational mixtites are the produce solely of regional conditions and are not connected with the development of the whole fold belt or with sizeable parts of it. Meanwhile the large volume and spread of $\alpha\beta$-mixtites confined to definite stratigraphic intervals and their extent in various tectomostratigraphic zones suggest that tectonogravitational mixtites reflect events of interregional significance.

Patterns of the Distribution of Tectonogravitational and Gravitational Mixtites

Tectonogravitational and related gravitational mixtites are associated (Leonov, Sokolov and Scherba, 1978) with several diagnostic groups of rocks such as ophiolites and flysch and molasse deposits. Most extensive are mixtites formed in the period of flysch sedimentation; mixtites associated with the ophiolites and molasse complexes are smaller in volume and extent, but are still significant. Tectonogravitational mixtites are not typical of other formations such as epiplatformal and miogeosynclinal deposits, and gravitational mixtites are encountered only sporadically. $\alpha\beta$-mixtites related to the above rock groups are not identical, their differences reflecting regular features of crustal development.

$\alpha\beta$-mixtites associated with ophiolites and found in many world regions (the Apennines, Turkey, the Caucasus, Oman, and so on), are (Sokolov, 1977) divided into two varieties—ophiolitic and ophiolitic-fragmentary.

The first, being of very early occurrence, is not widespread and is found in only a few places, such as the western Alps and the Apennines. This type is associated mainly with ophiolitic suites and is composed solely of fragments derived from rocks of ophiolitic association. Their inherent characteristics are occurrence at the base or in the lowest part of the Steinmann trinity, their insignificant volume and extent and the poor development of gravitational mixtites.

The second type is more widespread, being found in the Apennines, the Dinaric Alps, Turkey, the Lesser Caucasus, Oman, the Himalayas, and so on. Ophiolitic-fragmental mixtites are characterized by occurrence in various tectonostratigraphic zones; diversified detritus with the inevitable presence of fragments of ophiolites; spatial links with the ophiolitic association and particularly with ophiolitic melange; association with flysch-like deposits; great thickness and wide extent; association with gravitational mixtites, a younger age in comparison with the first type; and confinement to the inner zones of the Alpine belt.

Flysch-related $\alpha\beta$-mixtites are most widely developed and form the largest masses of mixtites. They are found in nearly all flysch troughs. The inherent features of these mixtites include the following: great extent and thickness; availability of diverse, often exotic, detritus; absence in most cases of fragment ophiolites; development at various stratigraphic levels, but mostly confined to Albian–Cenomanian, Campanian–Maastrichtian and Late Eocene–Early Oligocene periods; connection with the outer zones of the Alpine belt (especially the northern branch); widespread development of gravitational mixtites.

The molasse-related $\alpha\beta$-mixtites are found in the Pyrenees, the Alps, the Carpathians, the Apennines, North Africa, Sicily, Calabria, Turkey, the Caucasus, the Pamirs, and so on. Their specific features are usually the absence of detrital rocks of ophiolitic association, confinement to foredeeps (mainly in the southern branch of the Alpine belt) and Oligocene–Pliocene deposits.

An analysis of spatial, temporal and formational patterns of distribution of tectonogravitational mixtites over the Alpine belt indicates that all of them are closely interrelated. The early tectonogravitational mixtites are associated with the internal zones of the Alpine belt and associated with the ophiolitic suite. The later mixtites (Upper Cretaceous, Late Eocene–Early Oligocene) are confined to association with flysch deposits and are more often found in the flysch downwarps (Late Cretaceous in the southern and Late Eocene in the northern). The Neogene–Quarternary $\alpha\beta$- and β-mixtites are confined to the outermost zones of the Alpine belt and are associated with molasse deposits. In other words, the time-space interrelationships are not incidental, but reflect some paragenetic unity observed in a definite developmental stage of different parts of a fold belt.

However, in various regions tectonogravitational mixtites can have different compositions and be associated with different formations within one stage (Fig. 4). Thus, the Late Eocene–Early Oligocene $\alpha\beta$-mixtites occur towards the end of the flysch sedimentation stage on the southern flank of the Greater Caucasus and in the Ultrahelvetian zone of the Alps. In the Dinaric Alps and in Iranian Baluchistan they are included in the flysch deposits; the flysch stage continues through the Oligocene and, possibly, the Miocene. In the Subalpine chains in France mixtites form the lower horizons of the Alpine molasse and are found in the marginal zone of the platform, while in Bulgaria they originate in the zone of the 'stable' Rhodope Massif where no sedimentation took place for a long time. The same regular pattern may be traced back to other stages. Thus tectonogravitational mixtites of a single stage are produced in various tectonostratigraphic zones and in different phases of their development. The presence of tectonogravitational mixtites of differing compositions and associations with various formations of a similar age over vast areas of the fold belt confirms the idea of a unified process encompassing great territories and overlapping regional processes.

Time-space Regular Patterns of Formation of Nappes and Charriages

What is the significance of the identification of tectonogravitational mixtite formation stages and the establishment of spatial and formational patterns of their distribution? Tectonogravitational mixtites may serve as indicators of horizontal rock displacements and the stages

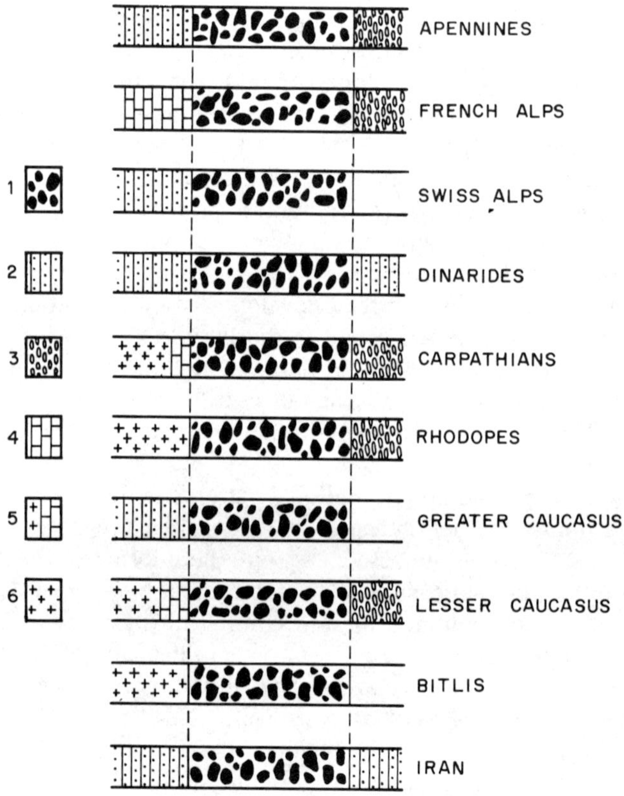

FIG. 4 Location of Upper-Eocene tectonogravitational mixtites in a profile in various regions of the Alpine belt.

1 – $\alpha\beta$-mixtites; 2 – flysch; 3 – molasse; 4 – epiplatform and miogeosynclinal facii; 5 – inner crystalline massifs with a sedimentary cover; 6 – granite-metamorphic foundation of inner massifs.

The figure shows that $\alpha\beta$-mixtites occur in various structural-facial zones of the Alpine fold belt and are formed in the Late Eocene irrespective of the previous development of this region.

identified are in effect stages of active horizontal displacements of crustal blocks that often result in the formation of nappes and charriages. Thus, the distribution of $\alpha\beta$-mixtites and their relations with definite formational groups reflect the specific manifestations of these displacements and their significance in the geological history of the Alpine fold belt.

Judging from the distribution of $\alpha\beta$-mixtites in time, the horizontal displacements of crustal blocks started in the Late Jurassic and continued up to the present. However, these displacements in space and time were uneven. The strongest compression was observed in short intervals (from 5 to 20 m.y.), separated by the relatively 'tranquil' epochs of horizontal displacement. Five stages, corresponding in general to Stille's orogenic phases, are identified as follows: Late Jurassic–Early Cretaceous (Late Cimmerian), Albian-Cenomanian (Austrian phase), Campanian–Maastrichtian (Laramide phase) and Neogene–Quaternary (corresponding to numerous minor phases of this period). Yet a study of tectonogravitational mixtites helped the author of this paper to unravel the significance of these phases (horizontal compression), to determine the length of individual pulses of horizontal displacements and to undertake the time-space correlation of not only tectonic movements in general, but also of dislocations related to the generation of nappes and charriages.

Horizontal rock displacements were observed in various areas of the Alpine belt and in different time periods. They were associated with various stages of geosynclinal development (Fig. 5).

The first displacements (the Late Jurassic–Early Cretaceous) occurred in the innermost zones and it is the ophiolitic suite that was mainly involved in the horizontal disruption. The second stage of major horizontal displacements, when systems of nappes were formed, took place in the Middle Cretaceous Aptian-Cenomanian). Similar to the first stage, these movements developed over the internal parts of the Alpine belt and in individual geosynclines. The composition of $\alpha\beta$-mixtites of this time and their relationships with the nearby formations revealed that areas with a mafic basement (oceanic crust) became much contracted and melanges were squeezed out among sialic blocks and thrust over them. Crustal parts with granite-gneiss basement were also involved in the movements and formed nappes. This is reflected in the composition of chaotic formations. In general,

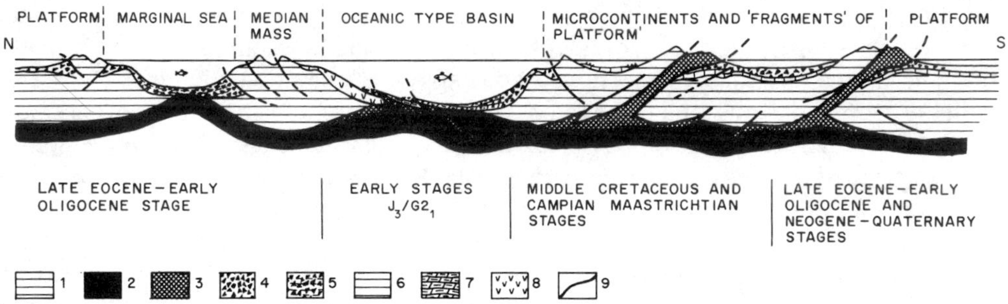

FIG. 5 Principal cross-section of the Alpine fold belt and location of tectonogravitational mixtites (stage-to-stage migration of $\alpha\beta$-mixtite formation zone is seen).

1 – earths crust with a granite-metamorphic foundation (platforms, median masses); 2 – basic and ultrabasic rocks; 3 – melange rocks of the ophiolitic suite; 4 – tectonogravitational and attending gravitational mixtites; 5 – flysch; 6 – shallow-water deposits of a sedimentary cover; 7 – deep-water deposits of a sedimentary cover in oceanic-type basins; 8 – volcanogenic rocks; 9 – overthrusts.

mixtites of Middle Cretaceous age indicate overthrusting (allochthonous covering) of basins with ocean-type crust and a stage of 'mature' ophiolitic melange.

The distribution of tectonogravitational mixtites in the subsequent stages (Campanian–Maastrichtian, Late Eocene–Early Oligocene) indicates that horizontal displacements occur in the more external (flysch) zones of the Alpine belt, including the Swiss Alps, the Carpathians, the Dinaric Alps, the southern flank of the Greater Caucasus, and so on. By that time areas with oceanic crust had actually ceased to exist, and this is confirmed by the absence (or rare occurrence) of ophiolitic rocks in the mixtites and the abundance of rocks derived from granite-metamorphic basement and its sedimentary cover. The dynamic process of disruption extended over more external zones, and involved mainly sialic crustal blocks.

Horizontal displacements, that resulted in the formation of nappes in the Neogene–Quaternary period, were manifest solely in the external molasse depressions found mostly in the southern fringes of the Alpine belt (North Africa, Calabrian-Sicilian arc, Turkey). In other regions molasse sediments contain mainly bodies of gravitational β mixtites, the wide occurrence of which points to the existence of a growing relief of emerging land and the existence of considerable gradients.

Therefore it is clear that horizontal displacements are discrete and their general trend is the shift (from stage to stage) of the horizontal compression from the internal zones with a basaltic basement to the external zones with a sialic basement.

As for β-mixtites, they were developing within the same stages and not only in the localities with the $\alpha\beta$-mixtites, but in many other regions as well. This proves that tectonic displacements may involve greater territories than those where nappes appear, and that tectonic displacements during a single stage result in several forms such as overthrusts and nappes in some regions, and uplift depression in others.

Horizontal displacements display yet another pattern. Beginning from the Middle Cretaceous and through the Early Oligocene $\alpha\beta$-mixtites (within one time interval) were generated in regions with varying geological histories and at various stages of development, such as both within flysch depressions, median masses and platforms margins. Compression extended over areas composed of ophiolitic rocks (oceanic crust) as well as massifs with granite-metamorphic basement. In other words, horizontal rock displacements affected different areas irrespective of their previous history and their former mobility.

Similar behaviour of crustal blocks of different structure and separated from each other by hundreds and even thousands of kilometres may obviously be attributed to a common process which is superposed on processes of narrowly regional significance.

Inasmuch as the formation of overthrusts and nappes in oceanic basements started earlier than in the continental crust and extended over blocks with granite-metamorphic basements only after the ocean-floor rocks were converted into tectonic melanges, the displacements in the lower parts of the granite-metamorphic crustal layer may be considered as the reason for the general deformation. This was established for some regions of the Alpine belt by Knipper (1978).

The same holds for the explanation of the spatial redistribution of regions where the greatest rock deformation is observed in one time interval or another. The first deformations will be recorded in the rocks composing the oceanic crust and occupying the central part of the Alpine geoscycline. With time sialic blocks become involved in the movement. At this time, however, the whole region is still heterogeneous and highly mobile, thus the formation of nappes occurs heterochronously, which is observed in the Middle Cretaceous. After the Albian–Cenomanian period the mobility of the central part of the Alpine belt drastically diminished, the areas of oceanic crust disappeared, sialic blocks were overthrust and formed a single sutured unit. The movement of such a unified and consolidated block took the form of tectonic overthrusts and nappes involving the destruction of $\alpha\beta$-mixtites in the marginal, still mobile, zones. The synchronous movement of various parts of the central blocks resulted in a simultaneous formation of mixtites throughout the whole stratigraphic level. At the molasse stage, when much of the alpine belt acquired stability, horizontal displacements, concentrated in the most marginal zones (bordering the platforms), which were the 'weakest'. I believe that this can explain the shift of zones of horizontal compression which takes place in the deeper crustal layers.

None of the above means that the formation of nappes and tectonogravitational mixtites is in any way related to previous development of the fold belt. The distribution of tectonogravitational mixtites demonstrates that the horizontal rock dislocations, contributing to the formation of nappes, are confined to a definite stage in the development of a fold belt, being transitional from a widespread development of basins with oceanic crust to the formation and consolidation of the granite-metamorphic layer. In the early stages no tectonogravitational mixtites are formed; they are also poorly developed during the orogenic stage.

TECTONOGRAVITATIONAL MIXTITES AND THE NATURE OF CRUSTAL HORIZONTAL DISPLACEMENTS

It is obvious that the fragmented materials of tectonogravitational mixtites are formed as a result of crushing, imbrication and melange formation of allochthonous rocks that were formed in the course of horizontal displacements of crustal blocks. It would be wrong, however, to think that these processes always proceeded in the same way. The data obtained on the geology of tectonogravitational mixtites show that the generation of chaotic layers went on under varying palaeotectonic and palaeogeographical conditions, and consequently the resultant manifestations were different. This paper will not dwell on the origin and formation mechanism of nappes, leaving that to special studies. Multiple publications have been devoted to this theme, including extensive books (Ruzhentsev, 1971; Belostotskiy, 1978). But studies of tectonogravitational mixtites give additional and very important information in this respect.

The analysis of available information indicates that there may be at least four types of tectonogravitational mixtites, differing in their horizontal displacements (Fig. 6). Let us examine every type more closely, using a particular region as an example.

Lycian-type (Fig. 6A)

The most typical examples of mixtites of this type are tectonogravitational mixtites confined to the Anatolian system of nappes, Lycian nappes and nappes formed by inner crystalline massifs (Bitlis). In all these regions outlining the southern border of the Alpine fold belt in Turkey, and the Aegean Islands tectonogravitational mixtites are similar in their position and appearance. Many geologists conducted their research here (de Granciansky, 1967, 1968, 1973; Bernoulli *et al*., 1974; Radelli, 1971).

When describing these regions in a general way, one can distinguish three groups of rocks: autochthon, intermediate complex and allocthon. Autochthonous formations from the Cenomanian to the Burdigalian inclusively are represented by platform-type carbonate deposits. The intermediate complex is an intricate package of tectonic imbrications and sheets, composed of various rocks from the Carboniferous up to and including the Upper Cretaceous. The intermediate complex and, partially, the autochthonous deposits are

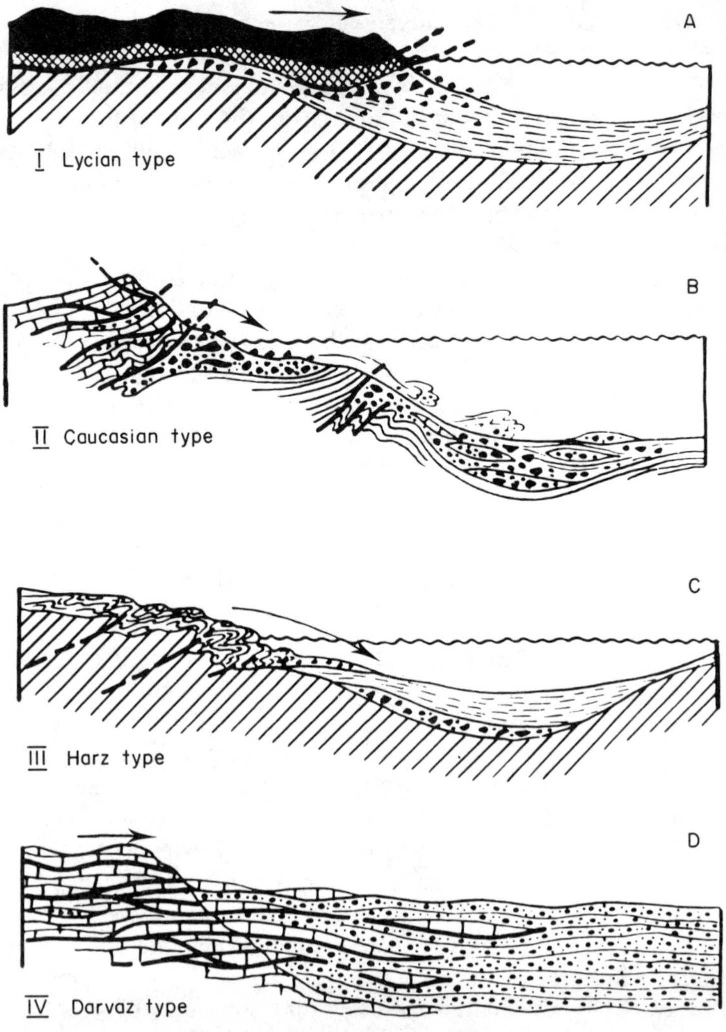

FIG. 6 Four ways of tectonogravitational mixtite formation in relation to forms of horizontal displacements (see details in the text).

overlain by a tectonic blanket of peridotites. Let us consider in more detail the structure of the intermediate complex and its relationship with the underlying and overlying formations.

The intermediate complex is represented by three series of rocks. The lower part is composed of terrigenous and carbonate marine deposits from the Carboniferous to the Cenomanian inclusive. It is overlain by the Cenomanian–Maastrichtian flysch, the upper part of which is represented by a chaotic complex, which I specifically referred to the tectonogravitational mixtites (Leonov, 1981). The authors, de Graciansky in particular, who studied these deposits describe this complex as a 'wildflysch'. Tectonogravitational mixtites are represented by the so-called Karabertlen Formation consisting of marls and limestones or argillites and sandstones with large fragments and plates of diabases, radiolarites, limestones and sandstones. Blocks are chaotically distributed, they are unrounded and the coarse-fragmented materials are unsorted, there is no internal stratification or lamination. Still upwards the 'wildflysch' horizon is replaced by an enormous breccia (300 m thick), composed of serpentine or 'tuffaceous' mafic material with random fragments and big plates of radiolarites, basalts, dolerites, ultramafic and metamorphic rocks. This breccia and underlying tectonogravitational mixtites are intensely tectonized. The transition between these two complexes is gradual and, as de Graciansky refers to it, is 'progressive'. It manifests in the changing appearance of the matrix, which is sandy-argillaceous below but acquires serpentine features up the section. However, the composition of blocks in either rock type is identical. The autochthons and intermediate complexes are overlain by the so-called peridotite cover and in some places even by an underlying cover of diabase. The rocks of these covers are found in the underlying breccias and in the 'wildflysch' horizons as fragments and blocks.

The lower breccia is referred to a category of tectonogravitational mixtites, namely formations which resulted from tectonic disintegration of nappes. This is confirmed by time association between chaotic formations and tectonic sheets as well as their nearly simultaneous origin. Breccias and overlying nappes extend over the faunally dated Cenomanian–Turonian deposits and are covered by Maastrichtian deposits. The cementing materials of the breccias include globotrumcates. Their tectonogravitational origin is also confirmed by the presence of rock lumps and blocks of the nappes overlying the breccias, as well as the upward transition of these nappes into a melange. The upper breccia, judging from the description and views of those authors who studied it, belongs to the category of chaotic complexes. Thus the following upward sequence is observed: autochthonous flysch sediments; tectonogravitational mixtites; melange (tectonic mixtite); covers of diabases and peridotites; the neoautochthon. However, one should mention two more features in the structure and relationship of the melange and the tectonogravitational mixtites which allow the recognition of an interesting pattern in their formation. First, the absence of any pronounced tectonic contact at the interface of the melange and the tectonogravitational mixtites. Second, the masses of tectonogravitational mixtites form sheets which may be likened to nappes of great extent, in that, neither along the trend nor across it are there any substantial facial changes, the size and composition of fragments being nearly the same everywhere. These two facts indicate, as already mentioned, that the formation of tectonogravitational mixtites is intimately connected with the tectonic disintegration of nappes, which is accompanied by the destruction of the nappe base and the formation of the tectonic mixture or melange. Later on the coarse fragmental material is moved over a short distance and is deposited directly at the nappe front as coarse breccias within the flysch. The 'progressive' contact between tectonogravitational mixtites and tectonic mixtites (melange) as well as similar mixtite composition over vast areas point at a continuous thrusting of nappes over their own debris. The contact between melanges and tectonogravitational mixtites is essentially of a tectonic nature, inasmuch as their origin is associated with the overthrusting of nappes, and the older rocks overlie the younger ones. But it manifests not as a discontinuity, but as a zone where fragmentary breccias with a sedimentary matrix are interbedded with breccias with a tectonic matrix. This process is only possible in the case of a rapid movement of the nappe, leaving no time for compensation by the destruction of its frontal parts (Leonov, 1972, 1981). It is also acceptable in view of the fact that the ophiolitic suite composing the nappes readily turns into a melange and forms a layer 'transitional' from the more or less monolithic allochthonous massif to the rocks underlying fragmental tectonogravitational breccias.

The Lycian-type mixtites are characterized by: the association with ophiolitic rocks and melange; spread of tectonogravitational mixtites over vast areas in the form of covers and the absence of inner facial variability across the trend; a gradual transition from tectonogravitational to tectonic mixtites (tectonic breccias and melanges). Such mixtites are usually associated with the early and middle stages in the development of geosynclinal areas and with areas of ophiolites or, in other words, with areas where the oceanic crust moved to the upper parts of the crust as a result of intense tectonic displacements. This type includes Campanian–Maastrichtian mixtites of the Lycian Taurus and Oman, possibly the Lesser Caucasus, the Albian–Cenomanian mixtites of Apusenides, the Apennines and some other regions.

Caucasian-type (Fig. 6B)

Tectonogravitational mixtites of this variety are usually generated at the margins of long-developed flysch depressions by means of a tectonic disintegration of adjacent uplifts (median massifs, microcontinents, and so on), compensating for the horizontal displacement by disintegration at the frontal escarpments. This type includes the Upper Eocene–Early Oligocene chaotic complexes of the southern flank of the Greater Caucasus, the Swiss Alps, Tithonian mixtites of Bulgaria, and so on. Let us briefly examine the tectonogravitational mixtites of the southern flank of the Greater Caucasus.

In the Caucasus, tectonogravitational mixtites extend along a narrow border between the flysch zone of the southern flank of the Greater Caucasus and the Transcaucasian median massif. The strip of outcrops of chaotic formations runs for over 200 km in the general Caucasian trend and has a maximum width of 5 to 7 km. In the north it is margined by an overthrust surface along which it is covered by the Cretaceous–Palaeogene flysch. In the south the horizons enclosing tectonogravitational and related gravitational mixtites are thrust over the molasse zone of the Transcaucasian median massif. Blocky layers contained in flysch deposits are confined to the upper flysch horizons in the southern flank. Chaotic rocks are of a heterogeneous composition. In the enclosing deposits, consisting of fine sandy clay and rarely of carbonate rocks typical of flysch and with varying stratification, there are inclusions of chaotic coarse fragmental materials. The inclusions can be found as lenses, layers and horizons of lumpy breccias, conglomeratic-breccias and associated conglomerates with rock fragments older than the matrix enclosing them. These ancient rocks occur in the form of individual randomly scattered fragments, mounds and sheets. Breccias are characterized by an abundance of poorly sorted and rounded debris. The size of the fragments varies from some centimetres to hundreds of metres and several kilometres in length. The thickness of lumpy breccias may be as much as 100–200 m, and the length, dozens and hundreds of kilometres. Rocks are intensively dislocated. The deposits described belong to a zone where flysch deposits accumulated from the Late Jurassic up to and including Eocene. It is stratigrafically related to the underlying deposits and belongs to the flysch formation of the southern flank. The age of the chaotic complex is Late Eocene–Early Oligocene.

The relations observed between fragmental breccias and enclosing deposits and the general palaeogeography of the basin shows that the mixtite formation can be attributed to the intensive disintegration of the source zone, collapse and sliding of coarse material off a tectonic, morphologically-pronounced, escarpment (Leonov, 1975, 1978).

The presence of tectonogravitational mixtites—complex indicators of horizontal rock displacements—is indicative (Leonov, 1981) of the fact that the Transcaucasian median massif was subject to horizontal dislocations in Late Eocene–Early Oligocene times. The inner structure of the massif (Leonov, 1970, 1974) as well as some structural features of the flysch display horizontal contraction, confirming this inference. There is no significant synsedimentary folding, while angular unconformities at the interfaces of deposit complexes of different ages, and in particular at the base of the mixtite complex, quickly disappear across the strike of the flysch zone, up to 15 km from the edge of the depression. Nevertheless in the Late Eocene strata there are no traces of tectonic overthrusting of the fragmental rocks or the flysch. Thus, structural data on the massif and the flysch as well as on the presence of tectonogravitational mixtites, which accumulated at the front of the escarpment by tectonic disintegration but which were not covered by overthrusts, permit the assumption that there was an equilibrium between the velocities of overthrusting and the destruction in the frontal part of the Transcaucasian massif. The mass movement within a narrow zone was compensated by the crushing of rocks and their redeposition as gigantic piles of fragmental materials in a flysch trough. The massif was not naturally a monolith, but consisted of tectonic imbricate slices and breccias (Leonov, 1974). Here the individual overthrusts slid together with the fragmental deposits into the flysch basin, acquiring the form of synsedimentational fragmental deposits into the flysch basin, acquiring the form of synsedimentational sheets.

The Caucasian-type tectonogravitational mixtites are characterized by:

(a) considerable masses of enclosing flysch-type deposits;
(b) two types of rock debris: basement rocks (granitoids, metamorphic schists) and debris of the sedimentary cover of platformal type (mainly limestones);
(c) considerable masses of gravitational mixtites;
(d) changing facies across the basinal trend, manifested by the decreased volume and size of coarse materials and gradual wedging out of coarse rocks.

Tectonogravitational mixtites of this type are mostly developed in the Alpine fold belt, but also occur in other fold systems (the Urals, the Tien Shan). They are mainly associated with the flysch stage of the geosynclinal development.

Harz-type (Fig. 6C)

The geology of the Harz Mountains has been studied by many research workers, particularly by Reichstein (1965, 1967, 1970), Lutzens (1959, 1973), Schwan (1974), Schwab (1976, 1977). Located in Central Europe, they comprise a range of mountains of moderate elevation. The Harz sediments were formed in the northern external part of the Variscan orogen referred to as the Rhenohercynian zone. This zone is bounded to the south by the Middle German crystalline swell and to the north by the Subvariscan zone. The sedimentary deposits under consideration are found in the eastern part of the southern edge of the Rhine depression belonging to externides of the Variscan orogen. Fragmental chaotic accumulations (mixtites) which are widespread here are up to 1200 m thick. They extend over dozens and hundreds of square kilometres. Among these chaotic formations, so far classified as olistostromes, several types were discovered which differ in types of debris, structure and composition, as well as the direction of sediment transport, and so on. One of them is the so-called 'Harzgerode olistostrome' occurring over an area of 400–500 sq. km, with a maximum thickness of 1200 m. The 'Harzgerode olistostrome' represents a complex of rocks with a chaotic structure and is referred to tectonogravitational mixtites (Leonov, 1981). It covers autochtonous deposits ranging from the Silurian to the Early Carboniferous, and is overlain by nappes of the Devonian age.

The complex has a mafic matrix and inclusions. The mafic matrix is composed mainly of pelites ranging from dark grey to black colour with admixtures of silt and sand distributed unevenly as small, well-defined lenses imparting a characteristic schlieren aspect to the whole mass. This lack of structure is one of the characteristic features of a matrix which is devoid of the slightest indication of primary sedimentary structures such as lamination, stratification or sorting. No fold structures can be seen in the matrix either, although tectonic processes have affected it. The main mass is composed of individual lense-type bodies with glide surfaces. These lenses have a scaly structure which is produced by multiple foliation surfaces.

The inclusions are represented by randomly-arranged fragments, lumps, blocks and gigantic sheets of different ages enclosed in a pelitic matrix. Sheets may cover an area of several dozens of square kilometres. They either form as accumulations of lumpy material, or as isolated blocks 'floating' in the matrix. Large sheets are divided into separate imbricate slices that are thrust one over another. The apparent continuity of the section sometimes conceals mixed rocks of various ages, while the more ancient horizons are either found higher up than the younger ones or are included among them. A feature inherent in the Harzgerode chaotic complex is the presence of faunal remnants of the same age in the matrix and as inclusions, which proves that mixtites are formed from the rocks of an originally single succession.

The origin of this type of chaotic complex may be reconstructed by relying upon the data on the structure of the Herzgerode mixtite complex and its relationship with nearby formations which are to be found not only in the works of German authors, but also in Leonov (1981). But in Carboniferous times flysch deposits accumulated within the Rhine. At the time of the most intensive tectonic movements caused by tangenal compression and, possibly, by asymmetric uplift of the Middle German swell, the borders of the rigid basement rocks and plastic Silurian–Devonian cover deposits were stripped and the latter became gravitational nappes, moving to the north through the Wippra zone and Lower Harz uplift in the direction of the Rhine flysch trough. While moving, the sedimentary rocks lost their uniformity the movement resulting in the formation of tectonic imbrications, the mixing up of individual elements of the formerly intact section, the formation of slates and disruption and partial loosening. Having reached the edge of the Rhine trough, the nappes composed of previously deformed rocks started moving into the flysch basin, but this time directed only by gravitational forces. Since rocks were sliding in water, their high thixotropy and continuing mechanical disintegration caused mudflows and even slurries. These processes (sliding, disintegration, dispersion) led to a complete loss of cohesion in friable rocks and the appearance of blocks, lumps and fragments in more compact rocks. But much of the rock still retains its cohesion and is found among the disintegrated mass in the form of gigantic plates. In moving, nappes caused the erosion of the Lower Harz uplift, fragments of which were entrapped and included in the chaotic complex. The disintegrated, mixed-up materials were deposited in a new basin as tectonogravitational mixtites.

The main features of tectonogravitational mixtites of the Harz type are:

(a) the dominant sandy-argillaceous or carbonate-argillaceous rock composition;
(b) the predominance of matrix over inclusions or their approximate parity;
(c) few or no interstratified normal sediments enclosing the bodies of mixtites;
(d) a total absence of sedimentary and fold structures and textures;
(e) separation of the matrix into pelitic and sand-silt components: silty-sandy sediment is distributed pelites in small schlierens;
(f) faunal remnants of the same age in the matrix and inclusions;
(g) the imbricate pelitic matrix.

One should also mention that such of tectonogravitational mixtites possess the characteristics of a nappe since both the matrix and the inclusions are older than the underlying strata and the mixtites are formed as a result of the transformation of a single rock-type which prior to disintegration made up the nappe. The mixtites also resemble melanges with the matrix and inclusions being mixed-up parts of a formerly unified succession disrupted by tectonic imbrication, movement and brecciation, and finally they appear as slumps. In other words, the Harz-type tectonogravitational mixtites reflect most vividly the unity and interrelation of tectonic and sliding processes.

The Harz-type mixtites were formed mainly in the flysch and molasse stages of geosynclinal development. An indispensable factor in their formation is believed to be not only a definite type of tectonic movement, but also the presence of considerable masses of plastic sandy-argillaceous or argillaceous-carbonate rocks conducive to the formation of nappes. This type may possibly include the Tertiary mixtites of Sicily, described by Marchetti (1957), who interpreted the origin of chaotic layers in this region in the same way, and 'imbricate clays' in the Apennines.

Darvaz-type (Fig. 6D)

Mixtites of this type have been most thoroughly studied in the Pamirs (Scherba, 1975; Lukyanov, Leonov and Scherba, 1975) and in the Crimea.

In the contact zone between the Neogene molasse and the Palaeozoic limestones in the northern Pamirs (the Darvaz range) crudely bedded, unsorted breccias are extensively developed in Neogene rocks. They either terminate against Palaeozoic limestones or overlie them discordantly. Palaeozoic rocks form a steep escarpment made up of several flatly arranged overthrust imbrications separated by tectonic breccias. The heavy brecciation accounts for nearly a half of the total volume of the rocks. Brecciation is manifest in two ways: as calcite veins cross-cutting the bedding and surrounding unmoved fragments from a single rock mass, and as limestone breccias with fragments which have rotated and shifted in relation to each other. Brecciated limestones and limestone breccias at the interface of plates merge into each other, which proves their tectonic origin. At the interface there are also gradual transitions of tectonic breccias that divide imbricate slices of the Palaeozoic basement into unsorted sedimentary breccias. Overthrust imbrications originating in the Palaeozoic rocks, like the breccias, gradually pass into the Neogene coarse fragmented horizon, becoming gigantic plates that extend over many kilometres. The gradual passing of tectonic breccias into sedimentary, and of tectonic plates into synsedimentary cover rocks, unambiguously points to the tectonic origin of mixtites in this region. Later on coarse fragmental materials move down the slope, though only for a few kilometres, by collapse-sliding processes.

Darvaz-type tectonogravitational mixtites are associated with the horizontal scaly foliation (imbrication) of the tectonic escarpment separating the areas of upward movements from the basin of sedimentation at the foot of the escarpment. Such an escarpment may result from a vertical fault or an overthrust. The necessary conditions for the formation of mixtites, however, are the vertical component of motion, on the one hand, and the subhorizontal tectonic fissuring of the escarpment on the other. The formation of coarse fragmental materials may be attributed to tectonic crushing and the squeezing out of breccias and large plates from the inner parts of the parent massif.

Mixtites of this type are characterized by:

(a) the presence of carbonate fragments;
(b) the availability of a really enormous (many square kilometres) but relatively thin (dozens, rarely, hundreds of metres) plates;
(c) the presence of thin breccia interbeds at the base of the plates;

(d) generally insignificant volumes of coarse fragmental materials in comparison to large plates and the separating foliated deposits;
(e) the absence of, or poorly developed, gravitational mixtites.

Tectonogravitational mixtites of this type occur mostly among molasse, and primarily coarse fragmental deposits. This type may also include chaotic complexes of the Crimea (Massandra suite), the Carpathians (Buchedji conglomerates) and the French Alps. The Darvaz-type mixtites may be also found as a component of the Caucasian-type mixtites.

Therefore it seems that notwithstanding the fact that fragmentary materials of tectonogravitational mixtites are formed as a result of a common tectonic rock crushing during horizontal displacements, tectonic disintegration acquires specific features depending on the conditions of horizontal displacements. In one case, tectonic imbrication and brecciation of large massifs with a granite-metamorphic basement and a sedimentary cover of a platform-type (Caucasian type) occur, in others, sedimentary rocks migrate from one basin to another, involving the crushing and mixing up of components which previously formed parts of an intact section (Harz-type), and so on. At the same time, each type of tectonogravitational mixtite is characterized by a set of features which may help to distinguish specific manifestations of horizontal displacements. Moreover, a definite pattern may be observed in a time sequence of various types of tectonogravitational mixtites and, consequently, in a sequence of horizontal displacements. Thus the Lycian-type tectonogravitational mixtites are most typical of the early and medium stages in development of geosynclincal areas. Their formation is associated with deformations in a basaltic basement, melange formation and its forced obduction onto the upper parts of the crust. The Caucasian-type is confined to a flysch stage, when the disruption extends over sialic blocks (median massifs, microcontinents). Flysch and molasse period is characterized also by the Harz-type, which may appear only after sufficiently thick layers of sandy-argillaceous and argillaceous-carbonate rocks have formed and a differential movement of crustal blocks with horizontal and vertical components has taken place. Finally, the Darvaz-type tectonogravitational mixtites are associated with vertical components of motion, with the formation of mountain massifs and molasse depressions, and with the tectonic 'extrusion' of breccias and plates from the escarpment formed by overthrusts or faults.

Such sequences may obviously be changed. For instance, the Darvaz-type mixtites may form earlier than the molasse, say, in a complex with the Caucasian-type, and the Caucasian-type mixtites may be generated at the early stages of the geosynclinal development, and so on. But, in general, the sequence observed reflects the evolution in horizontal displacements. The studies of tectonogravitational mixtites clearly show that horizontal displacements at various stages of the geosynclinal development manifest themselves in various forms, which depend on the 'maturity' of the earth's crust. This should not be neglected in studies of older fold belts.

To sum up it may be said that the study of chaotic complexes (tectonogravitational mixtites) helps to reconstruct the manifestations of horizontal displacements in the earth's crust and their evolution in time.

CONCLUSION

The present study was aimed at the identification of complex indicators of horizontal rock displacements and, consequently, the patterns in space and time observed in the formation of nappes and charriages, which is one of the pressing challenges of modern geotectonics. To summarize, there are five conclusions.

First, among the multitude of geological formations rock complexes were selected that are composed of gigantic chaotic piles of unsorted fragmental materials, all of which (melange, tillites, olistostromes, etc.) are classified in the present paper as mixtites. Specific features inherent in this group are defined.

Second, mixtites are subdivided into genetic types. Two types of process contribute to formation of any mixtite: those resulting in the clastic material formation and those resulting in its dislocation. Consequently, four monogenetic and four heterogenic types are identified. Among them new categories of chaotic complexes are distinguished—tectonogravitational ($\alpha\beta$) and gravitational (β) mixtites. Characteristics of these complexes are given.

Third, tectonogravitational mixtites with fragmental material resulting from tectonic disintegration of nappes and charriages and those with movement taking place in the form of collapse or slide emerge as a manifestation of a single overall sequence of events and the interrelation of tectonic and collapse-sliding processes. Tectonogravitational mixtites are complex indicators of horizontal displacements of rocks.

Fourth, the formation of gravitational mixtite is associated with collapse-sliding processes. The necessary factor is the adequate height difference between the original site of collapse or the slide and the depositional area of disintegrated material. It is this fact that allows mixtites to be considered as the complex indicators of vertical rock displacements.

Finally, identification of complex indicators of horizontal rock displacements helps to reveal important patterns of dislocations related to the nappe and charriage formation in the Alpine fold belt. The basic patterns are as follows:

(1) Horizontal dislocations of crustal blocks do not always result in the formation of clear-cut structures such as overthrusts or nappes. In many cases they cause only rock disintegration and crushing, and contribute to the formation of thick layers of tectonogravitational mixtites.

(2) Horizontal displacements occur at a definite stage of a geosynclinal development, being transitional from an extensive development of basins with a basaltic basement to the consolidation of the granite-metamorphic layer.

(3) Within a transitional stage (Late Jurassic–Neogene) horizontal displacements are discrete. Five time-limited (5–10 m.y.) stages of intensive tectonic movements (leading in many cases to the nappe and charriage formation) alternated with periods of relative rest from horizontal displacements.

(4) A general trend in horizontal displacements is expressed in shift with time of horizontal zones of compression from the internal areas with basaltic basement to the external areas with sialic basement.

(5) The development of nappes and charriages contributed to the shaping of tectonic relief.

(6) Tectonic movements at one and the same stage may manifest in different ways in various regions: in some areas nappes, while in others uplifts and depressions, are formed.

REFERENCES

Belostotskiy, I. I. (1978). *Structure and Formation of Tectonic Nappes*, Moscow: Nedra, pp. 1–195 (Russian).

Bernoulli, D., Graciansky, P.-C. de and Monod, O. (1974). The extention of Lycian nappes (SW Turkey), *Eclog. geo. helv.*, **67**, no. 2, 39–90.

Chumakov, N. M. (1978). Precambrian tillites and tilloids, *Trudy GIN AN SSSR*, issue 308 (Russian).

Graciansky, P.-C. de (1967). Existence d'une nappe ophiolitique à l'extrémité occidentale de la chaîne sud-anatolienne; relations avec les terrains autochtones (province de Mugla, Turquie), *C.R. Acad. Sci.*, **264 D**, 2876–2879.

Graciansky, P.-C. de (1968). Stratigraphie des unités superposées dans l'arc dinarotaurique, *Bull. Miner. Res. and Explor. Inst. Turkey*, no. 71.

Graciansky, P.-C. de (1973). Le problème des coloured melanges à propos de formations chaotiques associées aux ophiolites de Licie occiden tale (Turquie), *Rev. géogr. phys. et. géol. dyn.*, **15**, fasc.

Knipper, A. L. (1978). Ophicalcites and some other types of breccia accompanying pre-orogenic formation of an ophiolite complex, *Geotektonika*, no. 2, 50–60 (Russian).

Leonov, M. G. (1970). On the destruction of the frontal part of nappes, *Doklady AN SSSR*, **193**, no. 3, 669–71 (Russian).

Leonov, M. G. (1972). Flysch as a formation of the submarine slope, *Litologia i poleznye iskopaemye*, no. 2, 44–57 (Russian).

Leonov, M. G. (1974). Fault dislocations of the Dzirula massif. *Bull. Mosk. obschch. isp. prirody*, geol. sect., no. 6, 37–44 (Russian).

Leonov, M. G. (1975). Wildflysch of the Alpine area, *Trudy GIN AN SSSR*, issue 199, Moscow: Nauka, pp. 1–136 (Russian).

Leonov, M. G. (1978). Olistostromes and their genesis, *Geotektonika*, no. 3, 18–33 (Russian).

Leonov, M. G. (1979). Gravity and tectonogravity mixtites in the structure of fold belts, summary of Doctoral theses, Moscow, pp. 1–52 (Russian).

Leonov, M. G. (1981). Olistostromes in the structure of fold belts, *Trudy GIN AN SSSR*, issue 344, Moscow: Nauka, pp. 1–185 (Russian).

Leonov, M. G., Sokolov, S. D. and Scherba, I. G. (1978). Olistostromes of the Alpine fold belt, in *Tectonics of the Mediterranean Belt*, Theses of report, 32–33, Moscow (Russian).

Lukianov, A. V., Leonov, M. G. and Scherba, I. G. (1975). Olistostrome formation and the problem of pseudotillites, *Litologia i poleznye iskopaemye*, no. 4, 40–9 (Russian).

Lutzens, H. (1959). Die stratigrafische und tektonische Stellung der Wernigeröder Schicten nach Conodonten, *Geologie*, **8**, part 1, 71–92.

Lutzens, H. (1973). Zur Alterstellung der Oslistostrome und Gleitdecken im Hărz unter besonderer Berųcksichting der Initialmagmatite, *Z. geol. wiss*, no. 1, 137–144.

Marchetti, M. (1957). The occurrence of slide and flowage materials (olistostromes) in the Tertiary series of Sicily, in *Congr. Geol. Inter. XX sess*. Mexico, 209–225.

Radelli; L. (1971). Sur la tectonique de la chaîne anatolienne de Bitlis, in *Tectonique de l'Afrique*, Paris, vol. 6.

Reichstein, M. (1965). Motive und Probleme erneuter Deckenbanforstellungen für der Harz, *Geologie*, **14**, part 9, 1039–1076.

Reichstein, M. (1967). Die neuen Deckenbanvorstellungen im Harz (nach den neuesten Stand), Berlin: Vortz. deutsch. Akad. Wiss.

Reichstein, M. (1970). Deckenbaufragen im Harz und spätvaristische Strukturentwicklung, *Wiss. Zst. Univ. Halle*, no. I.

Ruzhentsev, S. V. (1971). Peculiarities of structure and mechanism of formation of thrust nappes, *Trudy GIN AN SSSR*, issue 223, Moscow: Nauka (Russian).

Ruzhentsev, S. V. and Khvorova, I. V. (1973). Mid-Palaeozoic olistostromes in the Sakmarian zone S. Urals, *Litologiya i poleznye iskopaemye*, no. 7, 21–2 (Russian).

Scherba, I. G. (1975). Olistostromes in the Neogene of the Darvaz Range, *Geotektonika*, no. 5, 97–108, Moscow (Russian).

Schermerhorn, L. J. G. (1966). Terminology of mixed coarse-fine sediments, *J. Petrol. Techn.*, no. 9.

Schwab, M. (1976). Beiträge zur Tektonik der Rhenohersynischen Zone im Gebiet der DDR mit besonderer Berücksichtigung der Verhältnisse im Unterharz, *Jahrb. Geol.*, vol. 5/6, 9–117.

Schwab, M. (1977). Structure and geological development of Harz in *Tectonics of the Variscides of Middle Europe and the USSR*, Moscow: Nauka.

Schwan, W. (1974). Flysch, Olistostrome und Gleitdecken im Harz, *Z. Dtsch. geol. Ges.*, **125**, part. 2, 253–267.

Sokolov, S. D. (1977). Olistostrome rocks and Late Cretaceous nappes in ophiolite zones of the Lesser Caucasus, *Trudy GIN AN SSSR*, issue 296 (Russian).

Global Correlation of Tectonic Movements
Edited by Yu. G. Leonov and V. E. Khain

4

Stages and Phases of Formation of Cenozoic Olistostromes in the Alpine Fold Belt

I. G. SCHERBA*

A number of recent works both by Soviet (Belostotskiy, 1970; Leonov, 1975, 1981; Sokolov, 1977; Scherba, 1975) and other (Elter and Trevisan, 1973; Caire, 1973; de Graciansky *et al.*, 1970) authors have showed that olistostromes are generally the result of active tectonic movements. Their formation is mainly related to nappe movements and, partly, to vertical breakdown of massifs of allochtonous materials. Proceeding from this, the present author will attempt to use olistostromes as a means of studying the long-standing geological problem of whether tectonic movements over vast territories are episodic or continuous, simultaneous or time-transgressive. With respect to the pre-orogenic stage in the development of the Alpine belt (Cretaceous–Eocene), Leonov (1975, and this book) showed that the movements which produce olistostromes take place episodically and synchronously. The author sets himself the task of tracing the history of the formation of olistostrome complexes in the Eocene–Quaternary time, that is, when the Alpine fold belt became a mountain system. A study of the distribution of olistostromes in Cenozoic sequences provided a convincing demonstration that Cenozoic movements were episodic and, what is especially important, that the formation of olistostromes over the territory of the Alpine fold belt took place within the same time intervals in a region from the Iberian Peninsula the west to the Pamirs and the Taurus mountains in the east, from the Alps and the Carpathians in the north to the Rif mountains of Morocco and the Tell Atlas in the south. This approach showed that it is quite possible to trace the phases of tectogenesis over the Alpine region established by Stille for the Western European branch of the Alpides.

Since the classification, genesis and morphology of olistostromes are disputed, the author considers it important to present his own approach to these chaotic formations. Olistostromes occur either as chaotic sedimentary breccias or as large isolated masses which slid into the sedimentary basin forming among its rocks strata of different thicknesses. Fragmentary materials observed in these strata were in some cases produced by denudation processes (collapse, downslope movement, landslides, and so on). These are true olistostromes, that is, landslide layers. In other cases, fragmentary materials were converted into sediments, as a result of tectonic disintegration of bedrock massifs. It is these massifs, referred to by Leonov (1981) as tectonogravitational mixtites, that are related to the activation of tectonic movements. This paper attempts to draw a difference between these

* Geological Institute, USSR Academy of Sciences, Moscow.

two categories of olistostromes and to establish the nature of their relationship with tectonic events taking place at the same time. A characteristic feature of the analysis is that in individual segments of the Alpine belt distinguished by a common structural pattern (Fig. 1), a study was made of the entire Cenzoic succession. In doing so, the author used the published data on the Apennines, the Dinarids, the Alps, the Pyrenees, North Africa and Turkey and added his personal observations in the Carpathians, the Crimea, the Caucasus and the Pamirs. The author has established that all these deposits include olistostromes and exhibit traces of sedimentary breaks, unconformities and other indications of tectonic activity, as well as of transgression or regression of sedimentary formations. This has allowed him to demonstrate a common rhythmic character of sedimentation. The typical sections were then compared to one another. Their age was correlated by using the scheme for comparison of standard scales of the Tethys and the Paratethys adopted at the meeting of the Neogene Commission held in Bratislava in 1977 (Table 1). As a consequence, six age levels within the Cenozoic era of the Alpine region have been identified and traced over all the territory under consideration where olistostromes appear to be most frequent and conspicuous. The six levels reflect regional rather than local formations of olistostromes. The author is aware that the boundaries of these levels are rather arbitrary and their ages rather inaccurate, which is due both to inadequate faunistic characteristics available for the sections and also to inadequate elaboration of Cenozoic stratigraphy. However, even assuming a stochastic distribution of olistostromes over the Alpine region, one can easily see a common rhythmic character in alternating sequences with and without olistostromes.

The limited scope of this article does not permit a uniform description of the 27 regions of the Alpine belt which were studied by the author; it is confined instead to examples of the regions with different styles of Alpine tectonics (the northern Apennines, the Rif of Morocco, the Carpathians, the Caucasus and the Pamirs). The author shows that it is possible to draw correlations among them and then to correlate them to other regions not described here.

The fold structures of the El-Rif (Figs 1, 2, Section 4), which were studied by Choubert and Faure-Muret (1971), and other French geologists, fall into three major tectonic zones separated by thrusts: an internal geanticlinal zone, a zone of the flysch-bearing ultra-Rifian troughs, and zones of an external trough of meso-Rif and pre-Rif.

Precambrian ultramafic metamorphic rocks as well as non-metamorphosed strata of the Palaeozoic and Mesozoic ages outcrop in the internal zone. They form a complicated fold structure known as the Sebtides composed during many phases, including the Late Miocene, a system of Palaeozoic nappes (Ghomarides) and a discontinuous belt of isolated uplifts of the Calcareous chains. These chains are composed of Triassic, Jurassic, Cretaceous and Palaeocene formations of platformal type deformed during the Early and Late Cretaceous, and at the end of the Eocene. Above the Palaeocene there occurs, with a slight unconformity, a sandy-argillaceous and occasionally conglomeratic complex with pebbles of Palaeozoic rocks attributed to the Upper Eocene–Oligocene. At the southern margin of the Calcareous range, within the lower (Eocene) portion of the section of this complex, one can find the slid masses of the Mesozoic rocks (Choubert and Faure-Muret, 1971). Their origin can be probably ascribed to destruction of the elevations and frontal portions of the nappes which moved in the Late Eocene. More common are Oligocene to Lower Miocene rocks. They overlie not only the Calcareous chains but also the Ghomaride nappes. It is suggested (Suter) that they formed after the overthrusting of these nappes and simultaneous movements in the Calcareous chains. The Oligocene and Lower Miocene rocks occur as facies of the Numidian Flysch, which is a very widespread formation reflecting a stage of great Oligocene subsidence in the Mediterranean area.

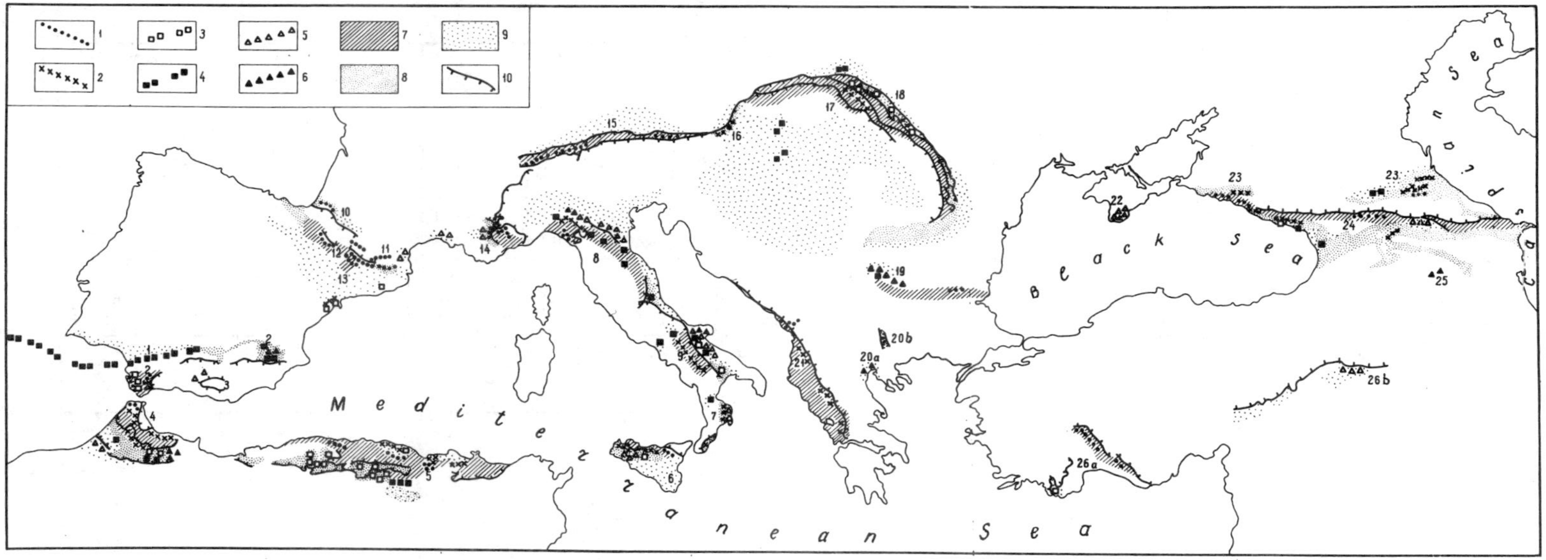

FIG. 1 Diagram showing Late Cenozoic olistostromes within the Alpine fold belt.

Olistostromes of: 1 – the end of Eocene; 2 – the end of Oligocene; 3 – the end of the Lower Miocene; 4 – the end of Middle Miocene to the beginning of Upper Miocene; 5 – the end of Upper Miocene to the beginning of Pliocene; 6 – the beginning of Pleistocene; types of rock: 7 – flysch; 8 – lower molasse; 9 – upper molasse; 10 – faults.

Figures relate to the numbers of sections in Fig. 2.

Table 1 Present Day State of Knowledge About the Correlation of Regional Stages of the Central and Eastern Paratethys with Standard Mediterranean Stages (adapted from materials of Project no. 25 'Stratigraphic correlation Tethys-Paratethys Neogene', 1977)

M.Y.	EPOCH		SUPER STAGES	STANDARD NEOGENE STAGES OF THE MEDITERRANEAN	BLOW 1969	MARTINI 1971	REGIONAL STAGES CENTRAL PARATETHYS	REGIONAL STAGES EASTERN PARATETHYS
	PLIOCENE		Rossellian	PIACENZIAN	N 19–21	NN 13–21	ROMANIAN	AKCHAGYLIAN
				ZANCLIAN			DACIAN	KIMMERIAN (CIMMERIAN)
5	MIOCENE	Late	Castellanian	MESSINIAN	N 18	NN 12	PONTIAN	PONTIAN
					N 17	NN 11	PANNONIAN	MEOTIAN
				TORTONIAN	N 16			KHERSONIAN (Sarmatian (Barbot))
10						NN 10		BESSARABIAN (Sarmatian (Barbot))
					N 15	NN 9	SARMATIAN	VOLHYNIAN (Sarmatian (Barbot))
		Middle	Cessolian	SERRAVALLIAN	N 14	NN 8		
					N 11–13	NN 7	BADENIAN	KONKIAN
						NN 6		KARAGANIAN
15				LANGHIAN	N 9–10	NN 5		CHOKRAKIAN
								TARCHANIAN
		Early	Girondian	BURDIGALIAN	N 8		KARPATIAN (CARPATHIAN)	?
					N 7	NN 4		
					N 6	NN 3	OTTNANGIAN	KOTZACHURIAN
20				AQUITANIAN	N 5	NN 2	EGGENBURGIAN	SAKARAULIAN
					N 4	NN 1	EGERIAN	CAUCASIAN
25	OLIGOCENE		↓	CHATTIAN ↓	P 21–23	NP25 / NP24	↓ ↓	

Oligocene sediments together with the underlying strata compose large nappes which are overlain by the Upper Oligocene–Lower Miocene Fnideq Formation which contains at its base the conglomerates from the rocks of the internal zone (Choubert and Faure-Muret, 1971). The frontal part of these nappes in the southern margin of the internal zone exhibits, among the flysch-like rocks of the Upper? Oligocene, bodies of olistostromes composed of large fragments of various rocks of the Rif internal zone (Choubert and Faure-Muret, 1971). In turn, a sequence containing Late Oligocene olistostromes is a part of the allochthonous complex in which it was involved during the Lower Miocene movements playing the major part in formation of the nappe structure in the internal zone. At the end of the Middle Miocene the previously formed allochthon in the inner zone was completely thrust over the ultra-Rif zone. As a result of intensive gravity sliding, the Middle Miocene strata with orbulinids were also transported into the same zone. Thereafter the Mesozoic and Palaeozoic complexes that were then exposed were overlain unconformably by Messinian deposits. At the end of Messinian times the early movements disturbed the contact zone between the Calcareous chains and the zone of flysch nappes. The unconforming Lower Pliocene sediments partially overlapped the Ghomarid nappes and were deformed at the end of the Pliocene.

In the ultra-Rif zone, all the Mesozoic and Tertiary rocks, which are mainly flysch, occur as large allochthonous massifs produced during Miocene times. The Oligocene deposits represented by a sequence several kilometres thick of Numidian Flysch are found in the Sugna and Beni-Inder nappes. The Beni-Inder nappe is largely composed of rocks of the upper part of the Numidian (the Upper Oligocene–Aquitanian), represented by micaceous and arkosic sandstones, grey marls with horizons of conglomerate consisting of rounded boulders of Ghomarid rocks, and Numidian Flysch which occur as a regressive sequence, resembling a molasse. The sandstones of the eastern portion of the nappe contain klippes of rose marls of the Upper Eocene in the Calcareous chains as well as fragments of the Cretaceous and Oligocene rocks in the ultra-Rif zone (Didon, 1977). The Oligocene olistostromes, again according to Didon, can also be found in the Sugna nappe among Numidian alternating sandstones and argillites. The olistostromes are represented by a chaotic mixture of sheared argillites enclosing blocks of Jurassic limestones (Celui-en complex) in unconsolidated sediments. All such olistostromes were formed in the pre-dorsal part in the ultra-Rif zone and were caused by Late Oligocene movements of the nappes in the El-Rif inner zone. At the end of the Middle Miocene they were tectonically transported into the inner parts of the ultra-Rif zone and the region of the marginal downwarp.

This marginal downwarp, which is well covered by flysh nappes, is 150 km wide in its inner part. This inner part, above the Triassic evaporites and dolerites forming the Palaeozoic platform cover, contains a deep-water, continuous succession of carbonate and clastic deposits (Jurassic–Eocene) which become shallower towards the outer margin. From the beginning of the Middle Eocene onward the axis of the downwarp shifted towards the outer margin. The succession of the inner portion of the downwarp is more clastic and from the Middle Miocene on it became both shallower and coarser. Conversely, the outer portion of the downwarp, called pre-Rif from the Middle Eocene on, subsided, and up to the Late Miocene pelagic marl sequences accumulate there. Today, all these sequences compose tectonic gravity complexes that were formed during several deformation phases, at the end of Oligocene–Late Miocene times.

In the inner zone of the downwarp, except for the meso-Rif region, continuous sedimentation took place from the Lias to the Early Miocene. In the meso-Rif region, which forms an anticlinal zone stretching along the outer margin of the inner zone of the downwarp, the

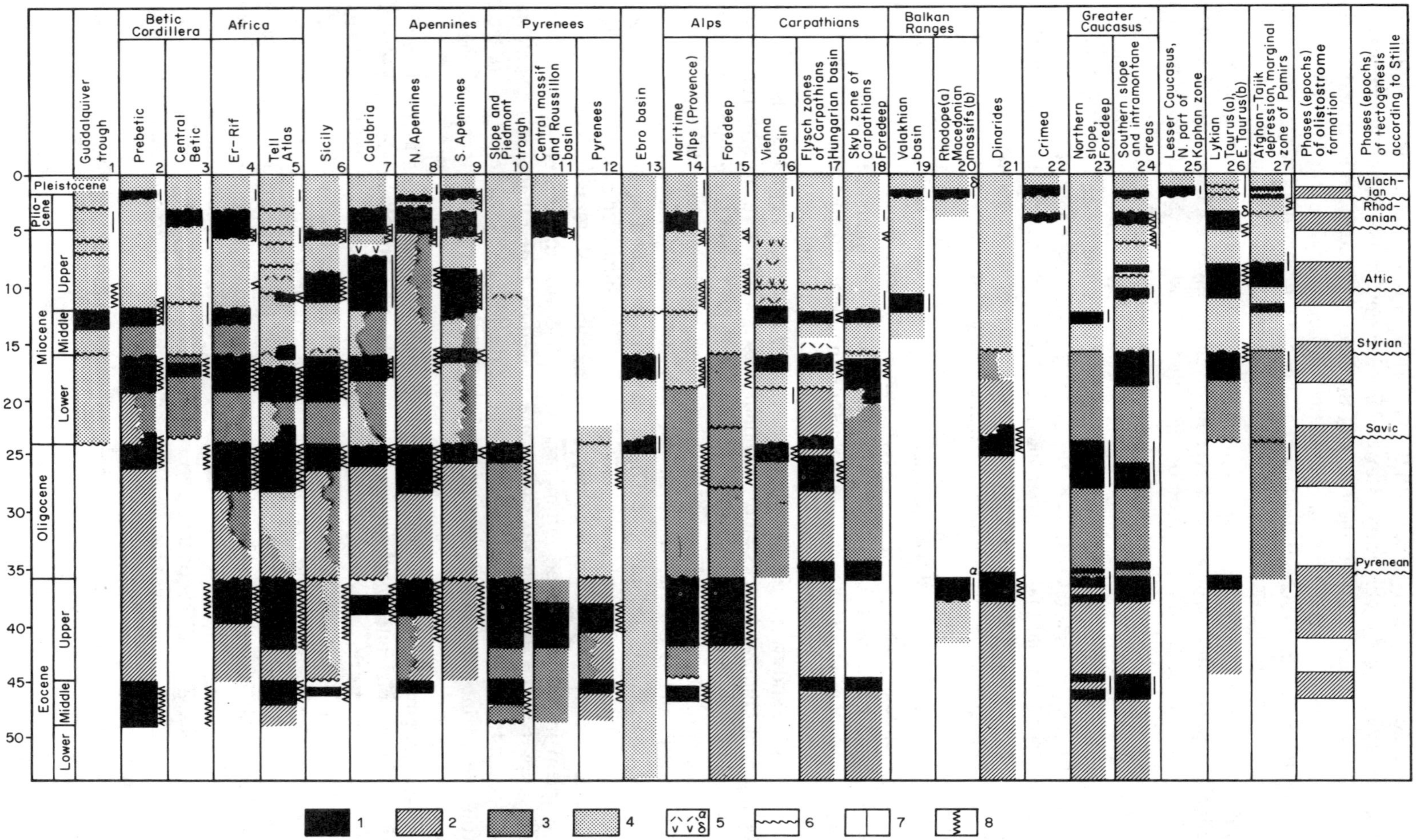
Guadalquiver trough 1
Betic Cordillera
Prebetic 2
Central Betic 3
Africa
Er-Rif 4
Tell Atlas 5
Sicily 6
Calabria 7
Apennines
N. Apennines 8
S. Apennines 9
Pyrenees
Slope and Piedmont trough 10
Central massif and Roussillon basin 11
Pyrenees 12
Ebro basin 13
Alps
Maritime Alps (Provence) 14
Foredeep 15
Carpathians
Vienna basin 16
Flysch zones of Carpathians Hungarian basin 17
Skyb zone of Carpathians Foredeep 18
Balkan Ranges
Valakhian basin 19
Rhodope(a) Macedonian massifs(b) 20
Dinarides 21
Crimea 22
Greater Caucasus
Northern slope, Foredeep 23
Southern slope and intramontane areas 24
Lesser Caucasus, N. part of Kaphan zone 25
Lykian Taurus(a), E. Taurus(b) 26
Afghan-Tajik depression, marginal zone of Pamirs 27
Phases (epochs) of olistostrome formation
Phases (epochs) of tectogenesis according to Stille
Pleistocene
Plio-cene
Miocene
Upper
Middle
Lower
Oligocene
Eocene
0
5
10
15
20
25
30
35
40
45
50
Valach-ian
Rhod-anian
Attic
Styrian
Savic
Pyrenean
1
2
3
4
5
6
7
8

FIG. 2 Position of olistostromes in the Late Cenozoic sections of the Alpine fold belt.

1 – olistostromes (the whole time interval of olistostrome occurrence in this region is hatched) ; 2 – flysch; 3 – lower molasse (grey clays, more seldom red clays, silts, sandstones with subordinate, occasionally predominant marls, and conglomerates often gypsum- and salt-bearing with pronounced large rhythmic patterns and mechanoglifs underlain by calcareous and flysch formations and overlain by upper molasse); 4 – upper molasse (conglomerates, sandstones, silts and clays which are often red, terrestrial, coal-bearing, with large rhythmic patterns, obliquely laminated); 5 – volcanism ((a) acid, (b) basic); 6 – unconformities in section; 7, 8 – tectonic environments: 7 – general uplift; 8 – general compression (formation of nappes, crumpling).

Numbers: 1 – Guadalquivir trough; 2–3 – Betic Cordillera: 2 – pre-Betic; 3 – Central Betic, 4–5 – North Africa: 4 – El-Rif; 5 – Tell Atlas; 6 – Sicily; 7 – Calabria; 8–9 – Apennines: 8 – North Apennines; 9 – South Apennines; 10–11 – Pyrenees: 10 – the northern slope and Piedmont trough; 11 – central massif and Raussilion basin; 12 – the southern Pyrenees; 13 – Ebro basin; 14–15 – Alps: 14 – Maritime Alps (Provence); 15 – pre-Alps; 15 – foredeep; 16–18 – Carpathians: 16 – Vienna basin; 17 – flysch zones of the eastern Carpathians, Hungarian basin; 18 – Skib zone of the eastern Carpathians, foredeep; 18–20 – Balkan Ranges: 19 – Valakhian basin; 20 – Rhodope massif (a), Macedonian massif (b); 21 – the Dinarides; 22 – the Crimea; 23–24 – Greater Caucasus: 23 – the northern slope, foredeep; 24 – the southern slope and intermontane areas; 25 – Lesser Caucasus, the northern part of the Kaphan zone; 26 – the Lykian Taurus (a), the Eastern Taurus (b); 27 – Afghan-Tadzhik basin, the marginal zone of the Pamirs.

first movements of the nappes, recognized by the appearance of olistostromes, occurred as early as at the end of the Oligocene. The end-Oligocene olistostromes were found (Lespinasse, 1977) in the front of the Loukkos nappe. The d'el Ouhahda white marls, of ages assumed to come between the Upper Oligocene and Lower Miocene, contain large slid blocks of Senonian marls which, according to the author, herald the first movements of the Loukkos nappe. Here, too, another rock sequence with olistostromes can be discerned, indicating subsequent movements of the nappe. Indeed, these formations with Upper Oligocene olistostromes are transgressively overlain by the Zoumi alternating sandstones and sandy marls, with an Aquitanian fauna, that cover the nappe. The Zoumi unit is overlain unconformably by the Dar el Qued sandy argillaceous rocks with a Middle Burdigalian fauna, laden with olistostromes whose fragments contain not only rocks of the Loukkos nappe, but also those of a more internal Tanger nappe (flysch, Upper Cretaceous and Eocene limestones, Triassic gypsum). These olistostromes owe their origin to the strong end-Early Miocene tectonic movements that produced an unconformity at the base of the complex with olistostromes.

The Early Miocene phase is the most important for the internal zone of the marginal downwarp. The Lower Miocene sediments, as well as Triassic and Cretaceous rocks, were imbricated and a complex tectono-sedimentary mixture of Lower Miocene age was produced.

The Lower Miocene strata are foliated, although this is not in general typical of the rocks of this age. At the end of the Middle Miocene the internal zone of the downwarp underwent a general uplift, causing a gravitational detachment of the Senhadja and Ouezzane nappes, their thrusting over the meso-Rif, and the formation of the true meso-Rif nappes. These nappes are transgressively overlain by the Tortonian marls of the Upper Miocene, on which Messinian rocks also rest transgressively. The two units were strongly deformed at the end of the Messinian. The olistostromes associated with the Middle and Later Miocene movements of the inner nappes can be readily observed in the external zone of the downwarp.

In the pre-Rif, among the Cenozoic marls olistostromes first appear at the level of the Burdigalian horizon of the Lower Miocene and in the lower parts of the Middle Miocene. They are widely spread in the inner portion of the pre-Rif and consist of the materials derived from internal nappes which moved into the region of meso-Rif and pre-Rif from the direction of Ketama and Tanger and even from the zones behind them. For the eastern pre-Rif they are observed (Feinberg and Leblanc, 1977) in front of the Aknoul nappe, at the base of the transgressive molassoid rocks of the Binet Series with a fauna of Langhian–Serravalian age represented by sandstones and conglomerates with marls and mudstones. The landslide packages consist of Numidian rock fragments of the Aknoul nappe, which moved into the sediments in the wake of Burdigalian movements.

The outer pre-Rif is composed of Miocene marly-clayey rocks and of Triassic and Jurassic salt-bearing formations. All these rocks, which suffered no deformation prior to the Middle Miocene, were redeposited as gigantic olistostromes buried among end-Middle-Miocene rock masses transgressively overlain by Upper Miocene (Tortonian) post-nappe marls. Choubert and Faure-Muret (1971) maintain that the magnitude of transport of gravitational materials into the downwarp, which underwent an intensive subsidence at the beginning of the Middle Miocene time, was so great that even quite 'trivial' Middle Miocene marls may be considered as products of redeposition of rocks derived from the nappes. Indeed, among the marls with pelagic fauna one can find interbeds of gypsum and Triassic ophiuroid fragments. Within the pre-Rif there also lies the Ouezan gravity nappe which is part of a tectonogravitational complex. Its constituent rocks belong to inner-Rif Ketama and Tanger complexes and represent a complex tectonogravitational melange of Eocene, Oligocene, Lower Miocene and the basal Middle Miocene sediments which were formed during its movement into the pre-Rif region. According to Choubert, all tectonogravitational events in the Middle Miocene took place during a short period of time equalling ~1 m.y. At the beginning of the Tortonian period the pre-Rif inner zone subsided and moved forward to encroach on the foreland.

The Messinian marly sands and calcareous sandstones occur with a local unconformity on the post-nappe Upper Miocene (Tortonian) strata, which overlie the Middle Miocene tectonogravitational mixtites. Within the southern Rif corridor Messinian rock types and Tortonian marls are overlain by the frontal part of the pre-Rif olistostrome. In the Gharb region this olistostrome was again redeposited and buried among the marls as a Tortonian olistostrome. Further to the southwest, near Rabat, one can observe (Feinberg, 1976) the interbedding of large allochthonous plates, the detached masses of the Middle Miocene olistostromes composed of sands and sandstones containing a Messinian and Early Pliocene fauna. The intensive Messinian movements also embrace the pre-Rif swells (Suter, 1978) which are occasionally allochthonous. Locally distributed Pliocene deposits rest unconformably on Messianian rocks and are overlain unconformably by non-deformed Villafranchian deposits. The Villafranchian time was marked by intense vertical orogenic movements.

Thus, in the complex El-Rif nappe structure the olistostromes consistently develop in different zones after the nappes have moved there. The tectonogravitational mixtites of the marginal downwarp consist not only of the detached nappes, but also of all autochthonous masses and those which were transported there. The relationship between the olistostromes and the nappe movements in El-Rif is quite obvious. These movements are attributed to the end of the Eocene, to the end of Oligocene, to the Burdigalian, to the end of the Middle Miocene, and to the end of the Messinian–Early Pliocene times. These were periods of active movements of the nappes, formation of unconformities and general uplift, separated by periods of transgression. The Oligocene transgression developed in all zones; the Aquitanian, Langhian, Tortonian and Messinian transgressions occurred in the outer zones successively involving other parts of the foreland of the epi-Palaeozoic platform into subsidence and forming a vast foredeep. An advancing wave of newly forming troughs and nappes moving into them together with the deposits of the increasingly external zones is the characteristic feature of the Late Alpine El-Rif fold structure.

During the Late Cenozoic period, the entire region of the Alpine ranges in the southern Mediterranean—the Betic Cordillera, Tell Atlas, Sicily, Calabria and the South Apennine (Figs 1, 2; Sections 1, 2, 3, 5, 6, 7, 9)—underwent development similar to that of the El-Rif. In analysing these areas the author used the reviews by Choubert and Foure-Muret (1971), and by Caire (1973) as well as the results of his own special investigations. The structure of this region is generally zoned. Everywhere one can observe an intrenal geanticlinal zone (the massifs of Cabilia, Calabro-Peloritains, the region of Alpujariides and Malagides) that represents the exposed part of the Variscian and pre-Variscian basement. On the external side, the massifs are bounded by the Calcareous chains which pass into flysch and molasse zones. The Late Cenozoic olistostromes, that resulted from the nappe movements, first appear in the Lutetian deposits of flysch troughs (Fenet, 1977; Bourgois and Chauve, 1971) of the Tell Atlas and Betic Cordillera. A transgressive sequence of the micaceous sandstones and siltstones (the Nimidian Flysch) appears in the Oligocene and Early Miocene in the geanticlinal and flysch zones and occasionally resembles either flysch or molasse. In Sicily, Calabria and South Apennine, the internal Numidian Flysh produces gravity flows of *argille scagliose* with blocks of Mesozoic and Lower Cenozoic rocks lining the West

Mediterranean basin (Caire 1973). These tectonized olistostromes are due to strong mass movement towards the pre-Mediterranean basin which formed at that time. The tectonogravitational formations, associated with the same process, are also common on the northern margins of the Cabil massif in the Tell Atlas orogen (Bouillin *et al.*, 1973). The internal olistostromes, at that time, appeared together with the external ones which occur among the Upper Oligocene and lower parts of the Aquitanian deposits in the flysch troughs of the Betic Cordillera, in the province of Campo-de-Gibraltar (Bourgois and Chauve, 1971), in Tell Atlas and Sicily (Caire, 1973), and were caused by nappe movements from geanticlinal zones towards the flysch troughs.

All over the region under consideration, the compressive phase at the end of Oligocene was followed by the Aquatanian transgression which changed, during a short interval at the end of the Burdigalian to the beginning of the Langhian period, to a stage of a great movement of nappes called 'debaile' by Choubert and Foure-Mouret (1971). Most of the olistostromes linked to this event are buried in the deposits filling Aquitanian foredeeps. In the Tell Atlas the olistostromes moved both into the foredeep and into the pre-Mediterranean basin. In the foredeep the advance of olistostromes continued up to the Middle Miocene (zones 9 and 10 of Blow) time, and was due to gravity sliding of nappes into the trough that was initiated in the Langhian period (Raymond, 1976). In the pre-Mediterranean basin, they are buried under Burdigalian zones 5–8 (Blow) and are transgressively overlain by Langhian conglomerates (Raymond, 1976).

The subsequent olistostromes appeared at the end of the Middle Miocene Serravalian in the Pre-Betic zone (Hoedemacker, 1973) and in the Miocene Tortonian interval in Sicily, Calabria, Tell Atlas and the southern Apennines (Caire, 1973). They owe their origin to the movements of the nappes into the external zones of the molasse troughs which were found in the Langhian period. In the southern Apennines olistostromes also occur among the deposits of Middle–Upper Miocene troughs situated in the area of allochthonous limestone sheets and are derived from their destruction (Iacobacci, 1965).

After a short period of build-up of transgressive formations at the beginning of Messinian times (diatomites of Tripoli and others) thick sequences of evaporites accumulated in the Mediterranean, which points to an abrupt shallowing of the basin at the end of Messinian times. At that time and at the beginning of the Pliocene a renewed movement of gravity nappes towards foredeeps started, involving both salt-bearing formations and the rock masses that underlie them. Olistostromes, composed of fragments of salt-bearing rocks, are widespread in the foredeep of Sicily (Catalano *et al.*, 1976) and on its southern shelf. In Calabria, olistostromes of this period are composed of *argile scagliose* (Perrone *et al.*, 1973), while in the southern Apennines they consist of limestone plates (Caire, 1973; Belostotskiy, 1970). The same time-interval witnessed gravity movements of large detached masses of the nappes in the Nevada-Felabrid complex in the central portion of the Betic Cordillera (Langenberg, 1973), associated with the spreading of these nappes during the anti-isostatic uplift of the Cordillera.

The latest large olistostromes in the southern Mediterranean region appeared at the beginning of the Pleistocene (Calabrian). In the foredeeps of Calabria and Sicily they are related to the tectonogravitational movements of the rock complexes composing the external troughs (Zanzucchi, 1959; Manfredini, 1978), while in the pre-Betics they are due to the gravity flow of the earlier nappes (Hoedemaeker, 1973).

The northern Apennines (Figs 1, 2, Section 8) are a classic region of olistostrome development of different ages, well described by Elter and Trevisian (1973). This territory differs from the previously described regions of the southern margins of the Mediterranean in that it includes an internal flysch-ophiolite zone which is a relic of the Jurassic Tethys.

From the northwest to the southeast the northern Apennines are divided into the following zones: Ligurian flysch-ophiolite, Toscanian geanticlinal zones, a zone of the external miogeosynclinal Marches trough and the Padana foredeep.

In the Ligurian zone flysch accumulated after the ophiolite complex had formed during the Cretaceous, Palaeocene and Early Eocene periods. This was interrupted by three long phases of deformation accompanied by the generation of olistostromes. The Cretaceous–Palaeocene (including the Middle Eocene) flysch and the underlying ophiolite complex form a complicated allochthon consisting of a recumbent fold in the upper part of the Bracco range on which other folds of internal origin are piled up. The allochthon is transgressively overlain by a complex of Oligocene Ranzano sandstones that fill the Tertiary molasse basin in the Ligurian Alps and the northern Apennines. These sandstones are products of erosion of the highlands formed during the late Eocene Ligurian phase. One can find there several bodies of slid breccias containing large blocks of ophiolite and metamorphic rocks which tended to become progressively exposed in the outer parts of the basin while the Ligurian nappes moved. At the end of the Oligocene, a tectonic nappe was thrust over Ranzano sandstones. The tectonic nappe is composed of the Palaeocene–Oligocene Canetolo complex (shales, carbonate and sandy flysch). In front of the nappe, in the Upper Oligocene–Lower Miocene, olistostromes were formed as a result of its destruction (Sistini, 1974).

The Toscanian geanticlinal zone exhibits the exposed metamorphic rocks of Variscan basement overlain by a Variscan molasse and by thin carbonates and carbonate-clastic sequence of Mesozoic and Lower Palaeocene ages. An intensely subsiding trough was formed in the Oligocene, along the western, northern and eastern margins of the zone. This trough contains a thick sequence of Macigno sandstones which is considered by some authors (Manfredini, 1978) to be the lower molasse of the Toscanian region similar to the Ranzano sandstones of the Ligurian region. In the upper portion of the sandstones and the overlying lower parts of the Cervarola marl-sandstone masses (Aquitanian), which mark a new subsidence stage of the Toscanian trough, one can find numerous bodies of olistostrome representing detached masses of the sub-Ligurian nappe which advanced at the end of the Oligocene and beginning of the Miocene into the Toscany region from the internal parts of Liguria. The total olistostrome mass is 1500 m thick. During Early Miocene the nappe continued moving and olistostromes travelled to the marginal portions of the trough. At the end of Early Miocene the olistostrome-containing rocks were tectonically overlain by the sub-Ligurian nappe.

The Marke external downwarp, bordered to the west by an uplift of the Briançon-type, is composed of miogeosynclinal carbonate-clastic deposits of the Jurassic–Oligocene periods which are underlain by Triassic evaporites and sandstones representing the upper molasse of the Variscan cycle. During the Early and Middle Miocene the downwarp subsided sharply, causing the flysch to accumulate and change towards the foreland into the Bisharo schlieren formation. At the end of the Middle Miocene during the Toscanian phase numerous bodies of olistostromes and detached plates of the Toscanian nappe, as well as nappes of the Ligurian region, appeared in the deposits of the downwarp and accumulated to form their present mass. The movements of these nappes were so vigorous that a region of tectonic denudation was formed behind Toscanian plates, while at their frontal part, Umbria, some of the Miocene flysch was squeezed out and replaced by the allochthonous masses.

The Po basin and the Adriatic margin subsided rapidly at the beginning of the Late Miocene. In the Tortonian time, according to Manfredini (1978), a thick turbidite mass (Picheno flysch) accumulated here and overlapped onto the slowly subsiding outer margin of the Marke downwarp. Within this area, Upper Tortonian and Lower Messinian deposits include olistostromes of Ligurian and Toscanian nappes, which were formed at that time near the outer margin of the Marke downwarp. The Upper Tortonian olistostromes represent large frontal landslides and detached plates of the nappes, which gravitationally moved into the troughs. Prior to the Late Messinian time, the trough became shallow, which caused tectonic movements and further gravitational transport of these nappes towards the northeast. The Upper Messinian–Lower Pliocene deposits, which rest uncomfortably on the Tortonian formations, together with olistostromes are represented by shallow-water carbonate-clay, occasionally salt-bearing, strata. Among them, in the low foothills of the Apennines, in the bends of the River Po there are olistostromes accompanying the allochthonous complex which moved from the Marke downwarp. According to Sestini (1974), the deposits attributed to the end of the Messinian–Early Pliocene and observed in the Padanian trough are 'parallochthonous'. They rest on the allochthonous plates and include the detached masses of these plates formed during insignificant movements of allochthon that took place at the boundary of the Miocene and Pliocene. The Messinian–Early Pliocene olistostrome masses are overlain by sandstones, conglomerates and clays (the upper molass of the Early to Middle Pliocene) that were slightly deformed at the end of the Middle Pliocene. The underlying older rocks are thrust over these conglomerates. The entire complex is transgressively overlain by Upper Pliocene marine clays and siltstones.

Thus, during the Oligocene, Miocene and Pliocene the northern Apennines witnessed a displacement of flysch and molasse depressions northwards and eastwards. This was accompanied by movements, in the same direction, of large allochthonous masses and olistostromes confined within the following age intervals: the beginning of Oligocene (Ranzano sandstones) in the Ligurian zone, the end of Oligocene to the beginning of Miocene (Macigno sandstones) in the Toscan region, the end of the Middle Miocene in the Marke downwarp, and the boundary of the Middle and Late Miocene and the end of Miocene to the beginning of Pliocene in the Po basin. The olistostromes formed at the beginning of the Oligocene, the end of the Oligocene to the beginning of the Miocene and at the end of the Middle-Miocene are associated with the Ligurian (Late Eocene to Early Oligocene), Toscan (Early Miocene) and Middle Miocene phases of nappe movements and general uplift. The frontal slides that occurred at the end of the Tortonian to the Early Messinian time were separated from the Middle Miocene nappes as a result of an abrupt subsidence of the Picheno flysch trough. The olistostromes of the end of the Messinian to the Early Pliocene were associated with the final horizontal movements in the Marke downwarp of the northern Apennines. Their internal parts underwent intense deformation and vertical faulting.

As in the case of the Sicilian-Calabrian and Betic-Rif arcs, the periods of intensive movements of nappes and the general uplift in the northern Apennines are separated by the stages of general subsidences and related accumulation of transgressive sedimentary strata. The Oligocene transgression occurred throughout the northern Apennines; the Aquitanian transgression which is less distinct manifests itself in the initiation of the trough, where fine detrital Chervarol rocks were deposited. In the Langhian time the Marke flysch downwarp had subsided, and at the end of the Tortonian to the beginning of the Messinian the Picheno flysch trough suffered subsidence. A large marine transgression is recorded in the Upper Pliocene deposits of the Po area. The presence of nearly coeval transgressive-regressive cycles in the regions under consideration points to common major rhythms in the entire region of the western Mediterranean. As a matter of fact this region is unique because of large-scale tectonogravitational movements producing olistostromes. This may be due to the fact that at the Late Alpine stage this region suffered not only the closure and deformation of relict geosynclinal troughs but also the generation of new troughs extending into the foreland. A constant coexistence of deep and wide troughs with zones of uplift and piling of allochthonous masses favoured tectonogravitational processes.

The northern branch of the Alpine belt is marked by considerably weaker tectonogravitational processes at the late Alpine stage. A vivid example of the northern Alpides is the Carpathian orogenic arc stretching for more than 1500 km.

The main structural features of the Carpathians are (Sandulesçu, 1978): internides consisting of the complexes deformed during the Cretaceous period; the externides occupying the region of the Carpathian flysch and deformed during the Oligocene and Miocene; the Penides separating them with their structures formed in the Late Cretaceous and at the beginning of the Miocene periods; the outer part of the marginal foredeep superposed on the foreland and deformed in the Miocene period; intermontane and inner molasse basins superposed on different parts of the internides (Figs 1, 2; Sections 16, 17, 18). All these, except for intermontane basins, have complex nappe structures. The nappes are mainly composed of sedimentary cover rocks, which formed complex multi-stage piles within their depositional basins and are now considerably compressed. The tectonogravitational nappes moving into the neighbouring tectonostratigraphic zones are typical only of the western Carpathians. The Carpathians differ essentially from the orogenic ranges of North Africa and the Apennines in not having been affected by great nappe movements.

A sequence of occurrence of Cenozoic olistostromes and their association with different types of movements can be best seen from the study of the externides, the foredeep and the eastern portion of the Hungarian (Transcarpathian) depression on the territory of the Soviet Carpathians (Fig. 2; Sections 17, 18). Here observations were made by the present author and documented by Kulchitskiy (1977) and by Gluschenko *et al.* (1980). The local analysis was supplemented by the information available on the non-Soviet territories (Fig. 2; Section 16). All the elements of the externides and the foredeep form cover nappes composed predominantly of the Cretaceous, Palaeogene and Oligocene flysch. The Cretaceous is marked by slight variations of the contemporanous deposits of different elements, whereas the Eocene displays a varied composition of rocks which testifies to the fact that at that time a rugged tectonic relief was being formed (Fig. 3). Underwater landslide layers are locally deposited during the second half of the Eocene and contain transported fragments of phyllites and carbonate rocks of the foreland (the Popel Formation of the Borislav-Pokut zone). The Oligocene deposits of the Carpathians are diverse and consist of sandstones and marls and clay, limestone, sandstone flysch that grades into bituminous shales of the Menilite formation in the inner parts of the externides. These rocks overlie conformably on Eocene rocks throughout the region—from the Krichev zone of the Transcarpathians to the Pokut zone of the cis-Carpathians—and contain at their base a layer of deep-water chert above which there are Menilite shales with Lower Oligocene deep-water fish fauna. The Oligocene time was apparently marked here by an abrupt subsidence of the flysch-bearing Carpathians that resulted in the smoothing of the rugged tectonic relief of the Late Eocene period. The lower cherty horizon is deformed by complicated synsedimentary folding of slump origin caused by the abrupt subsidence of the basin. The slump folds of the cherty horizon observed on the eastern margin of the basin in Pokutia and the East Skibas phyllites contain fragments which have been introduced by the sliding of foreland rocks (the Sandomir range). The overlying regressive complex of Oligocene deposit is distinguished by several facies caused by the basin differentiation during the general uplift. Thick bodies of tectonogravitational mixtites occur in the Middle Oligocene–Late Oligocene sandy flysch of the Silesian zone. The clastic products (Eocene marls, Cretaceous sandstones) of intra-

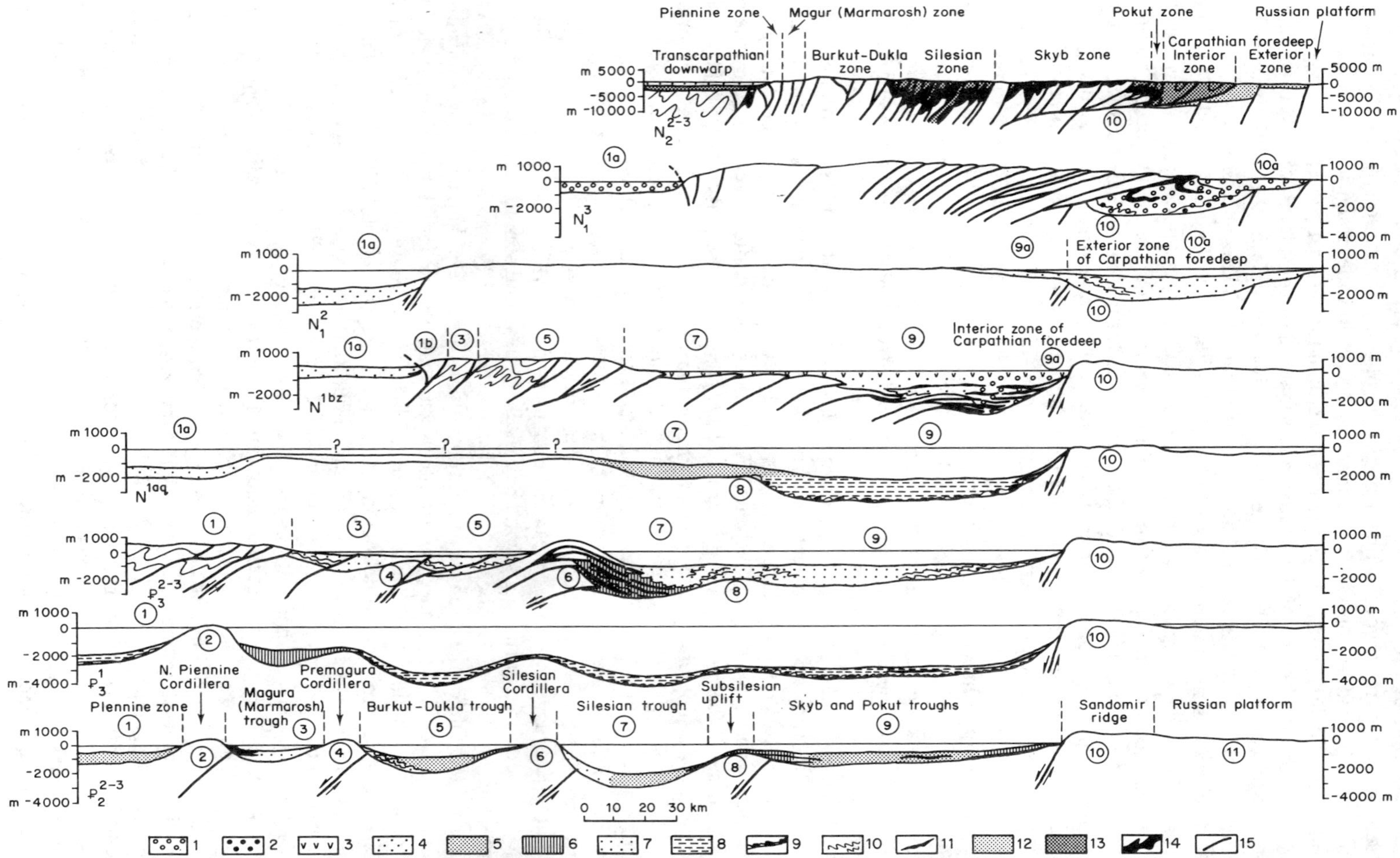

Piennine zone
Magur (Marmarosh) zone
Pokut zone
Russian platform
Transcarpathian downwarp
Burkut-Dukla zone
Silesian zone
Skyb zone
Carpathian foredeep
Interior zone
Exterior zone
N_2^{2-3}
N_1^3
Exterior zone of Carpathian foredeep
N_1^2
Interior zone of Carpathian foredeep
N^{1bz}
N^{1aq}
P_3^{2-3}
P_3^1
P_2^{2-3}
Piennine zone
N. Piennine Cordillera
Magura (Marmarosh) trough
Premagura Cordillera
Burkut-Dukla trough
Silesian Cordillera
Silesian trough
Subsilesian uplift
Skyb and Pokut troughs
Sandomir ridge
Russian platform
0 10 20 30 km

FIG. 3 Diagram showing the development of the flysch Late Alpine geosyncline eastern Carpathians (compiled by I. G. Scherba and M. A. Bear).

1 – coarse sand-conglomerate upper molasse; 2 – fan deposits in upper molasse; 3 – salt-bearing deposits, 4 – sandy and sand-argillaceous lower molasse; 5 – sandy argillaceous flysch; 6 – clay-carbonate flysch; 7 – sandy flysch; 8 – Menilite shales; 9 – cherty horizons; 10 – the slides in the basin sediments ('endo-olistostrome'); 11 – olistostromes with fragments of flysch nappes and rocks of the Sandomir range; 12 – deposits of the Middle–Upper Miocene cycle; 13 – deposits of the Lower Miocene cycle; 14 – deposits of the Oligocene cycle; 15 – faults.

Numbers in sections: 1 – the Pennine trough; 1a – Transcarpathian depression; 1b – Pennine zone; 2 – North Pennine Cordillera; 3 – Magur (Marmarosh) trough; 4 – Marmarosh Cordillera; 5 – Burcut-Dukla trough; 6 – Silesian Cordillera; 7 – Silesian trough; 8 – sub-Silesian uplift; 9 – Skiba and Pokut troughs; 9a – Interior zone of Carpathian foredeep; 10 – Sandomir range; 10a – exterior zone of Carpathian foredeep; 11 – Eastern European platform.

Carpathian origin are represented by detached nappe masses moving from the west through the sub-Silesian Cordillera (see Fig. 3). Upward in the Oligocene sequence, regressive features are more pronounced and coal and lignite intercalations appear, as for instance the Malovyzhenskian Formation of the Burcut-Duklyanskian zone and the upper parts of the Fusaru sandstones in the Silesian zone. The section of the sandy flysch is topped by the Vinetishu or Lopyanets Beds. These are represented by alternations of micaceous sandstones and clays with a considerable number of carbonates, which in general have an abundance of landslide layers consisting of autochthonous fragments. The age of these horizons, determined by nannoplanctons, is Early Miocene (Ştefănescu *et al.*, 1979). These beds as well as the upper parts of the Fusaru sandstones contain thick olistostromes at fronts of the Audia nappe in Romania and the Mogura nappe in Poland, the age of the olistostromes corresponding to the Upper Oligocene–Lower Miocene. They are tectonogravitational formations. The eastern parts of the basin also show regressive features which become more distinct up in the Oligocene section. The Menilite shales here are well developed and have intercalations of Kliwa sandstones which are fluxoturbidite deposits. As in the west, the Lopyanets Formation overlying them includes numerous slump sheets.

The regressive series of the Oligocene–Lower Miocene complex again changes into the fine bituminous shales of the Upper Menilite formation in the eastern part of the flysch-bearing Carpathians and into the fine sandy-clayey flysch of the Upper Krosno Formation in the Silesian zone (see Fig. 3). The Menilite shales are underlain by an upper cherty bed which, like the lower, includes synsedimentary slumps. These resulted from an abrupt deepening of the Krosno-Menilite basin which developed within the same boundaries as the Oligocene basin. The Upper Krosno and Upper Menilite deposits contain Aquitanian fauna (Ştefănescu *et al.*, 1979; Andreyeva *et al.*, 1981). In the Silesian and Skiba zones, the Krosno and Menilite rocks are conformably gradational into a considerably shallower Burdigalian grey sequence of sandstone-siltstone and marls with intercalations of gypsum (the Palyanitskian Formation, the Cornu Beds) in which there are sheets of underwater landslides of autochthonous rocks. In the Carpathian foredeep transition into Burdigalian deposits takes place across a level of erosion and is accompanied by a considerable change in rock composition. Intercalations and thick lenses of conglomerates that resulted from mud flows appear in the Palyanitskian Formation and in its age-analogue the salt-bearing Vorotytschenskian Formation. All these regressive sediments contain a great amount of tectonogravitational mixtites which are associated both with landslides at the front of the Skiba and Pokut nappes (plates of Menilite shales, Eocene marls and sandstones in the deposits of the Polyanitskian Formation on the Rybnitsa river and of the Vorotytschenskian Formation in the region of Rungursk Sloboda and Borislav) and a tectonic escarpment at the western edge of the Sandomir range (large and small blocks of the Jurassic limestones and phyllites widely occurring in the Pokut zone). The olistostrome masses here are a few hundred metres thick. The section of the Burdigalian deposits ends with a thick, up to 800 m, mass of shallow-water, frequently salt-bearing, deposits.

No equivalents of these deposits are known from the internal parts of the central sector of the flysch-bearing Carpathians which underwent intense compression at that time. However, in the west, where at that time the Magura nappe continued its movement, the upper Krosno Formation in the Silesian zone contains thick tectonogravitational olistostromes (Szymakowska, 1976). At the same time, a number of intermontane basins (Solotvin, Vienna) formed over the internide territories in the Transcarpathians. In the Solotvin basin of the Transcarpathian depression, fine sand-clay rocks of the Burkal Formation were at that time unconformably deposited over Palaeozoic, Mesozoic and Palaeogene complexes of the Penine zone. Up the section they grade into regressive masses of tuff and salt deposited at the end of the Burdigalian (Petrashkevich and Pishranova, 1967).* At their base on the eastern side of the depression, there are common coarse, dump-like Tereshul 'conglomerates' with fragments of Pennine rocks. These are probably tectonogravitational mixtites associated with back thrusting in the strongly folded Pennine zone and the destruction of the resulting tectonic escarpments. At that time the Lower Miocene deposits of the Transcarpathians were subject to deformations along fault zones.

The Middle Miocene deposits in the Carpathians occur only in the intermontane basins and in the foredeep which underwent local subsidence at that time. In the transcarpathian depression during the Carpathian period a rapid subsidence took place: the salt-bearing sequence changed to flyschoid clastics and argillites. Near the eastern side of the depression, it lies uncomformably on the steeply dipping Lower Miocene strata. The deposits of the middle portion of the Badenian is regressive, as indicated by the appearance of red shades, lignites, thick beds of sandstone and fresh-water fauna, all pointing to the shallowing of the basin, which reached its maximum in the Middle Sarmatian period. This uplift is associated with tectonic transport along the faults in the southern side of the Hungarian central Mountains. The aging of these events is about 13 m.y. (Hamor, 1980). Here, among Badenian and Sarmatian volcanic rocks replacing marine deposits, there are thick layers of redeposited Permian, Triassic and Palaeocene sedimentary rocks which are tectonogravitational formations associated with the movement on the Raba and Zagreb-Kul'ch fault zones (Biela, 1978).

The Carpathian foredeep, unlike the Transcarpathian depression, which continued to occupy the same position as in the Lower Miocene basin, in the Middle Miocene shifted from the geosynclinal position on to the platform. Here, the sand-clay deposits of the Carpathian and Badenian stages form the outer zone of the Carpathian foredeep. They overlie, transgressively and unconformably, the deposits of the Sandomir range and the margins of the Eastern European platform (Fig. 3). Up the succession they pass into sandy-argillaceous shallow-water Badenian deposits and finally into Lower Sarmatian (?) coarse Pystin conglomerates. These conglomerates, representing the deposits of the thick alluvial fans, mark upward movements at the end of the Middle Miocene and the beginning of the Late Miocene times. At the beginning of the uplift on the territory of Poland the nappes of the internal zone were displaced. This resulted in the appearance of olistostromes made up of Carpathian rocks intercalated with the autochthonous molasse formed as a result of the destruction of frontal portions of the nappes in the Lower Badenian (Konior, 1981). The overlying Sarmatian sandy-argillaceous deposits are separated from them by a surface of erosion. At the end of the uplift, prior to the Late Sarmatian, large-scale overthrusting of the internal zone molasse of the Carpathian foredeep took place upon the external zone and is manifest over the territory of Romania. The overthrust is concealed beneath the Upper Sarmatian–Lower Pliocene molasse (Sandulesçu, 1978).

The accumulation of the Upper Sarmatian–Lower Pliocene molasse marks the beginning of a new stage of local subsidence in the Carpathian foredeep, while the Transcarpathian depression did not enlarge. In the cis-Carpathians the sediments of that period occur only in Romania. In the Transcarpathians, the deposits of the Pannonian stage, that has a base close in age to the Upper Sarmatian of the cis-Carpathians, is represented by transgressive clays interbedded with andesitic tuffs. Pontian deposits contain abundant plant remains and fresh-water molluscs, indicating a considerable shallowing of the basin. At the end of the Pontian and prior to the Dacian, overthrusting and deformation of the Transcarpathian molasse took place along the northern margin of the Vyllan-Byckhor blocks. At the same

*The salt-bearing Tereblyan Formation may also be partly Langhian.

time, in the intermontane depressions and in the cis-Carpathians, a pre-Daccian unconformity was formed.

After the pre-Late Miocene uplift and deformation, local Late Pliocene lakes again developed in the Carpathians. The last overthrusts of the internal zone of the foredeep upon the external zone took place in the Romanian cis-Carpathians before the Pleistocene. These overthrusts are concealed beneath Eo-Pleistocene deposits. At that time, eruptions of basalt took place in the Transcarpathians.

The Pleistocene stage in the history of the Carpathians was truly orogenic. The mountain building was accompanied by erosion and formation of flat-lying complexes.

Thus we can distinguish six large tectonosedimentary cycles in the Late Alpine history of the Carpathians: Oligocene, Early Miocene, Middle–Late Miocene, Late Miocene–Early Pliocene, Late Pliocene and Pleistocene. Each cycle, except for the last, which is truly orogenic, started with a rapid subsidence and formation of sedimentary basins, which gradually shifted away from the centre, and ended in a general uplift. From the very beginning, the uplift caused the nappes to move and produce olistostromes of tectonogravitational origin. In the inner parts of the Carpathian externides, the earliest movements of the nappes started in the middle and finished at the end of the Oligocene or the beginning of the Miocene. At that time a thick mass of olistostromes was formed in the Silesian zone. Subsequent movements took place not only in the interior of the Carpathian flysch-bearing geosyncline, but also extended over the inner zone of the cis-Carpathian downwarp with detached masses appearing in Burdigalian deposits.

No olistostromes were formed in the flysch zones of the Soviet Carpathians which at that time underwent an intense compression, increasing the inclination of the earlier nappes. The compression produced steeper back thrusts in the Pennine zone at the margin of the Transcarpathian depression. Their movements gave rise to collapse and slumping. Uplift during the third tectonosedimentary cycle, which began in the Middle Badenian, was accompanied by minor overthrusts in the inner zone of the Carpathian foredeep and the formation of tectonogravitational olistostromes. In the Transcarpathians in the middle of Badenian the olistostromes formed in front of the Zagreb-Kultsch and Raba tectonic escarpments. No olistostromes formed during the pre-Dacian and Late Pliocene upthrusts and deformation accompanied by minor overthrusting within the Carpathian foredeep took place when the entire Carpathian territory emerged. Fold deformation at these stages took place under the condition of general compression with no tectonogravitational sliding of nappes. The upward movements in the Pleistocene orogenic stage did not produce considerable slide complexes although they included horizons of coarse gravels in the Tschop series of Lower Pleistoceneage.

The Alpine structure of the Greater Caucasus differs from that of the regions discussed above in that their nappe tectonics are on a smaller scale. The northern flank of the Greater Caucasus meganticlinorium has a gentle northerly incline and a foredeep trending into the epi-Hercynian platform. The southern limb of the meganticlinorium has nappes and southerly vergence develops in a series of flysch and molasse troughs which migrated southward during the Jurassic–Neogene period and formed on the site of a Hercynian basin. In the process, the Rioni intermontane area extended in the Oligocene–Neogene time to encompass the epi-Palaeozoic Transcaucasian median massif (Georgian block).

Late Cenozoic olistostromes occur on both the northern and southern flanks of the anticlinorium (Fig. 1, Sections 23 and 24). In the south they accompany the nappes, whereas in the north they were formed on the slope of a developing uplift. The author observed these olistostromes herself (Borukayev *et al.*, 1981) and knows them from the work of Keller and Menner (1945), Vassoevich (1949), Moskvin and Semikhatov (1956), and Leonov (1975). It is noteworthy that they are confined to approximately the same time-intervals as mentioned above. The olistostromes are widespread in the Upper Eocene flysch zone on the southern margin of the flysch basin (Leonov, 1975) bounded by the Georgian block uplift. Here the Jurassic–Eocene carbonate sequence is overlain by the Oligocene–Miocene lower molasse. In the western Caucasus the deposits of the Abkhazian zone (Georgian block) are separated from the Novorossiysk flysch synclinorium by the Akhtsu suture zone of Early Jurassic limestones (Fig. 4). The Eocene olistostromes are found here in the front of the Vorontsov flysch nappe (the Mamaya parautochthone) to the north of the Abkhazian zone which in the Mio-Pliocene was thrust over the Akhtsu zone. The olistostromes are confined mainly to the Middle and partly to the Upper Middle–Upper Eocene part of the Mamaya Group. The units or horizons, which are about 100 m thick, include blocks of different size (up to 2 km) composed of greenish-grey pelitomorphic limestones, marls, sandstones, red argillites and flints, and form several slides which differ in morphology and are separated by normal flysch strata. The clastics in olistostromes are derived from the rocks of the Abkhazian zone. They may have been supplied from the marginal parts of the Georgian block which in the Late Eocene was subject to uplift and was an overthrust of the Caucasus fold system. However, Leonov (1975) thinks that at that time the Georgian block or at least the Dzirulian massif suffered a northward overthrust on to the southern slope of the flysch basin.

Oligocene deposits unconformably overlay the Meso-Cenozoic rocks dislocated in the Late Eocene. They occur mostly within the Foredeep and in the intermontane molasse areas which were at that time subject to abrupt subsidence. These sediments, known as the Maykop Group (its lower and middle members are of Oligocene age), are conspicuously monotonous throughout the Greater Caucasus, where there was apparently a considerable subsidence at the beginning of the Oligocene. The Middle Upper Oligocene part of the Maykop Group, on the face of it very much like the Carpathian Menilite Group, is of a regressive character. Olistostrome horizons are very common at the base of the Maykop Group (the Khadum Horizon of the Lower Oligocene) and in the Middle Maykop member. They occur in the Adler and Gudauta depressions, on the northern slope of the Kura depression and in the foredeep.

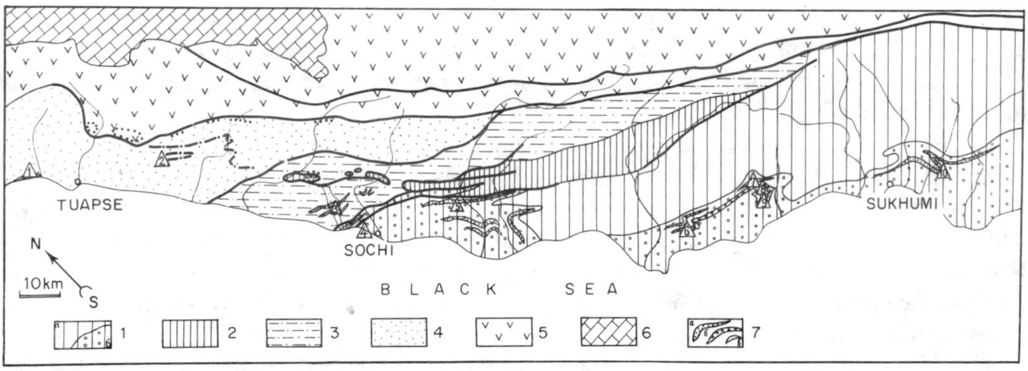

FIG. 4 Diagram showing Cenozoic olistostromes on the southern slope of the western Caucasus.
1–6 – tectonic zones: 1 – Abkhazian zone (a – Central Abkhazian uplift, b – Adler, Gudauta and Sukhumi molasse depressions); 2 – zone of Akhtsu; 3 – Tchvezhipsin zone; 4 – Lazarevskaya zone (flysch); 5 – zone of the Main ridge; 6 – Lagonakik zone of the northern slope; 7 – olistostrome horizons: a – Eocene–Oligocene, b – Kozachurian, c – Sarmatian.

In the Adler depression (Fig. 4), the Oligocene rocks are a part of a thick terrigenous series correlated with the Maykop Group. Its lower part is represented by interlayered grey shelly clays and brown sandstones (the Matsesta Group); the middle part comprises thick sandstones prevailing over clays (the Khosta Group); the upper, Miocene, part is composed of regularly alternating grey clays and brown sandstones (the Sotchi Group). At the base of the Matsesta Group (the Khadum Horizon) and in the Khosta Group (the uppermost Oligocene) slump sheets are very common and consist of chaotic fragments of allochthonous and autochthonous rocks. These horizons are thickest at the junction of the Gagra uplift of the depression basement, which has supplied most of the allochthonous material. The uplift consists of Upper Jurassic–Palaeogene limestones which have been crushed, indeed virtually pulverized, in the tectonic nappes. A specific morphology of the Gagra massif suggests that pockets of breccia and separate blocks were squeezed out from its interior. A similar process was described by the author in the Crimea and the Darvaz range (Scherba, 1975, 1978).

To the south of the Gagra uplift the Oligocene–Lower Miocene deposits known as the Maykop Group are very similar in their appearance. The lower part of the group was formed under conditions of abrupt subsidence of the large Maykop basin; the middle part (the uppermost Oligocene), composed of rather coarse clastics with sandstone intercalations with fossilized wood, is of a regressive character. As in the Khosta Group, in the Middle Maykop of the Gudauta depression there are slump sheets. These are made up of separate polygonal blocks (1.5–3 m) of grey Senonian limestones enclosed in an almost intact matrix and of beds with abundant limestone fragments from 5–10 to 20–30 cm diameter. Horizons of chaotic pieces of the Maykop Group are also common here. The entire complex of the chaotic sediments observed in the Upper Oligocene deposits of the Abkhazian zone is due both to the sliding of unconsolidated deposits of the Maykop basin and to the burial of the blocks of rocks of the Pre-Oligocene basement, which have been tectonically broken up. As a whole, they can be regarded as tectonogravitational formations.

In Azerbaidzhan a large overthrust of the Kakhetia-Vandam zone on to the deposits of the Kura depression has been noted. Here olistostrome horizons associated with the initial stages of this overthrust occur in the middle part of the Maykop Group in the Shemakha region as, for instance, at the village of Matrasa. They consist of blocks of Upper Cretaceous rocks of the Vandam zone (Kopp and Kurdin, 1981).

The Lower Miocene deposits completing the section of the Maykop Group as a rule gradually pass into the Middle Maykop Group. Their base is characterized by much finer rocks than the Underlying regressive Oligocene sediments, whereas the upper part becomes more sandy again.

In the Miocene deposits of the Abkhazian zone the main part of the collapse and slumping correlates with the Kotsakhur Interval of the section and occurs in the Gudauta depression. The deposits are comparatively thick clays, clayey and calcareous sandstones containing lenses of lignites, conglomerates and breccias. They pass, in the section, into the lowermost Miocene clayey sequence and, unlike the latter, have some regressive features. The submarine slumped rocks occur as packages (up to 50 m in thickness) of grey clays and coarse sandstones deformed by slip folds. Scattered among them are discordant polygonal blocks of various limestones, sandstone, porphyrites and flints derived from the Akhtsu zone and the Georgian block. Some klippen of Mesozoic limestone trend for 300–600 m, rising 50–70 m above the surface (Fig. 5). Fragments of these limestones occur in polygene conglomerates of Chokrak, overlain by fine Karagan, Konka and Sarmatian clays in which no slump sheets have been recognized.

At the beginning of the Middle Miocene, in the Tarkhan age, the entire territory of the Caucasus was subject to a considerable transgression which involved the previously uplifted inner sections of the Georgian block. Thus in the central part of the Kartalyk depression, Middle Miocene strata overlie Cretaceous limestones (Khain, 1978). The accumulation of the transgressive sequence was interrupted in the Chokrakian time. Minor shifts causing a disconformity at its base, established in the Abkhazian zone, took place at its lower boundary (Fig. 5). In Azerbaidzhan the Chokrakian rocks contain small olistostromes related to the redeposition of elevated reef of the sedimentary basin. The accumulation of the transgressive sandy-clayey sequence continued up to the Early Sarmatian to be followed by the shallowing of the basin.

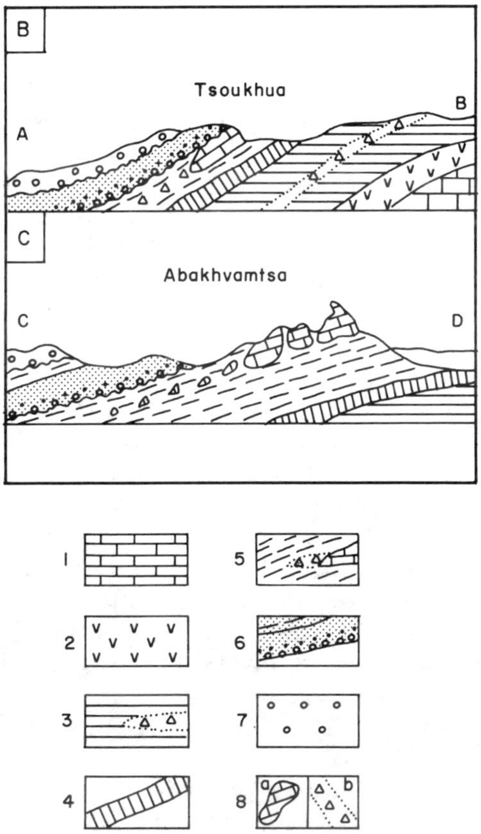

FIG. 5 Geological section of the Gudauta molasse depression via the large erratic of Cretaceous limestone in Kotzachurian deposits on the southern slope of Greater Caucasus.

1 – Cretaceous limestone; 2 – Eocene–Palaeocene marls; 3 – Oligocene clays with lenses of chaotic conglomerate breccia; 4 – Sakaraulian clays; 5 – Kotzachurian clays with exotic blocks of Cretaceous limestones; 6 – Chokrakian conglomerates; 7 – Middle Sarmatian conglomerates.

In the Abkhazian molasse at the Middle Sarmatian level there are horizons with olistostromes. The Middle Sarmatian is represented by a thick regressive sequence of conglomerates and sandstones which lies unconformably on different rocks ranging from the Lower Sarmatian to the Kotzakhuri Horizon. Thus at the base of the Middle Sarmatian bedded rocks unconformable lenses of coarse conglomerate-breccias with argillaceous cement can be detected. They contain large angular blocks (up to 100 m) of blue Eocene marls, sandstones and grits of the Kotzakhuri stage. Slumping was due not only to intense denudation but also to the growth of uplifts. At the southern border of the Rioni depression there are Upper Sarmatian products caused by landslide redeposition of the earlier complexes. These are horizons of conglomerates and sandstones with abundant redeposited faunas and random blocks of the Eocene andesitic basalt derived from the Adzhar-Trialeti zone that overthrusts the molasse from the south.

The pre-Pontian age (Late Miocene) of deformation can be clearly established in the inner zones of the Greater Caucasus anticlinorium in the eastern Caucasus. Folding and deformation of its Jurassic shales in the Main and lateral ranges also involved Sarmatian deposits which were later uplifted more than 4 km. In the northern part of the flysch zone, within the Belokan-Vandam anticlinorium, calcarenaceous Pontian deposits are observed to lie nearly horizontally on the deformed Miocene sequence. The Pontian transgression is recognized over the entire territory of the Caucasus. The accumulation of the thick productive sandy-clayey rocks of Pontian–Lower Pliocene age in the eastern Caucasus is terminated by a powerful fold-nappe deformation and general uplift. The regressive sequence in the northeast of the Kura depression is terminated by a broken formation (the uppermost Cimmerian) with numerous slump horizons of gravitational origin. At that time, throughout the southern flank of the Greater Caucasus, the flysch zone was powerfully overthrust upon the molasse basins; in some places (Tsiteli-Tskharo) the overthrusting resulted in the formation of olistostromes. These overthrusts and nappes overlie strata ranging up to and inclusive of the Lower Pliocene and are concealed beneath the rocks of the Akchagyl stage of the Upper Pliocene. Such structures are well demonstrated by the Baskal nappe transported from the Dibrar zone to the northern side of the Kura depression. It appears that the Vorontsov nappe in the western Caucasus must have moved at the same time. The southern border of the pre-Akchagyl deformations in the Kura depression is the Adzhichai overthrust. During the Akchagyl transgression the earlier elevated parts of the intermontane areas subsided. The transgression terminated in fold-nappe deformation which occurred mainly in the cis-Caspian part of the Kura depression and was accompanied by the formation of tectonogravitational olistostromes in the Apsheron and the lowermost Baku rocks.

At the same time olistrostromes were formed on the southern flank of the Greater Caucasus along the escarpment in front of the Main overthrust (Leonov, 1970). Their formation was favoured by the destruction of the tectonic escarpment.

To summarize, in the Eocene–Pliocene deposits of the southern flank of the Greater Caucasus there are seven olistostrome horizons: Middle–Upper Eocene, Lower Oligocene, Upper Oligocene, Lower Miocene (Kotsakhurian), Upper Miocene (Sarmatian), Middle Pliocene and Pleistocene. All these olistostromes were formed at the marginal part of a trough that gradually migrated to the south and filled up with flysch in the Eocene, and with molasse in the Oligocene, Miocene and Pliocene. Olistostrome horizons associated with flysch are clearly confined to the borders of formation complexes (Borukayev *et al*., 1981). The bulk of olistostromes occurring in the molasse are connected with regressive parts of sedimentary rhythms; thus all of them are associated in one way or another with changes in conditions of sedimentation.

All coarse clastics of the Eocene, Oligocene and Lower Miocene horizons are of tectonic origin. In the western Caucasus they are connected with tectonic breakdown of Mesozoic rocks at the edges of the Georgian block, which was by that time involved in uplift and overthrusting affecting the southern flank of the Greater Caucasus. Some clastics are due to the destruction of the frontal escarpment, which was caused by movement of nappes, while others were produced by the squeezing of breccia out of the interior parts of imbrications of basinal limestones. These horizons are tectonogravitational formations denoting epochs of growing tangential compression on the southern flank of the Greater Caucasus. Gravity mixtites are also widely developed among the olistostromes; these consist of autochthonous rocks and are, in this region, associated with reactivation epochs. The presence of abundant conglomerates and fragmental formations in the Middle and Upper Sarmatian sediments of the Rioni area is indicative of orogenic movements at the end of the Miocene and is mainly related to slumping on the slopes of growing orogenic uplifts. The materials of Pliocene olistostromes are derived from the flysch zone nappes, which apparently moved the greatest distance at that time. Their movement was largely due to gravity, but tangential compression was certainly also important since at that time the main phase in the formation of the Alpine fold structure in the Greater Caucasus was in progress.

On the northern flank of the Greater Caucasus olistostromes occur mostly in the Terek-Caspian foredeep (Fig. 2, Section 23), on the southern flank of which synsedimentary folding of the Front range and limestone Dagestan, having begun in the Cretaceous continued in the Late Cenozoic. Here Late Jurassic and Cretaceous sediments are represented by thin limestones of platformal aspect overlain by Eocene marls. On the flank of the foredeep in the Middle and Upper Eocene there are subaqueous slumps containing both chaotic autochtonous fragments and underlying Danian limestones. These horizons were formed by sliding on the flanks of the growing anticlines, in the crestal parts of which limestone and marl strata were being eroded while clastics piled up at the base (Moskvin and Semikhatov, 1956). Upper Eocene marls with olistostromes are transgressively overlain with the Maykop Group of which, at the bottom of the Khadum horizon, slides of argillaceous material are found which contain fragments of underlying marls. Up the succession, the chocolate clays of the Lower Oligocene pass into a sequence of interbedded Upper Oligocene clays and sandstones (the Mutsidakal-Miatlin Group of Dagestan, the Zelenchuck Horizon and the Middle Maykop of the West Kuban depression). The Terek-Caspian depression of this regressive sequence (about 600 m) comprises several olistostrome horizons a few dozen metres thick. These contain both small fragments and gigantic (up to several kilometres across) olistoliths of Upper Eocene clays and marls which moved as slides into unconsolidated sediments. These landslides comprise all of the Lower Oligocene and the uppermost Eocene rocks which were occasionally redeposited, and also include Upper Oligocene strata that directly overlie the lower horizons of the Eocene. In the West Kuban depression horizons of coarse conglomeratic breccias of slide origin occur in the Zelenchuk Group at the border of the Stavropol uplift that separates the two depressions.

No slide horizons occur in a thin argillaceous Upper Maykop Group that follows a regressive sedimentary sequence. Such horizons occur only in the upper part of the Upper Maykop (the Zurumakent Group of the uppermost Lower Miocene age) which also has regressive features. Up the succession, the subaqueous slumps of allochthonous and autochthonous material occur in the Chokrak deposits (Vassoevich, 1949) which are also characterized by abundant fluxiturbidite horizons. Clastic dykes widely developed in the pockets of clays point to irregularities in the Chokrakian sedimentation. The quiet argillaceous sedimentation of the Karagan and Konka was again interrupted in Sarmatian times when the basin gradually shallowed. Horizons of coarse conglomerates and fluxiturbides appeared. The next major transgressions took place in the Pontian and Akchagyl times. At their bases there are unconformities, though the movements that caused them produced no olistrostromes.

Moskvin and Semikhatov (1956) show that the subaequeous slumping on the northern flank of the Limestone Dagestan arose in the periods of the most intense growth of separate uplifts. These authors believe that the fact that subaqueous slumping in different parts of the Terek-Kuban foredeep developed at proximate stratigraphic levels 'points to the geologically simultaneous intensification of the upward movement in the North-East Caucasus' (Moskvin and Semikhatov, 1956, p. 83).

On the other hand, the olistostromes of the northern flank of the Caucasus were associated with sliding off the growing uplifts, and were coeval with the tectonogravitational mixites on the southern flank that occurred during the epochs of intense tangential compression. Both types of olistostromes correlate with regressive deposits of major sedimentary cycles. All these events indicate that in the Greater Caucasus the Late Cenozoic tangential compression and uplift were simultaneous.

Thus in the late Cenozoic Caucasus six tectonic sedimentary cycles can be recognized: in the Oligocene, the Early Miocene, the Middle–Late Miocene, the Early Middle Pliocene, the Late Pliocene and the Pleistocene. They began with general subsidence and ended with uplift and compression on the southern flanks of the mountains. The uplift and compression periods were accompanied by the formation of olistostromes in the middle and at the end of the Eocene, the end of the Oligocene, the end of the Early Miocene, in the Chokrak stage of the Middle-Miocene and the Pleistocene. Olistostromes connected with the rapid subsidence of the Maykop basin are found within Lower Oligocene deposits.

The Afghan-Tadzhik depression, the bordering Pamirs and the Tien-Shan mountains (Fig. 1, Section 27) are neotectonic structures in the epi-Palaeozoic orogenic region of Central and Eastern Asia. In the Tien Shan and the Pamirs, a complicated fold-nappe structure was formed during the Hercynian and Early Cimmerian phases of tectogenesis. The orogenic uplift of these structures during the Alpine epoch resulted in the Afghan-Tadzhik intermontane depression that was formed on the site of the epi-Palaeozoic platformal depression which was filled with thick continental molasse during the Oligocene–Quarternary age. The history of molasse accumulation, the formation of the fold structures in the area and its transformation into a mountain system were subjects of numerous investigations (Zakharov, 1958; Chediya, 1971; Byelskiy, 1978; and others). However, due to inadequate palaeontological information and disputed opinions on the stratigraphy of this molasse, the correlation of separate events and estimation of their age in the Afghan-Tadzhik depression are far less certain than those in the areas occupied by marine Upper Cenozoic rocks. This present paper utilizes the stratigraphic schemes of Melamed (1966) and Penkov (1974), supplemented and modified by the author, who compared the continental molasse of Tadzhikistan with the marine molasse of the Kopetdagh (Rastsvetayev and Scherba, 1980).

In the Afghan-Tadzhik area, thin salt-bearing, sandy-carbonaceous Jurassic–Eocene deposits pass into thin red lagoonal and continental rocks (Khanabad, Ispharine and Sumsar Beds) which apparently represent the lower molasse. At its base, in the middle part of the Rishta Formation, there are local unconformities. At the end of the accumulation of these rocks, apparently in the Khat time, tectonic movements took place which resulted in the uplift of the western part of the depression. Thin sandy-argillaceous deposits of from the uppermost Oligocene (?) to the lowermost Miocene (the Schurysay Formation) transgressively overlie the eroded Eocene and Oligocene rocks. They are represented by a transgressive-regressive sequence of thin clayey-sandy rocks deposited in a shallow marine basin. In the Middle Miocene or the uppermost Lower Miocene these give place to a thick formation of coarse continental molasse superposed with local unconformities on the underlying complexes. They begin with the sandy silt Baldzhuan Group which in the cis-Darvaz area is an essentially conglomeratic brick-red sequence and is succeeded by the predominantly conglomeratic deposits of the Kingou Group in the middle of the Middle Miocene. For the first time in the cis-Darvaz area olistostromes (Fig. 6) related to the intensification of differential movements along the edge of the mountain system against the valley appear within the upper Baldzhuan and lower Khingou groups, which are separated by an angular unconformity, probably at the Chokrak level. No olistostromic horizons have been found in the Upper Kingou Group (Karagan-Konka?) or the lower Tavildara Group (lower Upper Miocene). They appear among coarse conglomerates in the uppermost Tavildara Group both in the cis-Darvaz area (Fig. 7) and along the southern

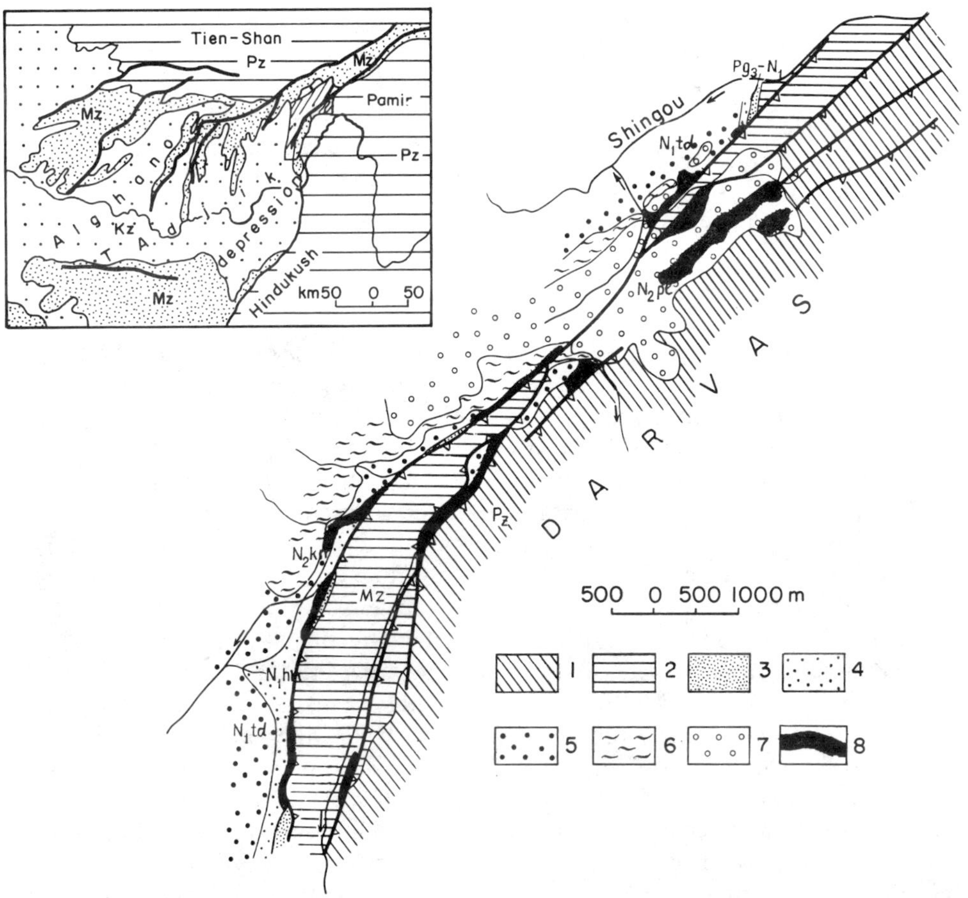

FIG. 6 Diagram showing olistostromes in Neogene molasse in the western margin of the Afghan-Tadzhik depression.

1 – Permian and Carboniferous; 2 – Trassic, Jurassic, Cretaceous, Palaeocene and Eocene; 3 – the end of Oligocene to the beginning of Miocene (Baldzhuan series); 4 – Middle Miocene (Khingou series); 5 – Upper Miocene (Tavildara series); 6 – Lower Pliocene (Karanak series); 7 – Upper Pliocene (Polisak series); 8 – olistostromes. The insert shows the position of the diagram.

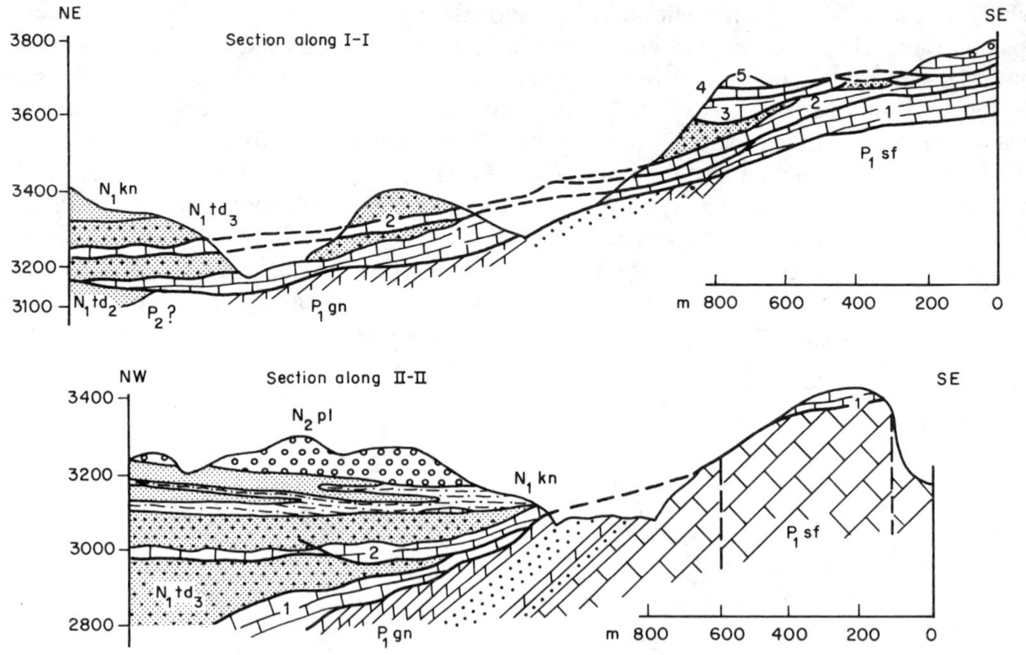

FIG. 7 Geological sections of the eastern margin in the Afghan-Tadzhik depression and the western slope of the Darvaz range in the region of the Vozgina river.

margin of the Tien Shan (Scherba, 1975, 1979). The olistostromes are thick (up to 100–400 m) elongated lenses of chaotic conglomeratic breccia interbedded with conglomerates. The breccias are composed of diverse angular blocks of different sizes originating from Mesozoic and Palaeozoic rocks exposed at the border of the uplift and contributing to the structure of major allochthons or steep fault zones. Within the Breccias are occasional large olistoliths—erratics of these nappes. It is considered (Scherba, 1975) that these braccias and large erratics are mainly a product of squeezing out of parts of the tectonically disrupted uplift block. The entire Upper Tavildara Group of apparently Late Sarmatian age was formed under the conditions of strong mountain-building movements. This group on the margins of the area is unconformably overlain by the alluvium of the lowermast Pliocene Karanak Group which differs from the adjacent formations in its mode of occurrence and distribution. The formation of olistostromes, which reflect the Late Sarmatian movements, in some places occurred at the time when this group started accumulating.

The Kuliab complex which overlies the adjacent rocks transgressively and with a sharp unconformity consists of coarse and poorly graded rocks. From palaeomagnetic and palaeontological data the complex is attributed to the Pliocene and its basal unconformity correlates with Middle Pliocene erosion of the cis-Caspian region. The upper Eo-Pleistocene section of the complex is also separated by an unconformity coeval with the commencement of the pre-Pleistocene movements. In the pre-Pleistocene section of the complex there are olistostrome horizons. In the cis-Darvaz region they occur as gigantic klippen of Permian limestones which are, like the Upper Miocene olistostromes apparently separated from the allochthonous complex. Horizons of chaotic breccias consisting of allocthonous rock fragments are recorded in the internal part of the area in front of the Babatag overthrust. These olistostromes are related to the final orogenic movements that produced a complex fold-overthrust structure of the Afghan-Tadzhik area. The overlying Pleistocene terrace deposits are hardly deformed.

Thus, in the 'pure form', mountain-building movements in the Afghan-Tadzhik area were followed by platformal conditions and where the folding occurred during the Hercynian and Early Cimmerian phases there was no gradual dying out of nappe overthrust movements and this essentially distinguishes it from the other parts of the Alpine geosynclinal region which have been described. The olistostromes in the Afghan-Tadzhik area occur mainly at the edges of the adjacent uplifts. They were mainly squeezed out from the earlier imbricate breccia massifs and large plates or olistoliths. Here one can observe direct transitions of the imbricate zone and breccia into a sedimentary complex where they are normally interbedded with conglomerates. These tectonogravitational formations originated prior to, or simultaneously with, unconformities and changes in conditions of sedimentation. Thus they correlate with the periods of active tectonic movements. These, predominantly vertical, movements caused horizontal flow of rock masses. The related olistostromes were formed in the Middle Miocene, the end of the Late Miocene and in the pre-Pleistocene times. The first phase of olistostrome formation is related to the obvious rise of the mountains which followed the moderate orogenesis at the end of the Oligocene to the beginning of the Miocene. This phase lasted for a long time. Upthrusting which began in the Chokrak was interrupted in the Karagan by a short period of denudation followed by strong upthrusting at the end of the Sarmatian epoch. In the pre-Pleistocene not only general upthrusting but also large-scale fold-overthrusting took place, accompanied by the formation of the second phase of olistostromes. Thus the olistostromes in the Afghan-Tadzhic depression correlate with regressive sediments and are connected mainly with orogenic movements.

It is noteworthy that the tectonic sedimentatary rhythms established here closely correlate with those of the classic Alpine region. One can distinguish (Rastsvetayev and Scherba, 1980) an Oligocene transgressive-regressive cycle (Khanabad, Isphara and Sumsar strata) that finished in the Late Oligocene uplift with the subsequent intense pre-Shurysai period of erosion. Late Oligocene–Early Miocene cycles (Shurysai Formation) began with a transgression at the end of the Oligocene and finished at the end of the Early Miocene with a general uplift and faulting affecting both sides of the depression. Middle–Late Miocene (Baldzhuan, Khungou and Tavildara groups) began with the denudation and uniform sedimentation of the entire area and finished at the end of Sarmatian time with a powerful orogeny accompanied by olistostromes. An early Middle Pliocene cycle followed and terminated with a Middle Pliocene uplift, dislocation and erosion. Then a Late Pliocene cycle finished with the formation of pre-Pleistocene olistostromes and the folding of all Meso-Cenozoic rocks in the area.

The data on approximate isochroneity of transgressive-regressive sedimentation cycles, obtained from the analysis of the Tertiary deposits in the Alpine belt, are in good agreement with the data on changes in the level of the World Ocean (Fig. 8; Vail and Mitchum, 1980). Considerable lowering of the World Ocean level (regressions) took place at the very end of the Ypresian age (49.6 m.y.) between the Bartonian and Preabonian (40 m.y.), in the middle of the Chattian of the Late Oligocene (29 m.y.), in the Serravalian of the Middle Miocene (13 m.y.), in the middle of Late Miocene, in the Sarmatian (9.8 m.y.), in the Messinian (6.6 m.y.) and at the end of the Pliocene (2.8 m.y.). The global changes in the ocean level which resulted in global unconformities at boundaries of the sedimentary cycles seem to be largely due to tectonic causes. The same conclusion is drawn by the Australian geologists who compare Cenozoic sedimentary cycles of New Zealand and Australia (Loutit and Kennett, 1981). It is noteworthy that the cycles they distinguished correlate well with the Alpine ones.

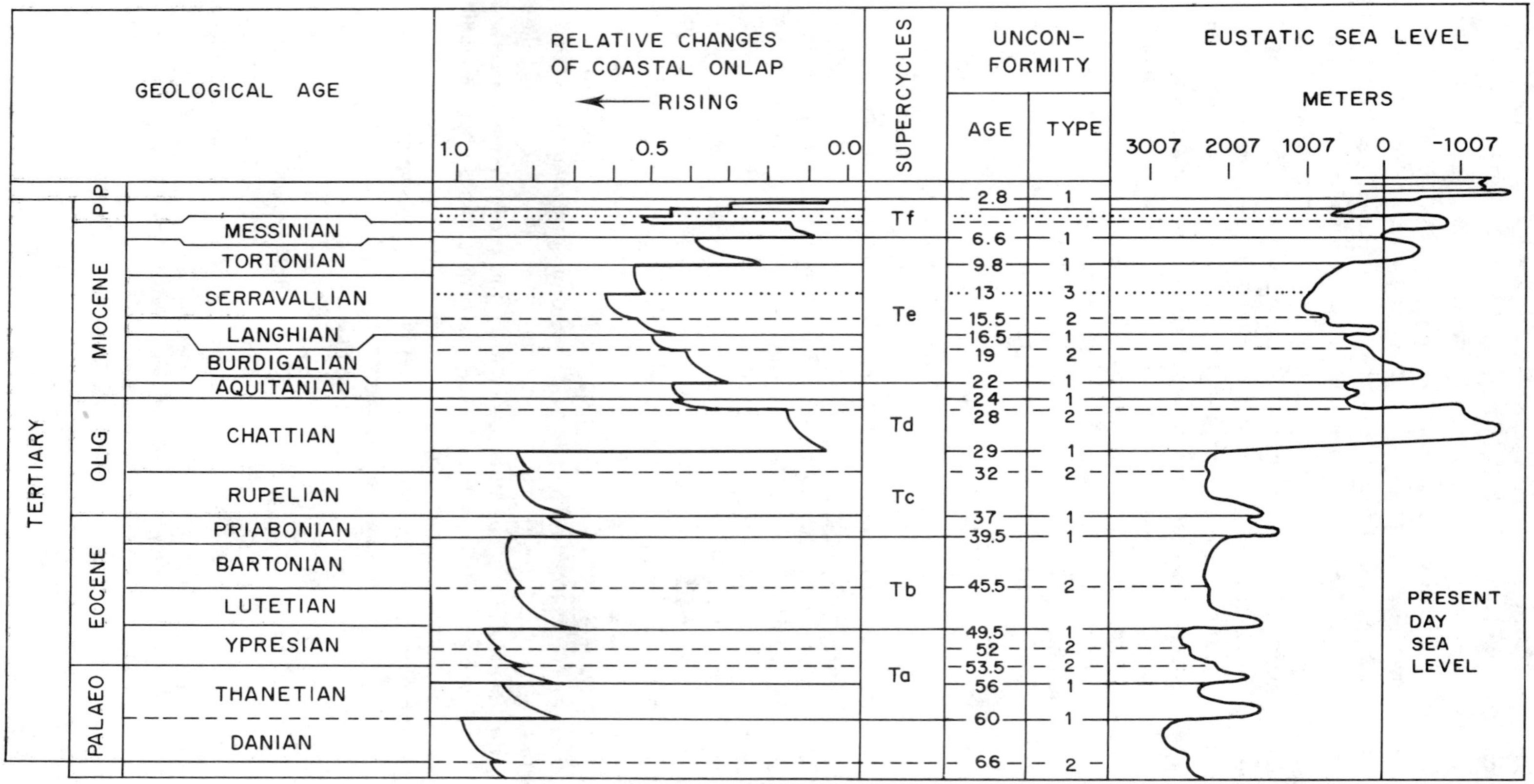

FIG. 8 Global cycles of relative changes in the level of the World Ocean in Cenozoic times (acc. to Vail and Mitchum, 1980, Fig. 1, p. 96).

CONCLUSIONS

Thus, in such widely separated and differently built segments of the Alpine region as the southern Mediterranean, the northern Apennines, the Carpathians, the Caucasus and the Afghan-Tadzhik area, the formation of olistostromes of tectonogravitational origin is discontinuous and is attributed to the same or closely similar time intervals (Fig. 2). These time intervals for the Mediterranean are: the Middle Eocene, the end of the Eocene to the beginning of Oligocene, the end of Oligocene to the beginning of the Miocene, the Burdigalian of Early Miocene or, rarely, of Middle Miocene, the end of the Messinian to the beginning of the Pliocene (Zanclian), the beginning of the Pleistocene (the Calabrian). In the Carpathians olistostromes are found in the Middle Eocene, the beginning of the Oligocene, the end of Oligocene to the beginning of the Aquitanian, the Burdigalian (Otnangian) of Early Miocene and the middle of the Badenian of the Middle Miocene. In the Caucasus these events occurred in the Middle Eocene, the Late Eocene to the Chadum Horizon of Lower Oligocene, the Kozachur Horizon of Lower Miocene, the Chokrak Horizon of the Middle Miocene to the Upper Miocene Sarmatian, the Cimmerian of the Middle Pliocene, the Akchagyl Horizon of the Upper Pliocene to the Apsheron Horizon of Eo-Pleistocene. In the Afghan-Tadzhik area these events occurred at the end of Middle Miocene (Chokrak) to the Sarmatian and at the Plio-Pleistocene boundary.

It can therefore be claimed that in all the regions discussed the phases or epochs of olistostrome formation are simultaneous and cover the Late (sometimes also Middle) Eocene to the Early Oligocene (42–36 m.y.), the Late Oligocene to the beginning of the Early Oligocene (30–24 m.y.), the end of the Early Miocene (16–19 m.y.) the end of the Middle Miocene to the Tortonian (Sarmatian) (15–19 m.y.), the latest Late Miocene to the Middle Pliocene (6–4 m.y.), the Plio-Pleistocene boundary (1.5–2 m.y.).

Olistostromes of each stage do not occur everywhere, but all their major occurrences in the discussed parts of the Alpine belt are confined to the above-mentioned time intervals. Moreover, it is to these and not to any other time intervals that all occurrences of tectonogravitational olistostromes, known to the author in many other parts of the Alpine belt, are confined. Hence the phases of olistostrome formation coincide over a large territory of the Alpine region from the Pyrenees to the Pamirs.

In addition to the previously described region, the Late Eocene–Early Oligocene phase manifests intelf in the Central Pyrenees and in the adjacent northern and southern troughs (Fig. 2, Sections 10, 11 and 12) (Meurisse, 1974; Kieken and Winnock, 1973; Choukroune and Séguret, 1973), in the Digne basin of the Maritime Alps, in the Ultradauphinoise and Ultrahelvetian zones of the western Alps (Fig. 2, Sections 14, 15; Gigot, 1973; Roux, 1973), in the Smolian and Khvoinen basins of the Central Rhodope (Fig. 2, Section 20; Moskovskiy and Ivanov, 1980), in the external flysch trough of the Dinarides (Fig. 2, Section 21; Belostotskiy, 1970), at the front of the Beyshehir trough in the western Taurus massif (Fig. 2, Section 26; Ricou *et al*., 1974) and in the south eastern segment of the Zagros (Stöcklin, 1968).

The Late Oligocene–Early Miocene olistostromes occur in the Ebro basin, on the northern slope of the Pyrenees (Fig. 2, Sections 13, 10; Cerdé and Santanach, 1974; Choukroune and Séguret, 1973), in the Gavrovo zone of the Dinarides (Fig. 2, Section 21; Richter, 1976).

The olistostromes of the end of Early Miocene to the beginning of the Middle Miocene occur in the Ebro basin (Fig. 2, Section 13; Canérot and Fondecave, 1975), in the Göcek window in the Lycian Taurus (Fig. 2, Section 26; Graciansky *et al*., 1970).

The olistostromes of the end of Middle Miocene to the Tortonian and Sarmatian stages of

Upper Miocene are known in the Stara Planina (Fig. 2, Section 19; Minchev, 1960) and in the western Taurus (Fig. 2, Section 26; Brunn, 1974).

The olistostromes of the end of the Messinian to the beginning of the Pliocene are known in the Russillion area of the Central Pyrenees (Fig. 2, Section II; Clauzon, 1974), in the Maritime Alps (Fig. 2, Section 14; Clauzon, 1972; Gigot *et al.*, 1974), in the Crimea (Fig. 2, Section 22; Scherba, 1978) and in the marginal fold zone of the eastern Taurus Massif (Fig. 2, Section 26; Ilhan, 1974).

The Pleistocene olistostromes are known in the cis-Balkanian region (Fig. 2, Section 19; Minchev, 1960), in the Sidirokastron basin of the Macedonian massif (Fig. 2, Section 20; Sotiriadis, 1973) and in the Kafan zone of the Lesser Caucasus (Fig. 2, Section 25).

All these olistostromes are associated with different manifestations of tectonic movements. In the Mediterranean region, they are formed during the movement of tectonic nappes into the troughs which enlarge at the expense of the foreland, the movements, becoming truly gravitational from the middle of the Miocene onwards. In many cases (the Lower Miocene of the Tell Atlas, the Middle Miocene of El-Rif) olistostromes obviously formed during short periods corresponding to 1–2 zones of Blow or 1–2 m.y. But the time intervals of one stage of olistostrome formation are usually longer. It is the time of general upthrusting and accumulation of regressive sedimentary sequence of a single tectonic sedimentation cycle and the beginning of subsidence related to the next cycle when the newly-formed escarpments contribute to the sliding of frontal parts of nappes and the formation of post-nappe olistostromes. Thus, if the beginning of olistostrome formation of one stage is synchronous with nappe movement, the end may, sometimes, include also the time of transition from one cycle to another. This systematic feature is characteristic of all the regions described here. It accounts for a stochastic formation of olistostromes both within a single region and within the entire Alpine area.

Indeed, the Late Oligocene–Early Oligocene and the Early Miocene Carpathian olistostromes are mainly confined to the beginning of general regression stages when major nappe movements took place. They appear in the middle to the end of the Oligocene and the beginning of the Burdigalian age, whereas the formation of the Mediterranean olistostromes of these stages took place at the end of the Oligocene to the beginning of the Aquitanian and in the Burdigalian to the beginning of the Langian. Conversely, Carpathian olistostromes of the beginning of Oligocene appear at the beginning of an abrupt subsidence which followed the late Eocene deformation, whereas in the Mediterranean and the Alps the olistostromes of this stage are mainly related to the nappe movements preceding the Oligocene subsidence. In the Middle-Late Miocene stage of the Carpathians, the Caucasus and the Afghan-Tadzhik area it was demonstrated that olistostromes, formed as a result of destruction of escarpments produced by vertical orogenic movements, do not form continuous Middle-Late Miocene sections as in Sicily, Calabria and the southern Apennines, but occur at two levels: in the Chokrak and at the end of the Sarmatian with short intervals during which uplift weakened and transgression renewed. The accumulation of olistostromes continued throughout the Lower Tortonian transgression in the previously discussed Mediterranean regions, where olistostromes accompanied gravity nappes. In El-Rif and Betic Cordillera the time of formation of olistostromes of this phase is only confined to the end of Middle Miocene when the nappe movement was over. On the whole, the Middle-Late Miocene stage of olistostrome formation is marked by the least distinct time boundaries. This is caused by regional differences in orographic and palaeogeographic conditions during transitions from the general compression to the uplift of orogenic zones. During the Late Miocene–Early Pliocene, in the Mediterranean, the formation of the olistostromes began together with the Late Messinian regression and is related to gravitational nappe movements which continued locally at the beginning of the Pliocene. The gravity flow of imbricated massifs, in the Maritime Alps and Pyrenees took place mainly in the Early Pliocene, and does not seem to continue anywhere until the formation of the pre-Pyachenzian unconformity. In the Caucasus and eastern Taurus the formation of olistostromes obviously continued until the end of the Early Pliocene and in the Middle Pliocene, which was due to considerable concurrent nappe movement.

Thus, the duration of a stage of olistostrome formation, which involves the maximum remote time of appearance of olistostromes and does not always correspond to the duration of formation of certain olistostromes, is from 1–6 m.y., the intervening (quiescent) periods being 3–9 m.y. From the middle (?) or the end of the Miocene such stages became more frequent, whereas their duration shortened. During the formation of olistostromes of a single stage most variable deformations take place (general uplift, accelerated uplift, nappe movement, formation of unconformities, erosion, abrupt subsidence), each being a shorter event than the formation of olistostromes. All these deformations are characteristic of the stages of olistostrome formation as the stages of prevailing tectonic movements. The deformations are not always accompanied by the generation of olistostromes, but it is remarkable that as a rule no olistostromes were formed in the intervals between the described stages.

All geological phenomena connected with olistostrome formation in the western part of the Alpine belt correlate well with Stille's orogenic phases: Pyrenean, the Savic, the Styrian, the Attic, the Rhodanian, and the Walachian (see Fig. 2). Thus in such a heterogeneous fold region as the Alpine, the orogenesis and overthrusting took place in pulses, their reactivation phases being persistent throughout the region from the Pyrenees to the Pamirs. This proves that the vertical and horizontal late Alpine movements are due to the same causes.

In the development of olistostromes and tectonogravitational nappes, the northern and southern branches of the Alpides differ fundamentally. Unlike the southern branch, where the movement of nappes and the accompanying olistostromes into the growing troughs proceeded throughout the Late Cenozoic up to the Pleistocene, the olistostromes of the northern branch are mainly confined to the Late Eocene, the Late Miocene and, to a lesser extent, the Early Miocene epochs. They occur mainly in the relict geosynclinal troughs which in the Middle Miocene slightly spread towards the foreland, thus restricting the scope of tectonogravitational processes. The presence of Late Miocene and Pliocene olistostromes in the Pyrenees, the Maritime Alps, the Crimea and the Afghan-Tadzhik area is due to gravity flow of earlier formed allochthonous massifs which underwent anti-isostatic uplift. The differences in the Late Cenozoic development of the northern and southern branches of the Alpine belt may be due to their different position with respect to the margins of the African and Arabian plates that were moving towards each other.

REFERENCES

Andreyeva, A. S., Gruzman, A. D. and Dosin, G. D. (1981). The problem on the Oligocene-Miocene boundary in the Pokutia-Bukovina Carpathians, summary of XII Congress, KBGA, Bucharest: Geology and Geophysics Institute, pp. 57–8 (Russian).

Belostotskiy, I. I. (1970). Zones of melange and chaotic structures, in *Chronicles of Structural Geology of Strongly Deformed Rocks*, Moscow: Nedra, pp. 170–92 (Russian).

Byelskiy, V. A. (1978). The most recent tectonics of the border zone between the northern Pamirs and the Tadzhik Depression, Dushanbe: Denish, p. 253 (Russian).

Boryukayev, Ch. B., Rastsvetayev, L. M. and Scherba, I. G. (1981). Mesozoic and Cenozoic olistostromes on the Southern flank of the western Caucasus, *Bull. Mosk. obshch. isp. prirody*, geol. section, no. 6, 32–44 (Russian).

Bouillin, J. P., Durand-Delga, M., Gelard, J.-P-., Leikine, M. *et al.* (1973). Les olistostromes d'âge miocène inférieur liés aux flyschs allochthone kabyles de l'orogène alpin d'Algérie, *Bull. Soc. géol. France*, **15**, nos. 3–4, 340–4.

Bourgois, I. and Chauve, P. (1971). Nouvelles données stratigraphiques et tectoniques sur les sierras du Lagarin et du Malaver (province de Cadix et Malaga, Espagne meridionale), *Bull. Soc. géol. France*, sér. 7, **13**, nos. 3–4, 321–7.

Brunn, J. H. (1974). Le probème de l'origine des nappes et de leurs translations dans les Taurides occidentales, *Bull. Soc. géol. France*, sér. 7, **16**, no. 2, 101–6.

Caire, A. (1973). In *Gravity and Tectonics*, eds. K. A. De Jong and R. Scholten, New York: Wiley, Chapter II, pp. 168–82.

Canérot, J. F. M.-J. and Fondecave, M.-T. W. (1975). Identification du Miocène inférieur-moyen dans la basse vallée de l'Ebre. Conséquences sur l'âge de la tectogenèse alpine aux confins des chaînes ibérique et catalane (Espagne), *C. R. Acad. sci.*, **280D**, no. 17, 1951–4.

Catalano, R., Renda, P. and Slaczka, A. (1976). Redeposited gypsum in the evaporitic sequence of the Ciminna basin (Sicily), *Mem. Soc. geol. Ital.*, **16**, 83–93.

Cerdé, M. E. and Santanach, P. P. F. (1974). Deslizamientos gravitacionales y olistostromas miocenos en el bajo Gaià y campo de Tarragona, *Acta geol. hisp.*, **9**, no. 4, 117–26.

Chediya, O. K. (1971). *The Southern Part of Central Asia during the Most Recent Period of Orogenesis*, vol. 1, Frunze; Ilim, p. 201 (Russian).

Choubert, G. and Faure-Muret, A. (1971). Mémoire sur la tectonique du continent africain, UNESCO, p. 534.

Choukroune, P. and Séguret, M. (1973). In *Gravity and Tectonics*, eds. K. A. De Jong and R. Scholten, New York: Wiley, Chapter II, pp. 154–67.

Clauzon, G. (1972). Sur la présence d'une brèche syntectonique d'âge miocène supérieur sur le flanc sud du Petit Luberon (Vauclux, France), *C. R. Acad. Sci.*, **275D**, no. 18, 1963–7.

Clauzon, G. (1974). La brèche d'Espira de l'Agly (Pyrénées Orientales): une brèche synorogénique d'âge messinien, *Bull. Soc. géol. France*, **16**, no. 3, 273–80.

Didon, J. (1978). Rôle des phénomènes de glissement et d'écoulement par gravité dans la mise du matériel flysch à la périphérie de l'arc Gibraltar, *Bull. Soc. géol. France*, **19**, no. 4, 765–71.

Elter, P. and Trevisan, L. (1973). In *Gravity and Tectonics*, eds K. A. De Jong and R. Scholten, New York: Wiley, Chapter II, pp. 183–95.

Faure-Muret, A. and Choubert, G. (1975). Proposition d'un nouveau modèle tectonique pour la Méditerranée occidentale, *C. R. Sci. France*, **280D**, no. 17, 1947–50.

Feinberg, H. (1976). Mise en place, au pliocène, d'une nappe de glissement à l'extrémité sud-occidentale de la chaîne du Rif (Maroc), *Bull. Soc. géol. France*, **18**, no. 6, 273–6.

Feinberg, H. and Leblanc, D. (1977). Preuves de l'âge anté-langien de charriages dans le Rif oriental (Maroc), *C. R. Acad. sci.*, **285D**, no. 16, (1977). 1387–90.

Fenet, B. (1977). La phase lutétienne est une phase majeure dans les zones externes de l'orogène nordmaghrébien. *C. R. Acad. sci.*, **285D**, no. 4, 283–6.

Gigot, P. (1973). Sur la présence de klippes sédimentaires et de mégabrèches dans la partie basale des formations continentales d'âge Eocène supérieur-Oligocène du bassin tertiaire de Digne, dans la région de Volonne (Alpes de Haute-Provence), *Bull. Bur. rech. géol. et minières*, **4**, sec. 2, no. 1, 17–25.

Gigot, P., Grandjacquet, C. and Haccard, D. (1974). Evolution tectono-sédimentaire de la bordure septentrionale du bassin tertiaire de Digne depuis l'eocène, *Bull. soc. géol. France*, **16**, sér. 7, 128–39.

Gluschenko, L. A., Zhugunova, T. F., Kuzovenko, V. V., Lozinyak P. Yu. and Temnyuk, F. P. (1980). Olistostromes found in the Oligocene deposits of the Krosno (Silesian) zone of the Ukrainian Carpathians, *Materialy XI Congressa KBGA (litologiya)*, Kiev; Naukova Dumka, pp. 55–64.

Graciansky, P.-Ch., Lorenz, J. and Magné, J. (1970). Sur les étapes de la transgression du miocène inférieur observée dans les fenêtres de Göcek (Sud-Ouest de la Turquiet), *Bull. Soc. géol. France*, sér. 7, **12**, no. 3, 557–64.

Hamor, G. (1980). Entwicklungsgeschichte, Paläogeographie und tektonische Phases des Badeniens in Ungarn, *Proceedings of the XI Congress of the Carpathian-Balkan Geological Association, Tectonics*, Kiev: Naukova Dumka, pp. 226–34.

Hoedemaeker, Ph. I. (1973). Olistostromes and other delapsional deposits and their occurrence in the region of Moratalla (Prov. of Murcia, Spain), Lieden: Rejkmuseum van geologie en mineralogie, p. 247.

Iacobacci, A. (1965). Frane sottomarine nelle formazioni geologishe. Interprotazione dei fenomeni olistostromici e degli olistolite nell—Apennino e in Sicilia, *Boll. Serv. geol. Ital.*, **86**, p. 65–85.

Ilhan, E. (1974). In *Mesozoic-Cenozoic Orogenic Belts*, ed. A. M. Spencer, London: Geological Society of London, pp. 234–47.

De Jong, A. and Scholten, R. (eds) (1973). *Gravity and Tectonics*, New York: Wiley.

Keller, B. M. and Menner, V. V. (1945). Palaeogene deposits of the Sochi region and related subaqueous landslides, *Bull. Mosk. obshch. isp. prirody*, geol. section, nos 1–2, 30–45 (Russian).

Khain, V. E. (1978). In Tectonics of Europe and its adjacent areas (explanatory notes to the International tectonic Map of Europe and its adjacent areas, Scale 1:2,500,000), Moscow: Nauka, Chapter III, 12, pp. 494–515 (Russian, English).

Kieken, M. and Winnock, E. (1973). Olistostromes dans le lutétien du Bas-Adour à l'avant du Front nord-pyrénéen, *Bureau rech. géol. et minières*, **4**, no. 1, 5–15.

Konior, K. (1981). Rola miocenu w budowie i tektogenezie brzeznej strefy Karpat obszaru Cieszyn-Wadowice, *Prz. geol.*, **29**, no. 1, 5–12.

Kopp, M. L. and Kurdin, N. N. (1981). Regional displacements observed in the South-Eastern Caucasus, *Izv. Vuz'ov, Geologiya i Razvedka*, no. 11, 30–9 (Russian).

Kulchitskiy Ya. O. (1977). Olistostromes, olistoliths and other subaqueous landslide phenomena found in the flysch of the eastern Carpathians, in *Geol. i polezn. iskopaem. Ukr. Karpat*, Ch. 2, L'vov: Vishcha shkola, pp. 44–54 (Russian).

Langenberg, C. W. (1973). Gravitational gliding in the northern Sierra de los Filabres (S.E. Spain), *Geol. en mijnbouw*, **52**, no. 4, 187–91.

Leonov, M. G. (1970). On the destruction of the frontal parts of overthrusts, *Dokl. AN SSSR*, **193**, no. 3, 669–71 (Russian).

Leonov, M. G. (1975). *Wildflysch of the Alpine Region*, Moscow: Nauka, Trudy GIN AN SSSR, issue 299, p. 138 (Russian).

Leonov, M. G. (1981). *Olistostromes in the Structure of Folded Regions*, Trudy GIN AN SSSR, issue 344, Moscow: Nauka Publishers, p. 173 (Russian).

Lespinasse, P. (1977). Les marnes à blocs de la région de Zoumi: leur signification dans l'évolution structurale des zones externes du Rif (Maroc), *Bull. Soc. géol. France*, sér. 7, **19**, no. 4, 781–7.

Loutit, T. S. and Kennett, I. P. (1981). New Zealand and Australian Cenozoic sedimentary cycles and global sea-level changes, *Am. Assoc. Petrol. Geol. Bull.*, **5**, no. 9, 1586–1601.

Melamed, Ya. R. (1966). Types of sections and stratigraphy of the Neogene deposits in the South Tadzhikistan, *Izv. Vuzov Geologiya i Razvedza*, no. 5, 10–20 (Russian).

Sandulesçu, M. (1978). In Tectonics of Europe and its adjacent areas (explanatory notes to the international tectonic Map of Europe and its adjacent areas, Scale 1:2500000), Moscow: Nauka, Chapter III.4, pp. 382–414 (Russian, English).

Meurisse, M. (1974). Sur les brèches vitrolliennes des Pyrénées centrales et orientales, *C. R. Acad. sci.*, **278D**, no. 12, 1545–7.

Minchev, D. (1960). The Sarmatian and Quaternary olistostromes in the region of Predbalkans near Cherven Bryag, Godishnik na Sofiyskiya Universitet, biol. geol. geogr. f-t, vol. III, part 2, geologiya, Sofia: Nauka i Izkustvo (state publishers), pp. 14–21 (Bulgarian).

Moskvin, M. M. and Semikhatov, M. A. (1956). The subaequeous slump dislocations in the Upper Cretaceous and Palaeogene deposits of Daghestan, *Izv. AN SSSR*, ser. geol., no. 10., 67–84 (Russian).

Moskovskiy, St. and Ivanov, Zh. (1980). Olistostromes of the Central Rhodope. Olistostromes in the Central Rhodope, Central and Eastern Stara Planina and in the Krayshtid Region (An Excursion Guide), Sofia, pp. 27–65.

Pen'kov, A. V. (1974). (Regional Palaeomagnetic Correlation of Continental Rock Masses (on the example of the Cenozoic molasse of South Tadzhikistan), candidate thesis, geol. miner. sciences, Moscow, p. 27 (Russian).

Perrone, V., Torre, M. and Zeppetta, A. (1973). Il Miocene della Catena costiera calabra. Primo contributo: zona Diamante-Bonifati-Sant' Agata d'Esaro (Cosenza), *Riv. ital. paleontaol. e stratigr.*, **79**, no. 2, 157–205.

Petrashkevish, I. I. and Pishvanova, L. S. (1967). Main stratigraphic units of the Transcarpathian Miocene and their possible analogues in the adjacent basins, *Voprosy geolog. Karpat. L'vov. Univers.*, v. 2, 99–114 (Russian).

Rastsvetayev, ?, and Scherba, I. (1980). Problems pertaining to global correlation of geological phenomena, *Trudy GIN AN SSSR*, issue 340, Moscow: Nauka pp. 111–33 (Russian).

Raymond, D. (1976). Evolution sédimentaire et tectonique du nord-ouest de la Grand Kabylie (Algérie) au cours du cycle alpin, *Ann. sci. Univ. Besançon Géol.*, no. 26, p. 47–76.
Biela, A., Hlyboke vrty v zakrytych oblastiach vmitozrych Zapadnych Kazpat. Reg. geol. Za'p. Karp't. Geologicky Ústav Dioniza Štúra, Bratislava, N10, 1978, p. 224.
Richter, D. (1976). Die Flyschzonen Griechenlands, III Flysch sowie spatund postorogene Serion in West-Griechenland zwischen Albanien und dem Golf von Patras, Tiel 2, *Neues Jhrb. Geol. und Paläontol. Abh.*, **151**, no. 2, 224–52.
Ricou, L., Argyriadis, I. and Lefévre, L. (1974). Proposition d'une origine interne pour les nappes d'Antalya et le massif d'Antalya (Taurides occidentales, Turquie), *Bull. Soc. géol. France*, sér. 7, **16**, no. 2, 107–11.
Roux, M. (1973). Sur la stratigraphie du bassin d'Eoulx-Brenon et l'âge Ludien supérieur des mouvements tectoniques à l'origine des fossés du Nord-Est du Var, *C.R. Acad. sci.*, **276D**, no. 26, 3413–20.
Scherba, I. G. (1975). Olistostromes in the Neogene of the Darvaz Range, *Geotectonika*, no. 5, 97–108 (Russian).
Scherba, I. G. (1978). The Pliocene-Quaternary olistostromes of the Crimea and the mechanism of their formation, *Bull. Mosk. obshich. isp. prirody*, no. 4, 23–34 (Russian).
Sistini, G. (1974). In *Mesozoic–Cenozoic Orogenic Belts*, ed. A. M. Spencer, Edinburgh: Geological Society of London, pp. 83–110.
Sokolov, S. D. (1977). The Olistostrome series and Late Cretaceous tectonic nappes of ophiolite zones of the Lesser Caucasus, *Trudy GIN AN SSSR*, issue 296, Moscow: Nauka. p. 170 (Russian).
Sotiriadis, L. (1973). Über festländische Olisthostrome im Nord Makedonien (Griechenland), *Neues Jahrb. Mineral. Monatsh.*, no. 1, 35–56.
Spencer, A. M. (ed.) (1974). *Mesozoic-Cenozoic Orogenic Belts*, Edinburgh: Scottish Academic Press, for Geological Society of London, p. 452.
Ştefânescu, M., Gheta, N. and Dicea, M. (1979). On the Oligocene-Miocene boundary in the external flysch zone of the Carpathian bend (between the Teleajen valley and the Dimbovita valley). A tentative solution by calcareous nannoplankton, *Rev. Roum. Géol., Géophys. et Géograph., Géologie*, **23**, no. 1, 89–94.
Ştefânescu, M. (1980). Slon olistostrome: lithological composition, geometrical position, tectonic significance, *Proceedings of the XI Congr. of Carpathian-Balcan Geological Association*, Kiev: Naukova Dumka, pp. 232–42.
Stöcklin, J. (1968). Structural history and tectonics of Iran, *Bull. Amer. Assoc. Petrol. Geol.*, **52**, no. 1, 79–98.
Suter, G. (1978). In Tectonics of Europe and its adjacent areas (explanatory notes to the International tectonic Map of Europe and its adjacent areas, Scale 1:2,500,000), Moscow: Nauka, Chapter III.2, 289–97 (Russian, English).
Szymakowska, F. (1976). Olistostromes in the Krosno beds (Polish Central Carpathians), *Ann. Soc. geol. Pol., Warszawa*, **46**, nos. 1–2, 39–54.
Vail, P. R. and Mitchum, R. M. (1980). Global cycles of sea-level change and their role in exploration, *Proceedings of the 10th Petroleum Congress*, vol. 2, Heyden, London: Philadelphia = Rheine, pp. 95–104.
Vassoevich, N. B. and Korotkov, S. T. (1935). Recognition of large subaqueous landslides ascribed to the Oligocene in the North Caucasus, *Trudy VNIGNI*, **52**, 19–21 (Russian).
Vassoevich, N. B. (1949). Middle Miocene subaqueous landslides of Daghestan, *Azerbaydzh. Neft. Knoz-vo*, no. 3, 13–17 (Russian).
Zakharov, S. A. (1958). Stratostructures of the Mezo-Cenozoic of the Tadzhik depression, *Trudy A.N. Tadzhik SSR*, **95**, Dushanbe, 223 (Russian).
Zanzucchi, G. (1959). Frane pleistocenione nell'Alta vall dell'Ofanto, *Boll. serv. geol. Ital.*, **80**, fasc. 2–3, 203–8.

Global Correlation of Tectonic Movements
Edited by Yu. G. Leonov and V. E. Khain

5

Summary of the Timing of Geotectonic Events during Cretaceous and Tertiary Times in the Balkan Peninsula

WERNER SCHWAN*

The aim of this report is to determine and summarize the timing of Cretaceous–Tertiary orogenies, especially in the area of the Balkan Peninsula. These investigations were performed for the IGCP project 107, 'Global correlation of epochs of tectogenesis'.

The Balkan Peninsula, referred to henceforth as the Balkan region, reveals an extremely variable geological development. Accordingly, its mountain structures have a complicated regional tectonic pattern, including the timing of the tectonic events. Before reviewing the timing of the orogenies, the geotectonic position and structural main features of the Balkan region will be considered.

In Triassic and Jurassic times, the Balkan region was largely orthogeosynclinal and/or marginal-oceanic and was subjected later on to several phases of Alpidic orogeny. Therefore the Balkan region belongs mainly to Neoeurope and is a part of the European plate. The Neoeuropean Balkan region is bordered by Mesoeurope to the northwest and by Eoeurope to the northeast. The Walachian spur was already consolidated in Palaeozoic times and projects as an isolated part of Mesoeurope from east to west into the Neoeuropean area of the Balkan region. The latter is in contact on its southern side with the African plate which is there flooded by the Mediterranean. The Adriatic spur, which was already consolidated in the Palaeozoic, defines the Balkan region on its southwestern side (Fig. 1).

The orogens of the Balkan region are located in the zone of the widespread, more than 10,000 km long Alpino-Himalayan bilateral mountain system. According to Kober (1931) and Stille (1924), this system has two belts—a northern and a southern. In the Balkan region, the southern belt is represented by the Dinarides-Hellenides and the northern belt by the Carpathides-Balkanides. The foreland for the southern belt is the African plate with Eoafrica. The foreland for the northern belt consists of Eoeurope and Mesoeurope as partial cratons of the European plate. Accordingly each of these belts verge towards the underriding foreland plates to the north and south. Between these two fold belts, marginal basins have developed within the Balkan region: the Pannonian basin in the northwest and the Aegean Sea in the southeast (Stegena *et al*., 1975; Boccaletti *et al*., 1976; Szádeczky-Kardoss, 1976; Schwan, 1976). Concerning the last mentioned area, a complete and still active island arc/marginal sea system exists which comprises the Hellenic deep-sea trench, the island arc with Crete and Rhodes, and the marginal Aegean sea with an active volcanic arc.

* Institut für Geologie und Mineralogie der Universität Erlangen-Nürnberg, Schlossgarten 5, 8520 Erlangen, Fed. Rep. Germany.

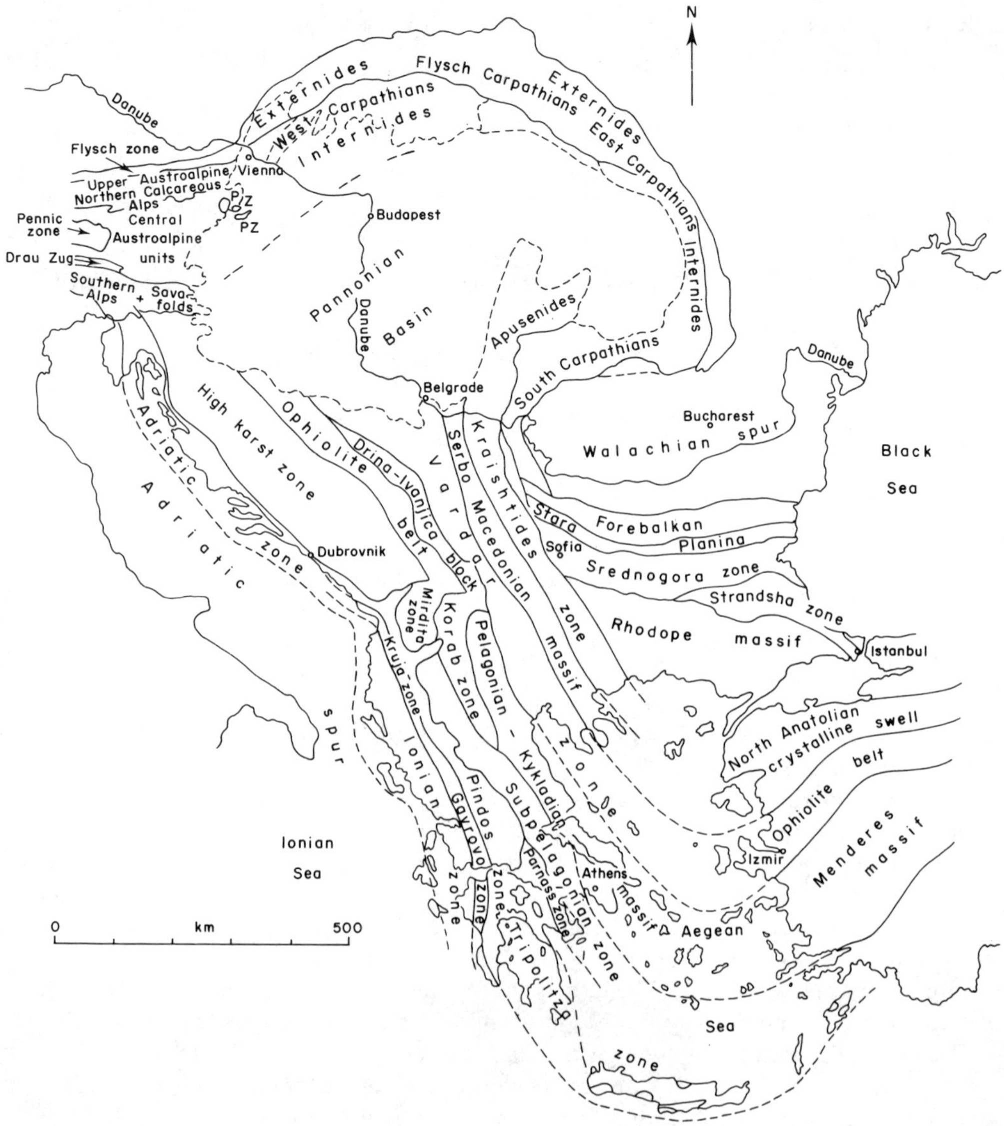

FIG. 1 Sketch of the geotectonic division of the Balkan Peninsula. Based mainly on Angenheister *et al*. (1972), Aubouin (1977), Bonchev (1970), Dimitrijević (1974), Hiessleitner (1951), Jacobshagen and Makris (1974), Janković (1977), Schwan (1976) and Stille (1953).

The regional palaeotectonic development and the evolution of the individual areas of the Balkan region up to its present stage cannot be investigated in this report. There are a number of kinematic interpretations which indicate the movement courses of macro-plates and micro-plates differently (Dewey *et al*., 1973; Boccaletti *et al*., 1974a; Biju-Duval *et al*., 1976; Hadzi *et al*., 1976).

Generally, existing interpretations support the idea of an European ocean and/or an orthogeosyncline in the Tethys area with at least one important active ridge and/or a rift zone, probably with branches, from which the oceanic crustal parts with continental micro-plates drifted partly to the north and partly to the south or in other directions. On the other hand, the European and African macro-plates were pushed towards the Tethys. This caused subduction, obduction, collision, or movements on transform faults. Due to the approach and enlargement of the European and African plates through oroplutogenic processes of 'sialization', the oceanic and/or orthogeosynclinal areas between these plates have gradually disappeared. In this way, the Balkan mountain belts also developed in Mesozoic–Tertiary time through the influence of several compressive orogenic processes. On the other hand, in Late Tertiary time, processes of extension of the marginal sea-floor spreading and/or of oceanization in a wider sense were effective in the rather stable 'intermediate massifs' and/or the marginal basins of Hungary and the Aegean Sea (Stegena *et al.*, 1975; Boccaletti *et al.*, 1976; Szádeczky-Kardoss, 1976; Schwan, 1976).

In the following text the main subject is the time of occurrence of the important structural events of the Alpinotype deformations and other essential processes in some parts of the Balkan region during Cretaceous and Tertiary times (Fig. 2). These events determine the main tectonic features of the region. The detailed time analysis of the areas concerned and their comparison may serve as a basis for interpretation of the kinematic and regional tectonic development of the individual parts in the frame of the Balkan region. The present investigation is based partly on the evaluation of literature and partly on the author's own experience in the field.

Firstly, the southern belt of the Alpino-Himalayan mountain system in the Balkan region, the Hellenides-Dinarides, will be described with regard to the timing of tectonic development. The following can be said about the Dinarides-Hellenides: above a basement which had in part already been subjected to the Hercynian (Variscan) folding, an eugeosynclinal-miogeosynclinal system ('couple' according to Aubouin *et al.*, 1960–63) again developed. Its Late Palaeozoic–Mesozoic–Early Tertiary sediments contain in their eugeosynclinal portion initial magmatic rocks which are typical of reactivation.

The Alpidic orogeny began in the inner zone of the Hellenides at the margin of the Rhodope Massif and in the Vardar zone at about the end of Jurassic time (Aubouin, 1960; Mercier, 1966a,b, 1973) or was already at the Dogger–Malm boundary (Kauffmann *et al.* 1976). Thereafter, the Vardar flysch developed in Upper Jurassic–Lower Cretaceous times (Aubouin, 1960). According to the rule of the shift of mountain-building processes, the Inner Hellenides were affected by compressive movements at the earliest point of time. This lasted between 160 and 120 m.y. during the Eohellenic tectogenesis and probably had several peaks of activity (Jacobshagen *et al.*, 1976b, p. 135) which correspond to the Late Cimmerian cycle. Intensified activity occurred during the Tithonian at the end of the Jurassic, during the 'révolution finijurassique' (Aubouin and Dercourt, 1975). During this process a nappe thrust of oceanic crust ('Eohellenic nappe') occurred, implying an obduction of oceanic basic rocks of the Vardar zone from the northeast to the southwest. The Neocimmerian phase especially affected Attica and Euboea (Albandakis, 1977). Approximately simultaneously with the Upper Jurassic orogeny, in the area of Chalcidice and in Macedonia (Kockel *et al.*, 1972) granites appeared over a period between 150 and 140 m.y. For the Pelagonian zone a somewhat younger tectonothermal Early Lower Cretaceous (Valanginian) event is indicated (Yarwood and Dixon, 1977). At the Jurassic–Cretaceous boundary an accelerated approach of the European and African plates and thus a reduction of the oceanic area of the Tethys between both plates has been recognized (cf. Dercourt, 1970; Bernoulli and Laubscher, 1972). This might be the cause of the Late Cimmerian tectogenesis.

A further orogenic activity is indicated for the Pelagonian and Subpelagonian zones during the Late Lower Cretaceous times, between the Valanginian and the Aptian (Bernoulli and Laubscher, 1972; Richter 1976, table 1). It can be correlated to the

OROGENES OF THE BALKAN PENINSULA

System	Series and/or stage	Age in m.y.	E- Alps	Hellenides	Dinarides	Carpatides	Balkanides	Anatol. orogenes	Ocean floor - spreading discontin.
Tertiary	Pliocene	5	Attic ph.						
	Miocene		Styrian phases	Attica, Cyclades, Kos; Externides		Moldavides		Outer Taurides Lycian nappe	Atlantic, Indian Ocean, Pacific ~ 10 m.y.
		23	Savic phases			Externides, Internides	Kraishtides		Atlantic, Indian Ocean, Pacific 27 - 23 m.y
	Oligocene	36	Pyrenean ph.; Illyrian ph.	Crete, Rhodes Internides, Vardar zone	Externides; Internides, Serb. zone	Externides W. Carpathians	Balkan and Forebalkan (Externides)	Outer Pontides Inner Taurides	N. Atlantic and Indian Ocean 42 - 38 m.y.
	Eocene	53.5				Apusenides S. + E. Carpathians	Srednogorie (Internides)	Middle and S. Anatolia	
	Palaeocene	65	Laramide phases	Internides, E. Aegean, Crete	Internides Serbian zone			Amanos	Atlantic and N. Pacific 65 - 60 m.y.
Cretaceous	Upper Cretaceous		Subhercyn. phases; Mediterr. ph.			Internides		Middle Anatolia	N. and S. Atlantic and Indian Ocean 81 - 77 m.y.
		100	2. Austrian ph.				Strandsha		
	Lower Cretaceous		1. Austrian ph.; Austroalpine ph.	Pelag. and Subpelag. zone	Internides	Internides, Apusenides	Forebalkan Ludogor. ph.	Outer Pontides	N. Atlantic, Arctic, Pacific 115 - 110 m.y.
		136							
Jurassic	Upper Jurassic		Late Cimmerian ph.	Attica, Euboea Eohellenic ph. Maced.	Internides, Vardar zone		Strandsha	Inner Pontides	N. Atlantic, E. Indian Ocean, E. Pacific 148 m.y.
		157							

Orogenic event without or with plutonism. The relative intensity is indicated by

Activity of Sea-floor spreading and/or readjustment of plate movements.

Austroalpine or Early Austrian phase. Apparently a development of flysch was also connected with this period of mountain-building. In the Pindos trough in front of the Eohellenic nappe the first Pindos flysch of Barremian–Lower Turonian age (Bachmann and Risch, 1976) developed. This flysch is more or less equivalent to the Bosnian flysch.

In the latest Upper Cretaceous times nappe movements in East Aegean areas and Crete, as well as processes of metamorphism and granitic intrusions, took place (Jacobshagen, 1976, table 1). Furthermore, tectonic movements at the Cretaceous–Tertiary boundary are indicated in the Inner Hellenides (Richter, 1976, table 1). In this area, the development of flysch began at the same time, and in some places probably even in the Upper Maastrichtian. The flysch sedimentation in the Pelagonian zone terminated in the Middle Eocene, probably because of orogenic movements. During the same period, intrusion of granites as well as extrusion of rhyolites and andesites took place in the neighbouring Vardar zone (Mercier, 1966; Boccaletti *et al.*, 1974b, p. 1079).

The end of the flysch sedimentation in the Inner Hellenides in the Upper Eocene is related to the Late Eocene main orogeny which above all is responsible for the structural formation of the Hellenic Internides. The central Hellenic nappes came into existence. The nappe transport of Crete also occurred mainly in this time (Baumann *et al.*, 1976), which is referred to as the Illyrian (Early Pyrenean) phase. The metamorphism of the Thessalian and the Cyclades crystallines as well as the granites of the Rhodope Massif are considered to be synchronous with the Late Eocene main tectogenesis (Jacobshagen, 1976, table 1).

The complete area of the Outer Hellenides was depressed in the Late Eocene (Priabonian; and widespread flysch sedimentation began in this area (Richter and Mariolakos, 1972). The thick olistostromes related to this point to the existence of the adjoining Pindos cordillera in the northeast (Aubouin, 1964; Richter, 1976; Doert, 1976). The flysch sedimentation lasted until the Aquitanian, and partly until the Burdigalian. Thereafter, the central Hellenic nappes which are more or less analogous to the Internides were thrust together with the Pindos nappe to the southwest over the Gavrovo-Tripolitza zone. The Externides, like the western Hellenic nappes, may have obtained their nappe and fold structures mainly during the period between the Lower and the Middle Miocene. On the Island of Levkas the front of the western Hellenic nappe is even thrust over Upper Miocene rocks (Jacobshagen *et al.*, 1976a, p. 357; Richter, 1976, p. 471). Thus it was again subjected to forward thrusting. Whether this movement should be attributed to the Attic or the Rhodanian phase is not yet clear. In the Upper Miocene, intense compression (Angelier, 1977), a rather young metamorphism of the Cyclades crystallines and an intrusion of acid plutons took place on the Cyclades islands, on the island of Kosi in Attica and in the Argolis (Altherr *et al.*, 1976; Jacobshagen, 1976, table 1). After the forward thrusting of the central Hellenic nappes in the Middle Miocene, Neogene basins collapsed in the central and southern Aegean region and in western Anatolia. During the same time and perhaps in kinematic connection with these events, the central Aegean and northwestern Anatolian volcanites (22–11 m.y.) erupted (Jacobshagen, 1976, p. 3).

Like the Dinarides, the Hellenides are a well-known example of orogenic activity which,

FIG. 2 Timing and regional occurrence of orogenic events in the mountain ranges of the Balkan Peninsula and the bordering eastern Alps and western Turkish mountain ranges as well as discontinuities in ocean-floor spreading. Dates based mainly on Andrusov (1960), Aubouin (1960), Blanchet (1974), Bonchev *et al.* (1969), Brinkmann (1976), Buday (1960), Ćirić (1963), Dewey *et al.* (1973), Hsü *et al.* (1977), Jacobshagen (1976), Jacobshagen *et al.* (1976a,b), Mercier and Vergely (1972), Sandulescu (1975), Schwan (1961), (1977), Sikoshek (1970), Stille (1924), (1953), Tollmann (1966), Vogt *et al.* (1971), (1974).

similarly to the flysch sedimentation, has shifted from the inner parts of a geosynclinal orogenic belt to its outer parts (Wunderlich, 1966). This was called 'polarité orogénique' by Aubouin (1960, p. 506). But it was not a continuous process; the shifting took place in specific stages and phases with intervening breaks. Thus in the case of the Hellenides, first one section, the Internides, were orogenically formed in the Late Eocene and thereafter the Externides were subjected to a separate orogenic shaping in the Middle Miocene. Therefore the different flysch complexes do not indicate time relations (Richter, 1976). However, there is an exception to such regular development of the outer zones as annexes to the inner zones. The exception is the Late Eocene Central Aegean orogenic belt which runs unconformably to the east-northeast and across to the main north-northwestern trend (Jacobshagen, 1976).

The Dinarides are located further along the north-northwestern trend of the Hellenides and both orogens show a far-reaching similarity. The Dinarides are also divided into an inner and an outer zone. The two mountain chains, the Hellenides and Dinarides, are separated by the important transverse Skutari-Peć fault which trends to the northeast.

The Dinarides also had a number of critical tectogenetic episodes (Cirić, 1963; Sikoshek, 1970; Blanchet, 1974). In the Inner Dinarides tangential movements occurred prior to the Jurassic–Cretaceous boundary and after the accumulation of a typical eugeosynclinal diabase-chert-formation and flysch-like sediments in Jurassic time: this points to Late Cimmerian movements. This orogeny was accompanied by regional metamorphism (Mercier, 1973; Boccaletti *et al.*, 1974b). As in the Eohellenic nappe, these movements caused a northeast to southwest overthrust of ultramafic oceanic rocks, again implying obduction. The Late Cimmerian fold structures are transgressed by Albian and Cenomanian sediments (Cirić, 1963, p. 576). In the Inner Dinarides, the Bosnian flysch developed in Cretaceous times and can be compared with the first Pindos flysch. Its materials, according to Rampnoux (1974, pp. 89, 91) are derived from the pre-existing mountain chains, the Palaeodinarides. Austrian movements prior to and at the boundary of the Lower and Upper Cretaceous are indicated for the Inner Dinarides, especially for the Kopaonik area (Cirić, 1963, p. 576; Andjelković, 1977). Thereafter, flysch developed in the Outer Dinarides during the Upper Senonian to Priabonian times, similarly to the Hellenides. However, in the former area it was already partly in the inner zone.

In the Serbian zone belonging to the Inner Dinarides the latest Cretaceous tectonic movements occurred in the Maastrichtian period (Cirić, 1963, p. 576; Blanchet, 1974, p. 151), belonging probably to the Early Laramide movements. The overthrust of the complex of the Serbian zone towards the southwest can be considered as post-Palaeocene and pre-Oligocene, implying the Illyrian or Pyrenean phase. The Outer Dinarides overthrusts also occurred after the Middle or Upper Eocene and thus represent Illyrian–Pyrenean movements. These phases (= first and second Pyrenean phase) were of major importance for the Dinarides and the Hellenides. The main orogenic movements in the Dinarides, generating intensive folding and horizontal thrusts, have to be attributed to these phases.

In the outermost strip of the Externides, in the Adriatic zone to the southwest, lies the Late Eocene and Oligocene flysch (Sikoshek and Medwenitsch, 1965, p. 348). In the Dinarides as well as in the Hellenides, the flysch is progressively younger, judging from its inner to its outer zones which confirm the rule. However, it should not be concluded that a long-term, continuous, and undifferentiated migration of orogenies took place in the course of time. This was emphasized by Blanchet (1974, p. 151).

The late tectonic stage is of post-Pyrenean age. As in the Hellenides, large Neogene molasse basins and horsts developed in Upper Oligocene–Miocene times. Such fault-block tectonics were accompanied by important and typical Late Tertiary acid to intermediate volcanism. Neotectonics, chiefly characterized by vertical movements and distension, may still express the activities of the Savic, Styrian, Attic and Rhodanian phases (cf. Cirić, 1963, p. 577).

Below, a summary of the timing of geotectonic development of the northern belt of the Alpidic orogenic system in the Balkan region, the Carpathides–Balkanides, will be given. First, the Carpathides will be studied. From the regional-geological point of view, they are divided into western, eastern, and southern Carpathian mountains. Also in this mountain range the longitudinal zones of the Internides and Externides have to be distinguished. The historic-tectonic features of the Carpathian mountains have been investigated by many scientists (Stille, 1953; Buday, 1960, 1961; Andrusov, 1960, 1963; Roth, 1964; Tollmann, 1955; Sikora, 1970; Mahel, 1974; Boccaletti *et al.*, 1974; Sandulescu, 1975).

The Carpathian geosyncline shows few clear indications of Late Cimmerian movements at the Jurassic–Cretaceous boundary. In places uninterrupted sedimentation since the Tithonian is indicated; elsewhere, however, there was a transgression of the Lower Cretaceous flysch, especially in the Beskides, and in the Romanian internal zone. Initial magmatic rocks occur within the flysch.

About the end of the Lower Cretaceous one main folding took place in the Carpathian mountains, namely during the first Austrian phase which was effective prior to the Upper Gaultian (Stille, 1953; Tollmann, 1966, p. 62). This phase had a very important effect on the Internides, especially on the western Carpathians, that is on the Pienides, the Romanian eastern and the southern Carpathian mountains where in the Banat area the eastern forward thrusting of the Getian nappes occurred, which is also the case for the Apusenides (Sandulescu, 1975; Mutihać *et al.*, 1977). This main phase has especially consolidated the inner zones of the Carpathian mountains and uplift and regression took place everywhere in this area. However, synorogenic plutons are not known in the Internides. The Early Austrian phase is limited to a short timespan in the High Tatra Mountains in that Aptian limestones are transgressed by Middle Albian sediments. Thus this phase can be attributed to the Aptian–Albian boundary (Andrusov, 1960). The diaphthorites of the Vepor crystalline in Slovakia with 107–106 m.y. point to an age corresponding to the Early Austrian phase.

In the north of the central Carpathian mountains, the flysch has an Upper Albian–Turonian age (Tollmann, 1966, pp. 67–8). The Late Austrian phase does not seem to have influenced this area. In the eastern Carpathian mountains and in the Apuseni Mountains the Cenomanian flysch was transgressive. The Carpathian flysch development also continued in the Turonian–Santonian.

The Subhercynian mountain-building has affected the western Beskides, the Klippen region and the Getian area. Therefore, according to Andrusov (1960), the Mediterranean phase in the Upper Turonian leading to the nappe development in the western central Carpathian mountains may be of relatively great importance. The stratigraphic fixing of this phase was possible on the basis of the angular unconformity of the Senonian and radiometric age dating of the diaphthoritic Vepor crystallines as 88–72 m.y. (cf. Tollmann, 1966, pp. 68, 78). Thus it could also be a Subhercynian movement of a younger age. In the Ilsede phase at the Lower–Middle Santonian boundary nappe movement with a northerly vergence took place in the Pienides (Andrusov and Bystricky, 1959). In the eastern and southern Carpathians in particular the Senonian is unconformable. In the Beskides the Senonian flysch is present. In the Apusenides and in the Banat subsequent magmatic rocks are present in the Senonian and Danian.

Laramide movements are recognizable from stratigraphic gaps and gentle folding (Roth, 1964, p. 337; Tollmann, 1966, p. 92). Orogenic deformations occurred in the southern

Carpathians ('finicrétacées', according to Sandulescu, 1975, p. 299), in the Apusenides, and in the Pienidian Klippen zone. The banatites and granites of the Zips area are related to these movements (Stille, 1953, p. 173).

In the Flysch Carpathians, the flysch development continued in the Eocene and Oligocene. In the inner zones, the Eocene shows an overlap. Pyrenean movements seem to have had only a minor influence (Roth, 1964, p. 338).

The Savic phase is the other main phase of the Alpidic Carpathian tectonics which affected the Externides and the Flysch Carpathians developed with their nappe structures. Thus, the western Carpathians were affected over their full width, as were the eastern Carpathians apart from some smaller marginal areas, and the Pienides. In the Hungarian massif intensified germanotype tectonic forces were active (Stille, 1953, pp. 202–3). An Early Savic phase (between Oligocene and Miocene) and a Late Savic phase Aquitanian–Burdigalian boundary) have been distinguished (Tollmann, 1966, pp. 117–18). The latter is the most important Neogene phase. The creation of the main structures of the Carpathian flysch zone can be attributed especially to this tectogenesis (Buday, 1960, 1961; Andrusov, 1963; Tollmann, 1966; Sandulescu, 1976, p. 349; Stegena *et al*., 1975, p. 76). The development of the Neogene Carpathian arc can be related to the intensive compression of the geosynclinal system. In the Miocene on the back of the arc, the Pannonian marginal basin was widened (Boccaletti *et al*., 1974a). After the Burdigalian transgression, producing considerable masses of conglomerates in the eastern and southern Carpathian mountains, and until the Helvetian, the salt deposits of eastern Galicia and Romania were developed.

The Styrian tectogenesis, especially the Early phase (Helvetian) led to the thrusting of the flysch nappe over the molasse (Buday, 1960) and during the Moldavian phase over the Lower Tortonian strata. During this process, overthrusts with a translation of 25 km came into existence at the outer margin of the eastern Carpathians (Roth, 1964). During the Late Styrian phase, the Hungarian Inner Carpathians and Transylvania underwent only Germanotype folding.

In the Helvetian, the andesitic-rhyolitic volcanism of Transylvania began, culminating in the Tortonian. In the Tortonian, the salt formation of western Galicia (Wieliczka) and western Podolia was deposited. In the Sarmatian brackish sediments were already present and in the Pliocene intensive volcanism which at first was post-orogenic and then terminal dominated. The Plio-Pleistocene volcanism was related to the extension in the Pannonian basin (Stegena *et al*., 1975; Boccaletti *et al*., 1976, p. 53).

The Attic folding at the Miocene–Pliocene boundary was unimportant. Limnic Pliocene sediments are up to several thousand metres thick in the southern marginal area of the eastern Carpathians. Intra- and post-Pliocene (Walachian) as well as intra-Pleistocene (Pasadenian) fold phases only weakly affected this area (Stille, 1953, p. 201).

The Balkanides, the second largest section of the northern belt of the Alpidic orogenic system in the Balkan region, will be described next. The southern Carpathian mountains continue to the southwest to the Carpathian-Balkan mountain arc which bends to the south and southeast. To the eastern trending part of this arc there adjoin the Balkan mountains of Bulgaria. The chronological development of the tectonics of the Balkanides has been investigated by Bonchev (1936, 1955, 1958, 1966, 1970), Stille (1940), Dimitrov (1958), Schwan (1961) and Foose and Manheim (1975).

The Balkanides originated in a geosyncline which developed between two Variscan cratons: the Rhodope massif to the south and the Walachian spur (Moesian plate) to the north. The Balkanides are divided into three longitudinal zones: (1) the Bulgarian central massif (Srednogorie), (2) the Balkans proper (Stara Planina), and (3) the Prebalkans. The division into Internides and Externides can also be ascertained. The Bulgarian central mountains represent the Internides, and the Balkans and Prebalkans the Externides.

In the Balkanides the following features are recognized: in the Sakar-Strandsha area in the eastern part of the Srednogorie area a special development took place in that, analogously to the Vardar zone, Late Cimmerian folding and regional metamorphism were active after the Jurassic mafic geosynclinal magmatism. Furthermore, Cenomanian strata are unconformable over folded Lower Cretaceous rocks, which confirms strong Austrian folding (Bonchev, Savov and Tschatalov, 1969; Savov, 1972; Gochev, 1976).

Generally, in the area of the Srednogorie, Upper Cretaceous sediments with geosynclinal magmatic rocks lie above Palaeozoic and Mesozoic sediments including those of Lower Cretaceous age. The Upper Cretaceous sediments, in particular, were regarded as an expression of orthogeosynclinal development. However, the volcanites which are often compositionally intermediate or felsic and sometimes mafic, as well as the coal seams of this period indicate that it was not a very typical orthogeosyncline. Boccaletti *et al.* (1974b) refer to a marginal sea basin with subduction from the south. Bogdanov *et al.* (1974) and Bogdanov (1977), however, support the development of the Tethys and the subduction from the north with respect to the Jurassic basites of the Strandsha area. The Laramide folding associated with intensive synorogenic felsic plutonism created Srednogorie. In the Early Tertiary later volcanites also developed there.

In the Stara Planina zone, as a result of pronounced depression, about 3000 m thick Upper Cretaceous and Palaeogene flysch was deposited above Palaeozoic and Mesozoic strata, especially in the eastern part. At the Middle–Upper Eocene boundary the Illyrian (= Early Pyrenean) folding which is the main orogeny in the Balkan zone (Stara Planina) affected the area. During this process, large folds with northerly vergence and, at the Stara Planina front line, nappes with a forward thrust displacement of 15 km developed, as, for instance, the central Balkan granite nappe (Bonchev, 1970; Schmidt and Schwan, 1960; Foose and Manheim 1975). The Upper Eocene (Priabonian) lies unconformably as a molasse facies.

In the Prebalkans two flysch sequences are found—the Tithonian–Berriasian flysch and the Lower to Middle Eocene flysch. The eastern zone of the Prebalkans was slightly affected by the Ludogorian (= Austroalpine) folding of Lower Cretaceous age, at the boundary of the Barremian and the Aptian. However, the main tectogenesis in the Prebalkan zone—similar to the case of the Stara Planina zone—took place during the Early Pyrenean (= Illyrian) phase after the deposition of a typical Eocene deep-water flysch which reaches a thickness of 1000 m. During this phase particularly, the large Jura-type north-verging anticlines of the Prebalkans were formed.

In the Balkanides the orogeny also shifted in two separate stages from the inner to the outer part, from the south to the north. It was the Laramide phase in the Internides in the south and the Illyrian in the Externides in the north. The Kraishtides which border the Balkanides in the west and partially cut them diagonally (Bonchev, 1958) were folded several times, chiefly at the boundary of the Oligocene and the Miocene (Savic phase).

In addition to the above historic-tectonic summary of the essential processes of mountain-building during the Cretaceous and Tertiary periods and other remarkable geotectonic events of this timespan in the area of the Balkan Peninsula, some comparisons with the mountain belts adjoining directly to the northwest and southeast can be made: The eastern Alps and the western Turkish mountain belts will be briefly touched on in order to determine on a broader basis the existence or lack of time coincidence between the tectonic events in the mentioned regions (Fig. 2).

Specific studies of the eastern Alps have been carried out by Tollmann (1966). The following orogenic phases in Cretaceous and Tertiary times are important for this area: the Late Cimmerian, the Austrian, the Subhercynian, the Pyrenean, the Savic and the Styrian phases.

As indicated by Brinkmann (1976) and other authors, the mountain belts of western Turkey are characterized by the following Cretaceous–Tertiary phases: the Late Cimmerian, the Austrian, the Subhercynian, the Laramide, the Pyrenean, and the Styrian.

DISCUSSION AND CONCLUSIONS

This brief preliminary summary is an attempt to present old and new data about geotectonic events which occurred in the Balkan region during Cretaceous and Tertiary periods.

With regard to the Alpidic tectonism in this rather large and complicated area of the Balkan Peninsula, different problems still exist today. On the one hand, the classification of orogenic processes in terms of certain phases is not always clear. Precise age dating is not possible where, for instance, there are larger stratigraphic gaps. On the other hand, events in different regions are sometimes not perfectly comparable or else cannot be correlated at all. Moreover, data are sometimes non-existent.

Most of the data which have been investigated and compared with regard to the Balkan region, however, reveal that in this region, as Blanchet (1974, p. 153) has already specifically pointed out for the Dinarides, the same geotectonic pulses and/or main critical time intervals during the Cretaceous and Tertiary periods can be recognized as having been determined globally and defined as orogenic phases (Stille, 1924). The important changes and/or gaps in geological records are related to these (Schönenberg, 1969).

A further comparison of the angular unconformities caused by the orogeny with the discontinuities of sea-floor spreading also reveals unambiguous time coincidences (cf. Fig. 2). This fact points to intimate kinematic relations between both processes as soon as their activities (change in direction and speed of plate movement and orogenic compression) took place in the same time interval (Schwan, 1977, 1980, 1985).

I believe neither that the orogenic pulses are only the expression of slow and rather uniform 'orogenic waves' migrating through the earth's cruse nor that there are resultant angular unconformities occurring irregularly as only incidentally exposed time marks along the course of such supposed tectonic movements. Rather, I feel that there are concrete timebound peaks of orogenic events in the earth's history and that therefore the Alpinotype orogeny was active not continuously but episodically during specific phases of deformation which were separated by fairly quiet stages and/or that the orogenic process proceeded regionally step by step in separate parts. This opinion has already been convincingly expressed by some authors and is supported by adequate data. In spite of such clear geotectonic principles which are upheld by new facts, the available data, although numerous, need further amplification and some more precise determination. This summary gives a rough outline but should be supplemented by further research in different countries of the Balkan Peninsula.

REFERENCES

Albandakis, N. D. (1977). The main phases of the Alpine orogenesis in Skyros, Euboea and Atticoboetia, in *VI Colloqu. Geol. Aegean Region*, I, 229–35.

Altherr, R., Keller, J. and Kott, K. (1976). Der jungtertiäre Monzonit von Kos und sein Kontakthof (Ägäis, Griechenland), *Bull. Soc. géol. France*, sér. 7, **18**, 403–12.

Andjelković, M. (1977). Tectonics of the Dinarides and Sumadides of Yugoslavia, in *VI Colloqu. Geol. Aegean Region*, III, 1243–51.

Andrusov, D. (1960). Die geologische Entwicklung der Klippenzone und der zentralen Westkarpaten, *Mitt. geol. Ges. Wien*, **51**, 1–18.
Andrusov, D. (1963). Les principeaux plissements alpins dans le domaine des Carpathes occidentales, in Livre Mém. P. Fallot, vol. II, 519–29.
Andrusov, D. and Bystricky, J. (1959). Über die Bedeutung der subherzynischen Faltungsphase im Gebiet der Westkarpaten, *Geol. Sbornik*, **10**, 324–6.
Angelier, J. (1977). Some observations on the tectonic evolution of the Hellenic Arc and the southern Aegean Region during Neogene and Quaternary times, in *VI Colloqu. Geol. Aegean Region*, III, 1279–1280.
Angenheister, G., Bögel, H., Gebrande, H., Giese, P., Schmidt-Thomé, P. and Zeil, W. (1972). Recent investigations of superficial and deeper crustal structures of the eastern and southern Alps, *Geol. Rdsch.*, **61**, 349–95.
Argyriadis, I. (1974). Sur l'orogénèse mésogéenne des temps crétacés, *Rev. Géogr. phys.*, sér. 2, **16**, 23–60.
Aubouin, J. (1960). Essai sur l'ensemble italo-dinarique et ses rapports avec l'arc alpin, *Bull. Soc. géol. France*, **2**, 487–526.
Aubouin, J. (1964). Reflexion sur le problème des flyschs et des molasses; son aspect dans les Hellénides (Grèce), *Ecl. geol. Helv.*, **57**, 451–69.
Aubouin, J. (1977). Brève présentation de la géologie de la Grèce, *Bull. Soc. géol. France*, sér. 7, **19**, 6–10.
Aubouin, J., Brunn, J.-H., Celet, P., Dercourt, J., Godfriaux, I. and Mercier, J. (1960–63). Esquisse de la géologie de la Grèce, in Livre Mém. P. Fallot, vol. II, 583–610.
Aubouin, J. and Dercourt, J. (1975). Les transversales dinariques dérivent-elles de paléofailles transformantes? *C. R. Acad. Sci. Paris*, **281D**, 347–50.
Bachmann, G. H. and Risch, H. (1976). Ein oberjurassisch-unterkretazischer (eohellenischer) Flysch in der Argolis und der Bau der Lighourion-Mulde (Peloponnes, Griechenland), *N. Jb. Geol. Paläont., Abh.*, **152**, 137–60.
Baumann, A., Best, G. and Wachendorf, H. (1977). Die alpidischen Stockwerke der südlichen Ägäis, *Geol. Rdsch.*, **66**, 492–522.
Bernoulli, D. and Laubscher, H. (1972). The palinspastic problem of the Hellenides, *Ecl. geol. Helv.*, **65**, 107–18.
Biju-Duval, B., Dercourt, J. and Le Pichon, X. (1976). From the Tethys ocean to the Mediterranean seas: A plate tectonic model of the evolution of the western Alpine system, in *International Symposium on the Structural History of the Mediterranean basins. Split (Yugoslavia)*, Paris: Editions Technip, 143–64.
Blanchet, R. (1974). De l'Adriatique au Bassin Pannonique. Essai d'un modèle de chaine alpine, *Mém. Soc. géol. France*, **120**, 172 pp.
Boccaletti, M., Horváth, F., Loddo, M., Mongelli, F. and Stegena, L. (1976). The Tyrrhenian and Pannonian basins: A comparison of two Mediterranean interarc basins, *Tectonophys.*, **35**, 45–69.
Boccaletti, M., Manetti, P. and Peccerillo, A. (1974a). Hypotheses on the plate tectonic evolution of the Carpatho-Balkan arcs, *Earth Planet. Sci. Letters*, **23**, 193–8.
Boccaletti, M., Manetti, P. and Peccerillo, A. (1974b). The Balkanids as an instance of back-arc thrust belt, *Geol. Soc. Am. Bull.*, **85**, 1077–83.
Bogdanov, B. (1977). Metallogeny of Srednagora zone in the context of plate tectonics, in *Metallogeny and Plate Tectonics in the Northeastern Mediterranean*, Janković, S. (ed.), Fac. Mining and Geol., 493–504.
Bogdanov, B., Dachev, H. and Vulchanov, A. (1974). Metallogeny of Bulgaria in the context of plate tectonics, in *Problems of ore deposition*. 4. *IAGOD Symposium, Varna*, Vol. II, 435–44.
Bonchev, E. (1936). Versuch einer tektonischen Synthese Westbulgariens (Bulgar. mit deutscher Zusammenfassung), *Geologica Balkanica*, **2**, 5–48.
Bonchev, E. (1955). *Geologija na Bulgarija*. 264 pp.
Bonchev, E. (1958). Über die tektonische Ausbildung der Kraishtiden, *Geologie*, **7**, 409–19.
Bonchev, E. (1966). Revue générale de la structure géologique de la Bulgarie, *Bull. Strashimir Dimitrov' Inst. of Geol.*, K. H., **15**, 5–24.
Bonchev, E. (1970). Über einige Gesetzmässigkeiten in der Rhegmagenese der Balkanhalbinsel, *Rev. Bulg. Geol. Ges.*, **31**, 2, 135–45.
Bonchev, E., Savov, S. and Tschatalov, G. (1969). Über die Schollengliederung des Strandsha-Antiklinoriums, *Bulg. Acad. Sci., Bull. Geol. Inst., Series Geotect.*, K. H., **18**, 143–57.
Brinkmann, R. (1976). *Geology of Turkey*. 158 pp.

Buday, T. (1960). The development of the West Carpathians during the Neogene, in *Tectonic Development of Czechoslovakia*, 199–206.

Buday, T. (1961). Der tektonische Werdegang der Neogenbecken der Westkarpaten und ihr Baustil, *Geol. Práce, Zosit*, **60**, 87–106.

Cirić, B. (1963). Le développement des Dinarides yougoslaves pendant le cycle alpin in Livre Mém. P. Fallot, vol. II, 565–82.

Dercourt, J. (1970). L'expansion océanique actuelle et fossile; ses implications géotectoniques, *Bull. Soc. géol. France*, sér. 7, **12**, 261–317.

Dewey, J. F., Pitman III, W. C., Ryan, W. B. F. and Bonnin, J. (1973). Plate tectonics and the evolution of the Alpine system, *Geol. Soc. Am. Bull.*, **84**, 3137–80.

Dimitrijević, M. D. (1974). The Dinarides: a model based on the new global tectonics, in *Metallogeny and Concepts of the Geotectonic Development of Yugoslavia*, Fac. Mining and Geol. (Belgrade Univ.), 141–78.

Dimitrov, S. (1958). Über die alpidische Regionalmetamorphose und ihre Beziehungen zu der Tektonik und dem Magmatismus in Südostbulgarien, *Geologie*, **7**, 560–8.

Doert, U. (1976). Ergebnisse kleintektonischer Untersuchungen zum Bau und zur Stellung der Olonos-Pindos-Zone in der mittleren Peloponnes, *Z. Deutsch. geol. Ges.*, **127**, 417–28.

Foose, R. M. and Mannheim, F. (1975). Geology of Bulgaria: a review, *Am. Assoc. Pet. Geol. Bull.*, **59**, 303–35.

Gochev, P. M. (1976). L'évolution géotectonique du mégabloc bulgare pendant le Trias et le Jurassique, *Bull. Soc. géol. France*, sér. 7, **18**, 209–16.

Hadzi, E., Pantić, N., Aleksić, V. and Kalenić, M. (1976). Un modèle préliminaire de l'évolution tectonique de la péninsule balkanique dans le cadre du développement de la Méditerranée entière au cours du cycle alpin, *Bull. Soc. géol. France*, sér. 7, 2, **18**, 199–203.

Hiessleitner, G. (1951). Der dinarische Ophiolithzug in den Ostalpen, in Verh. Geol. B.-A., Sonderheft A, *Geol. F*ührer zu den Exkurs, 102–4.

Hsü, K. J., Nachev, I. K. and Vuchev, V. T. (1977). Geologic evolution of Bulgaria in light of plate tectonics, *Tectonophys.*, **40**, 245–56.

Hynes, A. J., Nisbet, E. G., Gilbert-Smith, A., Welland, M. J. P. and Rex, D. C. (1972). Spreading and emplacement ages of some ophiolites in the Othris region (eastern central Greece), *Z. Deutsch. geol. Ges.*, **123**, 455–68.

Jacobshagen, V. (1976). Thesen zum Bau und zur Entwicklung der Helleniden des griechischen Festlands, Unveröff. Ms., 8 pp.

Jacobshagen, V. and Makris, J. (1974). Zur Geodynamik Griechenlands und der Ägäis, *Nachr. Deutsch. geol. Ges.*, **9**, 78–85.

Jacobshagen, V., Makris, J., Richter, D., Bachmann, G. H., Doert, U., Giese, P. and Risch, H. (1976a). Alpidischer Gebirgsbau und Krustenstruktur des Peloponnes, *Z. Deutsch. geol. Ges.*, **127**, 337–63.

Jacobshagen, V., Risch, H. and Roeder, D. (1976b). Die eohellenische Phase. Definition und Interpretation, *Z. Deutsch. geol. Ges.*, **127**, 133–45.

Janković, S. (1977). Major alpine ore deposits and metallogentic units in the Northeastern Mediterranean and concepts of plate tectonics, in *Metallogeny and Plate Tectonics in the Northeastern Mediterranean*, Janković, S. (ed.), Fac. Mining and Geol., 105–71.

Kauffmann, G., Kockel, F. and Mollat, H. (1976). Note on the stratigraphic and palaeogeographic position of the Svoula Formation in the innermost zone of the Hellenides (northern Greece), *Bull. Soc. géol. France*, **18**/2, 227–32.

Kober, L. (1931). *Das Alpine Europa*. 310 pp.

Kober, L. (1933). *Die Orogentheorie*. 184 pp.

Kockel, F., Mollat, H. and Walther, H. W. (1972). New facts and ideas on the innermost zones of the Hellenides, *Z. Deutsch. geol. Ges.*, **123**, 349–52.

Mahel, M. (1974). Introductory articles, in *Tectonics of the Carpathian Balkan Arcs*, 7–49.

Mercier, J. (1966a). Paléogéographie, orogénèse, métamorphisme et magmatisme des zones internes des Hellénides en Macédoine (Grèce). Vne d'ensemble, *Bull Soc. géol. France*, **7**, 382.

Mercier, J. (1966b). Sur l'existence et l'âge de deux phases régionales de métamorphism alpin dans les zones internes des Hellénides en Macédoine centrale (Grèce), *Bull. Soc. géol. France*, sér. 7, **8**, 1020–49.

Mercier, J. (1973). Plissements symmétamorphiques d'échelle Kilométrique, d'âge jurasique supérieur-éocrétacé dans les Hellénides internes (Macédoine, Grèce), *C. R. Acad. Sci.* **276D**, 2249–52.

Mercier, J. and Vergely, P. (1972). Les mélanges ophiolithiques de Macédoine (Grèce): Décrochements d'âge anté-crétacé supérieur, *Z. Deutsch. geol. Ges.*, **123**, 469–89.
Mutihać, V., Preda, I. and Conovici, M. (1977). L'évolution et la structure géologique des montagnes et du plateau de Mehedinti (Roumanie), in *VI. Colloqu. Geol. Aegean Region*, Abstracts, 113–14.
Rampnoux, J. P. (1974). Contribution à l'étude géologique des Dinarides: Un secteur de la Serbie méridionale et du Monténégro oriental (Yougoslavie), *Mém. Soc. géol. France*, **119**, 100 pp.
Richter, D. (1976). Das Flysch-Stadium der Helleniden—Ein Überblick, *Z. Deutsch. geol. Ges.*, **127**, 467–83.
Richter, D. and Mariolakos, I. (1972). Paläomorphologie und eozäne Verkarstung der Gavrovo-Tripolis-Zone auf dem Peloponnes (Griechenland), *Bull. Geol. Soc. Greece*, **9**, 206–28.
Roth, Z. (1964). Die Tektonik des Westabschnittes der Äusseren Karpaten in der CSSR, *Z. Deutsch. geol. Ges.*, **116**, 312–41.
Sandulescu, M. (1975). Essai de synthèse structurale des Carpathes, *Bull. Soc. géol. France*, sér. 7, **17**, 299–358.
Savov, S. (1972). Strukturelle Beziehungen im Strandsha-Antiklinorium (SE-Bulgarien) *Z. Deutsch. geol. Ges.*, **123**, 337–42.
Schmidt, K. and Schwan, W. (1960). Tektonische Untersuchungen in der Nordost-Randzone der Zentralbalkan-Decke, *Abh. Deutsch. Akad. Wiss. Berlin, Kl. Chemie, Geologie, Biologie*, **19**, 35 pp.
Schönenberg, R. (1969). Hans Stille's Auffassung vom Grenzbereich Orogenese-Epirogenese in heutiger Sicht. *Beih. Geol. Jb.*, **80**, 155–66.
Schwan, W. (1961). Deckenfragen im Balkan. Mit einer Einführung in die geotektonischen Verhältnisse Bulgariens, *Geol. Rdsch.*, **51**, 181–218.
Schwan, W. (1976). Geokinematische Faktoren in Inselbogen/Randmeer-Systemen, speziell im Helleniden-Ägäis-Raum, *Z. Deutsch. geol. Ges.*, **127**, 105–24.
Schwan, W. (1977). Höhepunkte der Geodynamik bei alpinotyper Orogenese und bei Ocean-floor spreading bzw. Plattenbewegungen, *Z. Deutsch. geol. Ges.*, **128**, 143–52.
Schwan, W. (1980). Geodynamic peaks in alpinotype orogenies and changes in ocean-floor spreading during Late Jurassic–Late Tertiary time, *Am. Assoc. Pet. Geol. Bull.*, **64**, 359–73.
Schwan, W. (1985). The worldwide active Middle/Late Eocene geodynamic episode with peaks at ±45 and ±37 m.y. B.P., and implications and problems of orogeny and sea-floor spreading, *Tectonophys.*, **115**, 197–234.
Sikora, W. (1970). Tectonic phases in the geosyncline of the polish outer Carpathians, *Sympos. Beograd-Bor*, 65–83.
Sikoshek, B. (1970). Kurze Übersicht der orogenetischen Phasen in Jugoslawien, *Sympos. Beograd-Bor*, p. 45.
Sikoshek, B. and Medwenitsch, W. (1965). Neue Daten zur Fazies und Tektonik der Dinariden, *Z. Deutsch. geol. Ges.*, **116**, 342–58.
Stegena, L., Géczy, B. and Horváth, F. (1975). Late Cenozoic evolution of the Pannonian basin, *Tectonophys.*, **26**, 71–90.
Stille, H. (1924). *Grundfragen der vergleichenden Tektonik*. 443 pp.
Stille, H. (1940). Magmato-tektonische Verhältnisse Bulgariens im Lichte allgemeinerer Erfahrungen, *Z. Bulgar. geol. Ges.* **11**, (Festschrift Stephan Bonchev), 1–27.
Stille, H. (1953). Der tektonische Werdegang der Karpaten, *Beih. Geol. Jb.*, vol. 8, 239 pp.
Szádeczky-Kardoss, E. (1976). Plattentektonik im pannonisch-karpatischen Raum, *Geol. Rdsch.*, **65**, 143–61.
Tollmann, A. (1966). Die alpidischen Gebirgsbildungs-Phasen in den Ostalpen und Westkarpaten, *Geotekt. Forsch.*, **21**, 156 pp.
Vogt, P. R. and Avery, O. E. (1974). Detailed magnetic surveys in the Northeast Atlantic and Labrador Sea, *Geophys. Res.*, **79**, 363–89.
Vogt, P. R., Johnson, G. L., Holcombe, A. L., Gilg, J. G. and Avery, O. E. (1971). Episodes of sea-floor spreading recorded by the North Atlantic basement, *Tectonophys.*, **12**, 211–34.
Wunderlich, H.-G. (1966). *Wesen und Ursachen der Gebirgsbildung*. 367 pp.
Yarwood, G. A. and Dixon, J. E. (1977). The Mesozoic and Tertiary evolution of the Pelagonian zone of Mainland Greece—progress, possibilities and problems, in *VI Colloqu. Geol. Aegean Region*, 49.

Global Correlation of Tectonic Movements
Edited by Yu. G. Leonov and V. E. Khain

6

Synchronism and Heterochronism of Compressive Tectogenesis—An Example: the Central and Southeastern European Alpine Chains

MIRCEA SANDULESCU*

INTRODUCTION

The analysis and verification of Stille's philosophy of tectogenic phases is still going on and has new implications with respect to plate-tectonic concept.

The discussions of the episodicity of the deformations in an orogenic belt were long, hard and partly inefficient. In fact the analysis of the evolution of deformed areas shows clearly that the major tectonic compressive structures were generated during relatively short periods of time—compared with the duration of the geosynclinal evolution. Even if the stress field never vanishes, the long periods of its very low intensity are separated by short periods of active deformation (Stille's 'phases' or tectogenic moments of deformation).

Why tectogenic? We accept that the whole process through which a mountain (fold) chain is generated is an orogenic process. The objections against this term are rather of a semantic nature. In fact Stille's orogenic cycles remain the major basis for dividing the geotectonic history of the earth (Alpine, Variscan or Hercynian, Caledonian, Cadomian and so on). The orogenic processes show two main aspects: tectogenesis and morphogenesis. That is why deformation moments are called tectogenic moments or tectogenesis.

Tectogenic movements generating the main tectonic structures of an orogenic belt are compressive. They are accompanied by variable contractions of the crust. The distensive moments in an area precede or succeed the tectogenetic ones. Thus the spreading and/or rifting processes on the Tethys realm preceded the tectogenic compressive period of the Alpine region, whereas taphrogenic processes follow the compressional tectogenesis of this area.

The criteria which allow us to recognize the compressive tectogenic moments are based on various structures which are, in order of importance: nappes, imbrications and thrusts and folds.

Basement shear and obduction nappes indicate important crustal contractions. Cover detachment nappes could also represent major contractions. From the latter group are excluded the gravitational gliding nappes, which may be produced without contractions of the crust and may develop intermittently over a long time interval. Imbrications, thrusts and

* Institute of Geology and Geophysics, str. Cararrseloes 1, 78344 Bucharest, Romania.

folds may precede or accompany the generation of nappes or they may develop independently. The obduction and cover detachment nappes are related to crustal consumption (subduction). The basement nappes occur in continental areas and produce 'epidermal' contraction.

Finally, the actual orogenic belts are zones of accretion (agglutination) of several deformed ranges generated by compressive tectogenesis and corresponding to areas of oceanic and continental crustal contraction.

GENERAL STRUCTURE OF THE CENTRAL AND SOUTHEAST EUROPEAN ALPINE CHAIN

The Alpine symmetry in central and southeastern Europe is mirrored across the oceanic Tethys suture zone. It divides the Carpatho (p.p.)-Balkan branch (peri-European) from the Dinaro-Hellenic branch (peri-Apulian). In this scheme the Carpathians—as well as the Alps—do not occur entirely in the peri-European branch because the Tethyan ophiolitic suture divided the Carpathians into two parts: the main part, outward of the suture in a peri-European position and the inner part (Inner or Western Dacides corresponding to the Austroalpine), inward of the suture. It is still a problem whether the latter belongs to the Apulian block or to an intermediate zone; the general pre-Alpine history of the Inner Dacides, as well as that of the Alps, pleads against its 'Africanisation' (Săndulescu, 1980).

South of the Danube (Belgrade), the Tethyan suture is represented by the Vardar zone. In the Carpathians, it was recognized in the Transylvanides, which correspond to the Liguropiemontais (Săndulescu, 1975, 1980) (Fig. 1). The Vardar zone runs between the Serbo-Macedonian or the Rhodopian massifs and the inner Dinaric (and Hellenic) units, namely the Pelagonian, Korab and Golija massifs (Aubouin *et al*., 1970; Aubouin 1973; Grubič, 1980). The Transylvanides separate the Inner or Western Dacides (northern Apusenides and Central Western Carpathians) from the Median Dacides (Central Eastern Carpathians, Getic and Supragetic nappes). The relations and connections between the Vardar zone and the Transylvanides are complex and several models have been proposed (Andjelkovici and Lupu, 1967; Săndulescu, 1975, 1980; Chanell *et al*., 1979; Bleahu, 1976; Durand Delga, 1980 among others). It should be stressed that the southwestern part of the Transylvanides (Metaliferi Mountains and their continuation below the Pannonian Depression, (cf. Visarion and Săndulescu, 1979) has the same vergence (away from the European margin) as the Vardar units. In the Central Eastern Carpathians obducted nappes—the Transylvanian nappes—cropped out, proceeding from the eastern zone of the Transylvanidian domain, their vergence is towards the European margin. The 'oceanic' southern Strandjides seem to have a similar position; they might proceed from the Vardar zone and are obducted on the European margin (Gočev, 1979). On the other hand, also from the Vardar zone, there proceeded the allochthonous ophiolites which at present crop out in the Pelagonian and Golija units as well as in the Serbian (Zlatar), Mirdita and Subpelagonian units (Aubouin *et al*., 1970; Aubouin 1973, 1977). Thus, in some traverses, the Tethyan suture shows a bilateral structure (like the Eastern Carpathians-Metaliferi Mountains and the Strandja-Hellenides traverses), even if the divergently vergent structures are of different ages.

The overthrust units originate from the split European margin which outcrops in the Carpathian, Balkan, Serbo-Macedonian and Rhodopian massifs. Thus the Tethyan suture is bordered first by a group of basement nappes: the Median Dacides (Central Eastern Carpathians, Getic and Supragetic nappes) in the Carpathians, which correspond to the Inner Balkanides, Luznica-Kraijšte, Serbo-Macedonian and Rhodopian units, in the Bal-

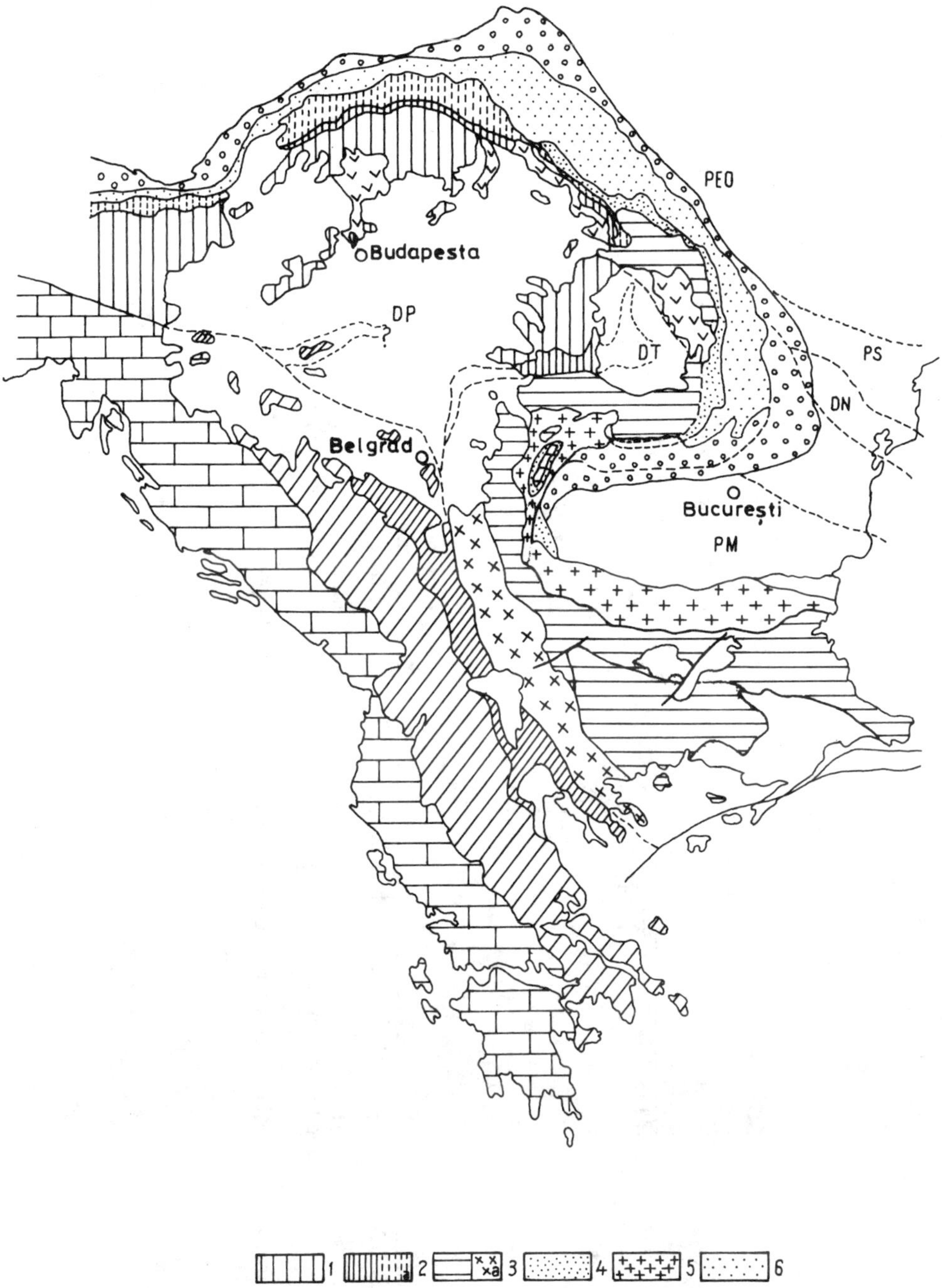

FIG. 1 General tectonic sketch of central and south-east Alpine Europe.
1 – Inner Dacides; 2 – Transylvanides, Pieneides and adjoining flysch nappes (a); 3 – Median Dacides and Serbo-Macedonian Massif (2); Rhodope and Inner Balkanides; 4 – Outer Dacides; 5 – Marginal Dacides (Danubikum) and Outer Balkanides; 6 – Moldavides; 7 – foredeep; 8 – Vardar zone; 9 – Inner Dinarides; 10 – Outer Dinarides; 11 – young depressions and volcanics.

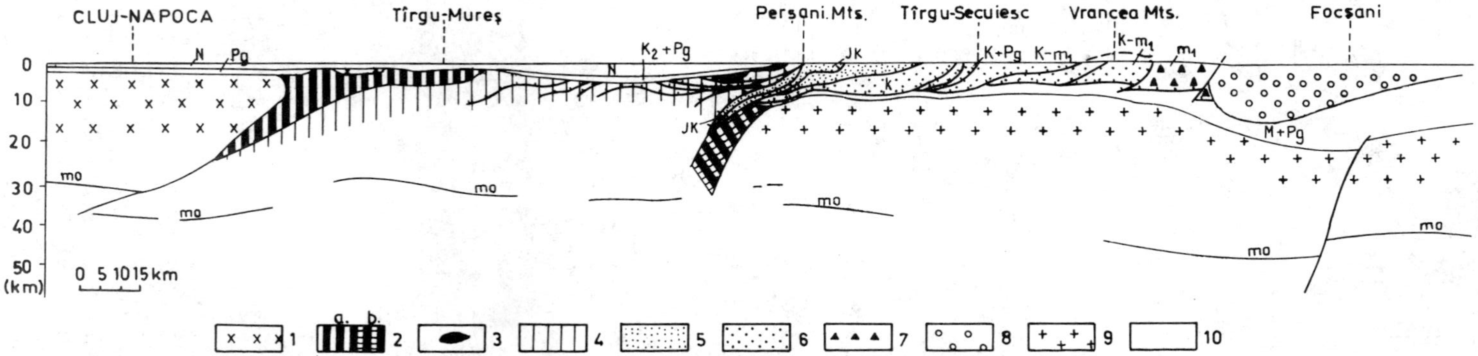

FIG. 2 General structural cross-section in the Romanian Carpathians (acc. to Săndulescu *et al.*, 1981).

1 – Inner Dacides; 2 – oceanic (a) or thinned (b) crust in the suture zones; 3 – Transylvanian nappes; 4 – Central Eastern Carpathian nappes; 5 – Outer Dacidian nappes; 6 – Moldavides; 7 – Subcarpathian nappe; 8 – foredeep (Focşani depression); 9 – foreland basement; 10 – platform cover (M + Pg), post-nappe covers (K2 + Pgl,Pg), Transylvanian Depression (N).

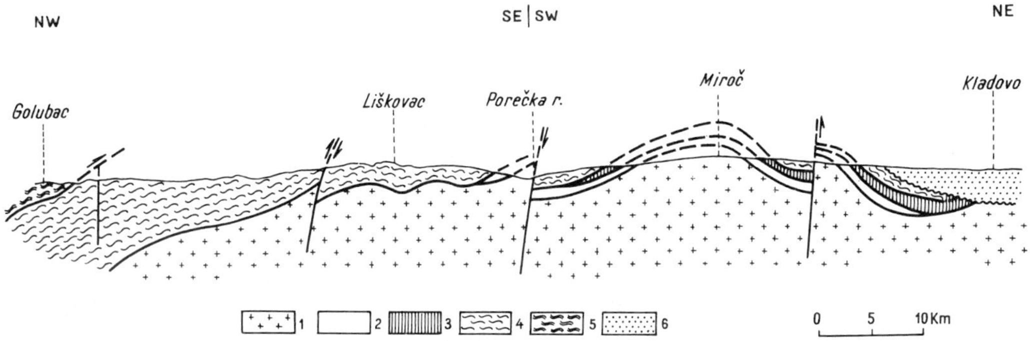

FIG. 3 Structural cross-section in the Southern Carpathians (acc. Grubiç, 1974, simplified). 1 – Danubikum (Marginal Dacides); 2 – Outer Dacidian nappes (Severin nappe); 3 – Infragetikum; 4 – Getic nappe; 5 – Supragetic nappe (Morava); 6 – Tertiary.

kan peninsula (Săndulescu, 1975, 1980; Grubič, 1980; Gočev, 1979). These are units consisting of pre-Alpine crystalline formations as well as Upper Palaeozoic and/or Mesozoic sedimentary formations; Permo-Carboniferous acid volcanic rocks and, locally, Jurassic alkaline magmatic formations. A group of nappes—the Outer Dacides—is pushed out from an intracontinental rift area; with mafic, locally ultramafic, Jurassic rocks and Tithonian–Eocretaceous flysch forming the major part of these units. Overthrust in front of the Median Dacides, these units also extend between the Inner and the Outer Balkanides, namely between the Srednagora and the Stara Planina (Săndulescu, 1980) (Fig. 2).

In the Southern Carpathians and the Balkanides the outermost deformed units—the Danubian and the Outer Balkanides (Stara Planina and Prebalkan) respectively—represent an extension of the foreland (the Moesian Platform) into the orogenic belt (Fig. 3). Pre-Alpine crystalline rocks (metamorphites and granitoids) are covered by Upper Palaeozoic–Mesozoic (Danubian) and even Palaeogene (Outer Balkanides) sedimentary formations (Mahel, 1974; Năstăseanu, 1975; Săndulescue, 1975, 1980; Gočev, 1979).

In the Eastern and Western Carpathians the outermost nappes—the Moldavides—are cover nappes with flysch formations (Cretaceous–Lower Miocene) and locally also with molasse (Lower and Middle Miocene), foreland is deeply underthrust below these Moldavidian nappes (Fig. 4).

A peculiar group of units—the Pieniny Klippen belt and the Magura nappe (Group)—develop in the northern Eastern Carpathians between the Median and Inner Dacides; they extend into the Western Carpathian in front of the latter. The Pieniny Klippen belt is a substitute, or a part, of the Transylvanides; the Magura nappe may have a similar position, but it groups only Senonian–Palaeogene flysch formations sedimented before the overthrusting in a trough superposed on deformed units possibly belonging to the suture area. The overthrust of the Magura nappe (at least) has tectonically overlapped the western continuation of the Median and Outer Dacides in the Western Carpathians. Thus this nappe contacts the Moldavides directly (Fig. 1).

The internal (western and southern) margin of the Transylvanides is represented in the Carpathians by the Inner (Western) Dacides, which group together the northern Apuseni and the Central Western Carpathian units, partly covered by the Pannonian Depression. The Inner Dacides, which prolong in the Austroalpine nappes of the Eastern Alps, are built up of several nappes with pre-Alpine crystalline rocks (metamorphites and, in the lower units, also granites) and/or Permian–Mesozoic sedimentary formations (Andrusov *et al.*, 1973; Mahel, 1974; Ianovici *et al.*, 1976).

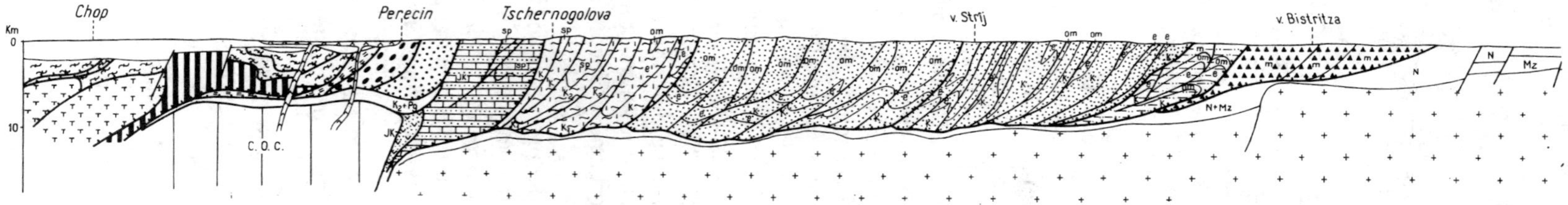

FIG. 4 Geological cross-section in the Ukrainian Carpathians (acc. to Krugliov, Smirnov *et al.*, modified by Săndulescu) showing the complete overthrusting of the Central Eastern Carpathians (COC) by the Pienides and the Inner Dacides.

In the Dinaro-Hellenic belt outward of the Vardar zone follow the Pelagonian, Korab and Golija units which have a continental basement. The ophiolites cropping out in the Pelagonian and equivalent units as well as in the neighbouring units—Subpelagonian, Mirdita and Serbian (Zlatar)—are in an allochthonous position; they are underlain by Triassic–Jurassic and overlain by Cretaceous–Eocene rocks. The outer Dinarides, or Hellenides, consist of cover nappes built up of mainly carbonate rocks (with flysch of different ages at the top) of Triassic–Eocene age and even up to Lower Miocene in the most external units (Aubouin, 1973, 1977; Grubič, 1980; Cadet *et al*., 1980).

The molasse troughs or superposed depressions are in different positions and exhibit different ages of their infillings as a result of differences in age of tectogenesis of the overlapping units. The best developed foredeep is found in the Carpathian area; it has an inner deformed limb and an outer non-deformed limb postitioned over the foreland. The filling of the Carpathian foredeep is of Sarmatian–Lower Quaternary age. The Balkan foredeep is poorly developed and is effectively restricted to the Varna basin. The Dinaric foredeep may lie below the Adriatic Sea.

TECTOGENIC HISTORY

The main compressive tectogenic deformations took place between the end of the Jurassic and the Upper Miocene, at different moments and in different ranges of the central and southeastern European Alpine chains. Before this large tectogenic period, in the geodynamic evolution of the Alpine realm, distensive processes predominated. They determined the spreading of the Tethyan ocean as well as the rift processes inside the two continental margins (Argyriadis *et al*., 1980; Săndulescu, 1980, 1981).

In the analysed area the oldest important tectogenesis with consequent crustal contraction occurred at the end of the Upper Jurassic. Nevertheless, older angular discordances are known on the European continental margin. In the earliest and the latest Lower Jurassic as well as at the base of the Tithonian, in different zones, the effects of such early tectogenesis were found. These may be local or may involve large areas, but they have never caused crustal contraction (consumption). The areas where such early tectogenesis is best manifest are the Median Dacides and their equivalents, more precisely the proximal part of the European margin against the Tethyan ocean.

Partly similar disconformities also occur in the Inner (Western) Dacides which in the Carpathians represent the internal margin of the Transylvanidian ocean. Such an event did not occur on the Apulian continental margin (Aubouin, 1973; Argyriadis *et al*., 1980), which stresses the difference between the Inner Dacidian–Austroalpine area and the 'normal' Apulian margin.

Among the early discontinuities, the most widespread known on the European margin is that of the earliest Lower Jurassic. It may be correlated with the synchronous deformations which occur in the Northern Dobrogea-Southern Crimea intracratonic geosyncline which is situated inside the European continent and represents the *locus tipicus* for the Old Cimmerian tectogenic 'phase'. There, this tectogenesis determined crustal contraction, accompanied by the obduction of intracontinental (rift) basalt rocks, which is not the case in the Carpathian or Balkan areas (Fig. 5).

The end-Jurassic tectogenesis affected the 'Almopias ocean' (Boillot, 1977) representing the southern and southwestern parts of the Vardar zone and determined the large obduction of the ophiolitic complexes onto the Apulian margin, namely the Pelagonian (and equivalents) and Subpelagonian (and equivalents) areas (Aubouin *et al*., 1970; Aubouin, 1973, 1977). Subsequent northerly and northeastwardly subduction indicates crustal contraction (Figs 6, 7).

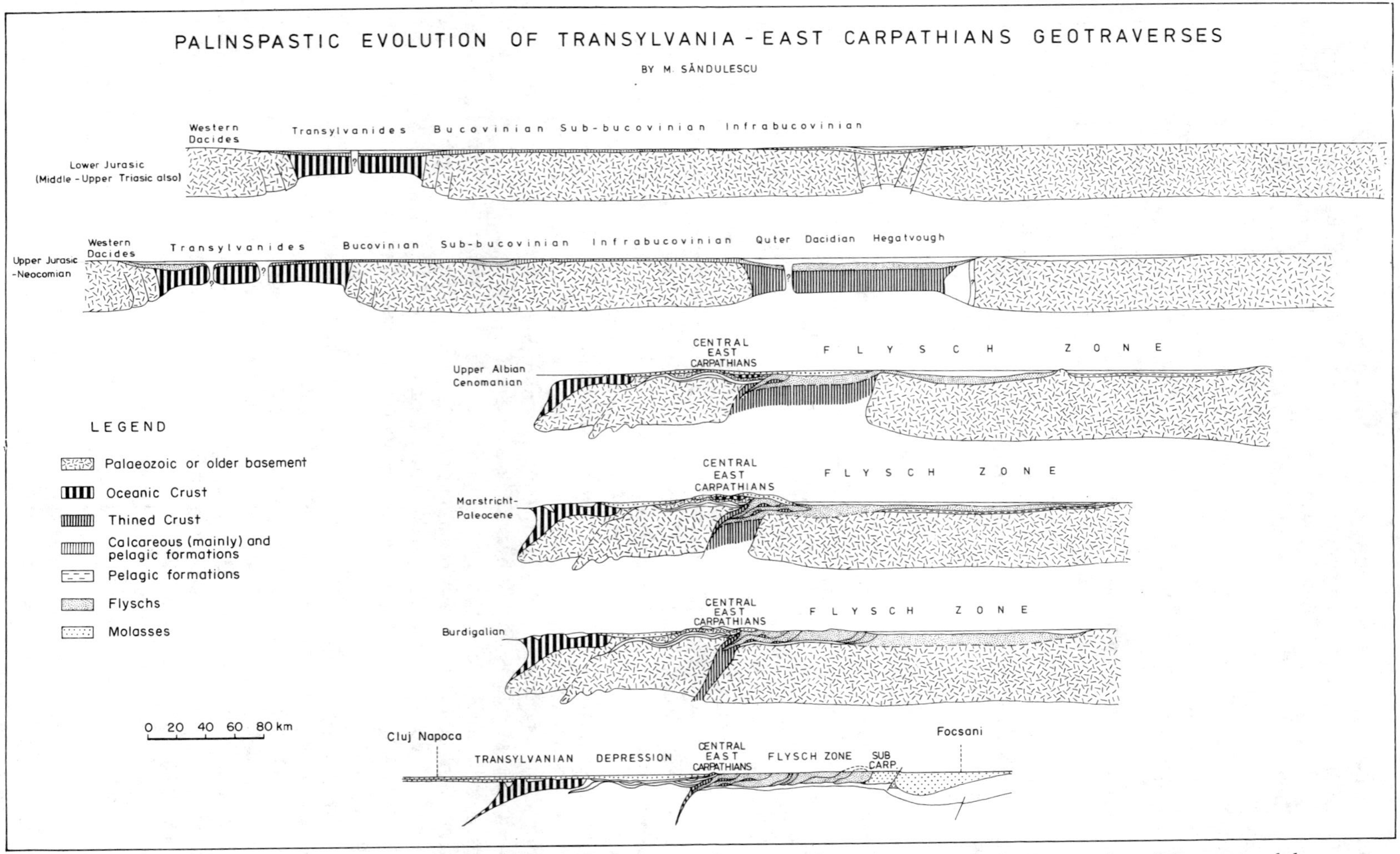

FIG. 5 Palinspastic evolution of the Transylvanian-eastern Carpathian geotraverse showing the successive contraction of the crust and the differences in the age of the main tectogeneses.

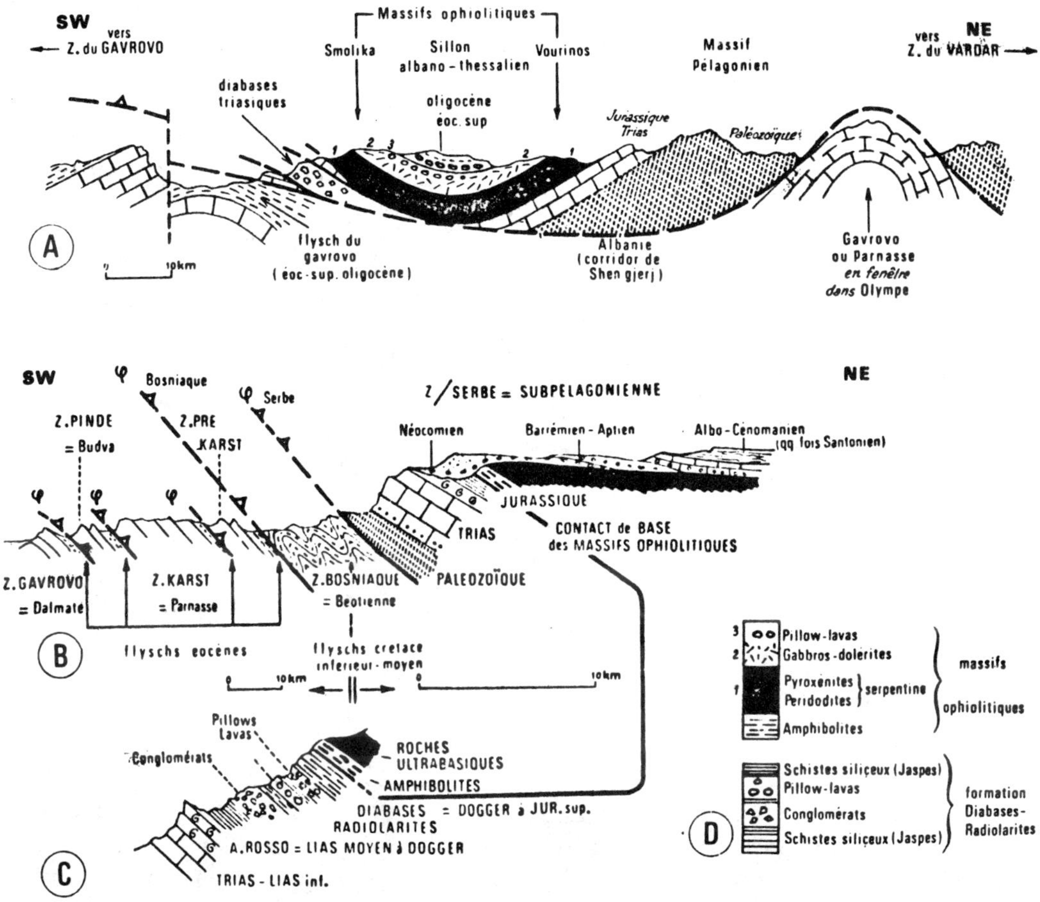

FIG. 6 End-Jurassic overthrust of the tethyan ophiolites above the Pelagonian Massif)acc. to Aubouin, 1973). A – North Greece cross-section, B – Yougoslavian cross-section.

The end-Jurassic compressive tectogenesis seems to be determined by the motion of the Apulian margin (block) eastward. This could explain the areal distribution of the compressive tectogenesis of this age and its absence from other areas. On the European margin Upper Jurassic–Neocomian continuous sequences are common in the flysch troughs (Outer Dacides) as well as in the pelagic and neritic sedimentary areas (Median Dacides). Even in the Transylvanian nappes and in the Metaliferi Mountains (both belonging to the Transylvanides) Tithonian–Neocomian sequences are continuous (Săndulescu, 1975; Lupu, 1976). In turn in several localities (cf. above) Tithonian rocks lie unconformably over older formations. This fact suggests a certain heterochronism between the two margins or, more exactly, between the obduction processes of the Vardar ophiolites and the weak deformation of the European margin. At the same time, it should be emphasized that a main tectogenic compressive moment in one area (Dinarides–Hellenides) may represent early tectogenesis (less important) in another (Carpathians). It should also be stressed that along the same major zone (Vardar and Transylvanides, namely the Tethyan ocean) compressive tectogenesis may occur with unequal intensity. In the case of the end-Jurassic tectogenesis this situation is determined by movements of the Apulian block.

The Mesocretaceous tectogenic 'phases' are manifested well on the European margin

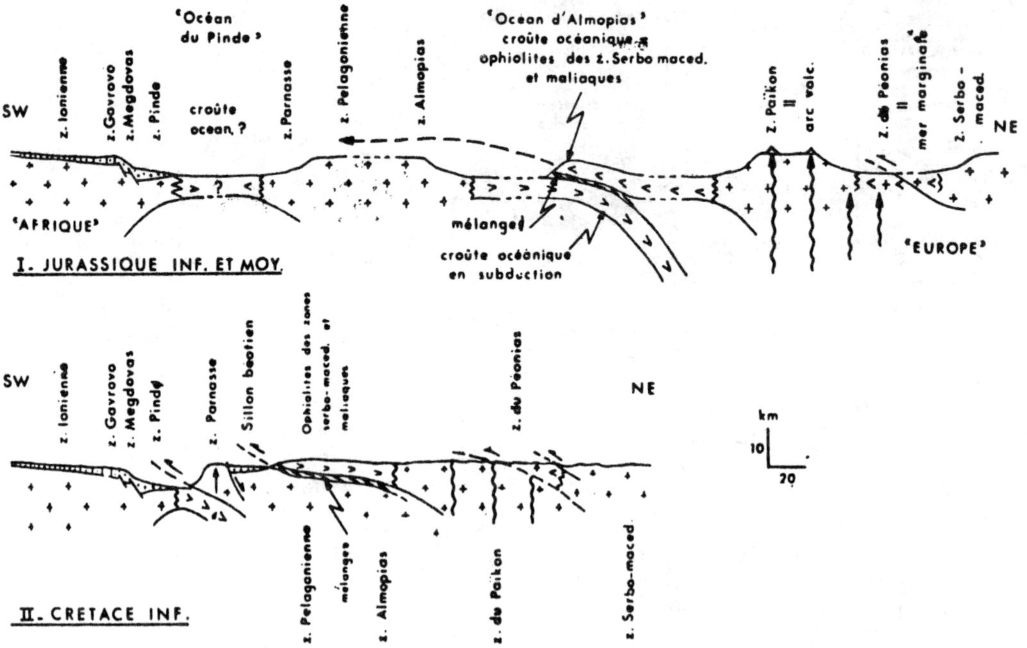

FIG. 7 Schematic evolution of the tethyan ophiolites (oceanic crust areas) in the Hellenides (acc. to Boillot, 1977).

(Median Dacides) and also in a part (at least) of the Tethyan ocean (Transylvanides). The Mesocretaceous compressive tectogenic time is longer in respect to other tectogeneses and it seems to have several rapidly repeated climaxes. It is rather a tectogenic interval than a tectogenic moment.

An obvious example of Mesocretaceous tectogeneses is presented by the Central Eastern Carpathian nappe system and the superposed Transylvanian nappes There the basement shear overthrusting occurs at least during the Albian, possibly starting also in the Aptian time. The nappes underthrust each other successively, from the outside towards the inside (namely starting with the Infrabucovinian nappes and ending with the Bucovinian). The obductive overthrust of Transylvanian nappes started in the Late Barremian or in the Aptian and ended in the (Late ?) Albian time.

The whole nappe structure of the inner Eastern Carpathians, namely the Central Eastern Carpathian nappes together with the superposed Transylvanian nappes are overlapped by a post-nappe cover which starts in the Vraconian or Cenomanian. This cover follows a very active but short erosion period proved by the sealing of the contours of the overthrust planes of practically all nappes, even of the deepest Infrabucovinian nappes. These erosion events have removed from the nappe anticlines several hundreds or even thousands, metres of crystalline schists and sedimentary rocks of the superposed nappes, as well as an important part of the uppermost units represented by the Transylvanian nappes.

The Mesocretaceous tectogeneses were active in the Southern Carpathians as well. They were described as the First Getic 'phase' (Codarcea, 1940) younger than the Lower Aptian. Supragetic and frontal Getic overthrusting took place during this 'phase' synchronous with the equivalent units of the Eastern Carpathians: older in the outer part and younger in the inner one, but both developing between the Lower Aptian and Late Albian.

According to Gočev (1979) compressive basement shear nappes as well as obduction nappes of Mesocretaceous age occur in the Inner Balkanides: the Morava nappe (equivalent of the Supragetic nappes) and the southern Strandjides (similar to the Transylvanian nappes) were generated before the Upper Cretaceous. Supposing that the latter proceeded from the Vardar zone (possibly from the Peonias zone?) it would be necessary to assume the occurrence of Mesocretaceous compressions there. Even if a different model is accepted in the Strandja, such as Tollmann's hypothesis of a tectonic window, it is also Mesocretaceous in age (Tollmann, 1965).

The above comments show that the Mesocretaceous tectogeneses were active along the European margin and generated the splitting of its outermost part (basement shear nappes) and also the obduction of the neighbouring oceanic zone. Concomitantly, on the Apulian margin the tectogenic regime was rather quiet. In turn, in the Inner Dacides area (inner margin of the Transylvanidian ocean), Mesocretaceous discontinuities (intra-Aptian, intra-Albian) without crustal contraction are known.

The areal distribution of Mesocretaceous tectogeneses was controlled by the southward motion of the European margin, which like the Apulian in the end-Jurassic time determined the obduction of the oceanic crust, but unlike the latter was itself deformed as well. Apart from the obductions of the oceanic crust (Transylvanian nappes), during the Mesocretaceous time of tectogenesis other different deformations in the oceanic area occurred and generated the Albian wildflysch and/or the olistostrome formations found at present in the Metaliferi Mountains (Lupu, 1976; Bleahu *et al.*, 1981); these are symmetrical with respect to the Bucovinian wildflysch which received blocks of Transylvanian nappes lying on the other side of the ocean.

Very complex in its development, the Mesocretaceous tectogenic interval is mainly represented on the European margin and decreases gradually towards the Apulian, where it is practically absent.

The Mediterranean or pre-Gosau tectogenesis is manifest in the Inner Dacides and their equivalents (Austroalpine nappes). Important nappe structures accompanied by crustal contraction are known there (Andrusov, 1965; Andrusov *et al.*, 1973; Mahel, 1974; Ianovici *et al.*, 1976; Bleahu, 1976; Patrulius, 1976).

In the Central Western Carpathians and in the Apuseni Mountains (northern Apusenides) the north and northeast vergent nappes are covered by the Gosau Formation the lowermost levels of which are Lower Senonian (Coniacian) in age. Thus the overthrust developed in the Turonian time or at the end of it. The area involved in this nappe structure represented in fact the inner margin of the Transylvanidian ocean.

On the other side—European margin—there are zones where pre-Senonian discontinuities are known, but zones with continuous Upper Cretaceous sedimentation also frequently develop. Even where discontinuities do occur, they correspond to weak deformations without crustal contraction.

Inside the Tethyan ocean, already (partly) squeezed, the pre-Gosau tectogenesis seems to be also present locally, namely in the Metaliferi Mountains (Lupu, 1976). The amplitude of the structures there is not well known and they may be mainly imbrications rather than overthrusts.

In the Dinaro-Hellenic area, the Apulian margin had no pre-Gosau (Mediterranean) tectogenesis generating nappes with the resultant crustal contraction. From this point of view, the differences between this margin and the Inner Dacidian-Austroalpine area are significant.

The areal distribution of the pre-Gosau tectogenesis shows that the most important effects were on the inner continental margin of the Transylvanidian and Liguropiemontais segments of the Tethyan ocean as well as in a part of these segments. It looks like a belated reflection of the Mesocretaceous tectogenesis developed on the other side of the ocean.

'Laramidian' tectogeneses involve several tectogenic climaxes which occur between the Lower Senonian and the Palaeocene. The most common seems to be the Early Laramidian tectogenesis which developed as an intra-Senonian deformation (intra-Campanian or intra-Maastrichtian).

Along the Carpathians and the Balkan the European margin is strongly affected by these deformations. Basement shear nappes (Getic and partly Supragetic overthrusts, Srednegorie overthrust), cover and cover-obduction nappes (the Outer Dacidian nappes) reflecting an important crustal contraction were generated at that time (Codarcea, 1940; Mahel, 1974; Săndulescu, 1972, 1975, 1980; Năstăseanu, 1975; Pop, 1973; Stefănescu, 1976; Gočev, 1979). These 'Laramidian' tectogeneses are progressively younger from the Carpathians towards the Balkans and the more distal Pontides. Thus in the Eastern Carpathians the Outer Dacidian nappes, namely the Ceahlău, are overlapped by a post-nappe cover starting at least in the Maastrichtian or Campanian (Stefănescu, 1976), while in the Srednegora overthrust also involves the Maastrichtian (Mahel, 1974).

'Laramidian' tectogenesis also occurs in the Transylvanides (Ianovici *et al*., 1969, 1976; Lupu, 1976; Săndulescu, 1975, 1980; Bleahu *et al*., 1981) and the Pieniny Klippen zone (Andrusov, 1965; Mahel, 1974; Birkenmajer, 1970) resulting in nappes and/or imbrications. In the Transylvanides (Metaliferi Mountains) the 'Laramidian' as well as the pre-Gosau tectogeneses show vergence towards the Inner Dacides, namely the inner continental margin of this segment of the Tethyan ocean. This is opposite to that of the Transylvanian nappes (Mesocretaceous in age). Thus the Transylvanides have a pseudo-bilateral structure.

In the Inner Dacides the 'Laramidian' deformations generated only weak and large folds accompanied by faulting. The same situation is encountered in the Central Eastern Carpathians where the post-Mesocretaceous cover, together with the overlapped nappes, was largely folded and/or faulted; retrothrustings also occur (Fig. 8).

Therefore, in an east-west traverse in the Romanian Carpathian zones with intensive 'Laramidian' deformations (Outer Dacides, inner Transylvanides) alternated with zones where weak post-nappe deformations took place (Central Eastern Carpathians, Inner Dacides). In the western Carpathians 'Laramidian' deformations show a normal orogenic polarity between the Central Western Carpathians (with Mediterranean tectogenesis) and the Flysch zone (with Neogene tectogenesis). In fact the possible prolongation of the Central Eastern Carpathians (with Mesocretaceous tectogeneses) is hidden by the Magura nappe overthrusting. Thus, alternating zones with 'Laramidian' tectogenic and weak deformations may also have occurred in the Western Carpathians. Along other traverses, such as the Southern Carpathians-Apuseni Mountains, the main 'Laramidian' deformations are divergent (Getic and Supragetic nappes in respect of the Metaliferi Mountains) and only inwardly (Inner Dacides in the northern Apuseni) associated with a weak post-nappe deformation.

In the Dinaro-Hellenic area the 'Laramidian' tectogenesis is either absent or, if it exists at all, is local and weak. The same situation occurs in the Vardar zone. It seems to be determined by the active motion of the Carpatho-Balkan foreland and the passive position of the Apulian at that time.

The general and specific features of the Cretaceous tectogenic period (including the end-Jurassic ones) may be summarized as follows:

(a) some of the Cretaceous tectogeneses—namely Mesocretaceous and 'Laramidian'—are in fact a group of rapidly repeated tectogenic climaxes;

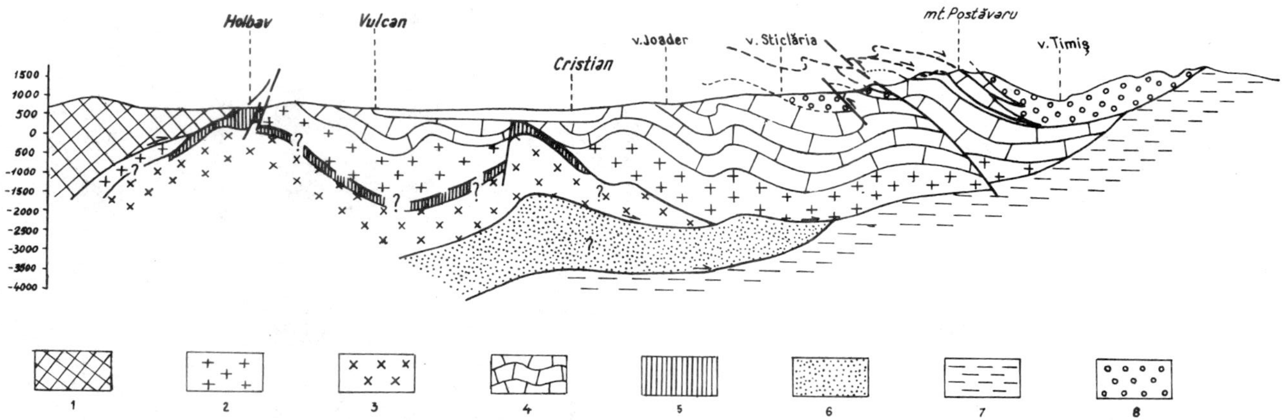

FIG. 8 Superposed tectogeneses—a cross-section in the Braşov mountains (eastern Carpathians) showing retrothrusting. 1 – Supragetic nappe; 2 ± 3-Getic metamorphic formations; 4 + 5 – Getic sedimentary formations; 6 + 6 – Outer Dacidian nappes; 8 – post-nappe cover (Upper Cretaceous (acc. to Săndulescu, 1975).

(b) during all the tectogeneses of this period, obduction nappes were generated, generally contemporaneous with continental shear nappes;
(c) Cretaceous tectogeneses involved the Tethyan oceanic area and also the proximal parts of its continental margins, but not with the same intensity simultaneously;
(d) at the end of the Cretaceous tectogenic period the squeezing of an important part of the Tethyan oceanic crust took place;
(e) in the present-day Alpine area of Europe no contemporaneous oceanic spreadings existed during the Cretaceous tectogeneses;
(f) except at the Apulian margin, the Cretaceous tectogenic period involved important parts of the Alpine area in central and southeastern Europe, as well as in the other European Alpine systems, in crustal contractions;
(g) during the Cretaceous tectogenetic period, the main tectogeneses alternately shifted from one part of the Tethyan ocean to the other: at the end-Jurassic over the Apulian margin, during the Mesocretaceous on the European margin and neighbouring oceanic area, the pre-Gosau tectogenesis on the inner margin of the Transylvanidian-Piedmontais segment of the Tethyan ocean and the 'Laramidian' tectogeneses mainly on the European margin and the remnants of the Tethyan ocean.

The Pyrenean tectogenesis involved important parts of the Apulian continental margin as well as some areas on the European.

In the Dinaro-Hellenic area important nappe overthrusts were generated after the Middle Eocene and before the Uppermost Eocene or Oligocene (Aubouin *et al*., 1970; Aubouin, 1973, 1977; Cadet *et al*., 1980). *En bloc* basement shear nappes and/or cover nappes are related to the tectogenesis proceeding from the Apulian continental margin (Pelagonian and equivalents, Bosnian and Beotian flysch, High Karst and equivalents, Pindus and equivalent units).

While the Apulian margin was involved for the first time in important overthrusts on the European margin, namely the Balkans, during the Pyrenean tectogenesis, older overthrust planes were again moved (Srednegorie) (Mahel, 1974; Gočev, 1979). However, the amount of Pyrenean overthrusts seems to be smaller in the Balkan as compared to the Dinaro-Hellenic area (Fig. 9).

An important Pyrenean overthrust is considered along the Eastern Alps, accompanied by crustal contraction (Oberhauser, 1968, 1978; Trümpy, 1973, 1976; Tollmann, 1978; Mahel, 1974; Debelmas *et al*., 1980 and others): the Austroalpine unit was thrust *en bloc* over the eastern Alpine flysch units. Therefore, taking into account the correspondence between the Austroalpine and the Inner Dacides (Central Western Carpathians), the *en bloc* Pyrenean overthrusts could be placed on the frontal part of the latter. No evidence of such an age of this possible process exists, except for the scarcity or the general absence of the Oligocene sedimentation in the Pieniny and Magura areas. In contrast, the intra-Eomiocene (Old Styrian) overthrust of the Pieniny Klippen belt and of the Mangura Group is well developed. Several solutions may be acceptable: (1) the *en bloc* overthrust of the Inner Dacides (Central Western Carpathians) above a part of the Pieniny zone is restricted in the Carpathians to the deep levels and surfaces only in the Alps; (2) there were two different moments of overthrusting along the same contact, one Pyrenean in the Alps and the other Old Styrian in the Western Carpathians. The second possibility is, in fact, a variant of the first: while during the Pyrenean tectogenesis the overthrusting was mostly active in the Alps and decreased progressively in the Carpathians, during the Old Styrian tectogenesis the overthrust occurred in the Carpathians (Western Carpathians) and decreased westward, in the Alps.

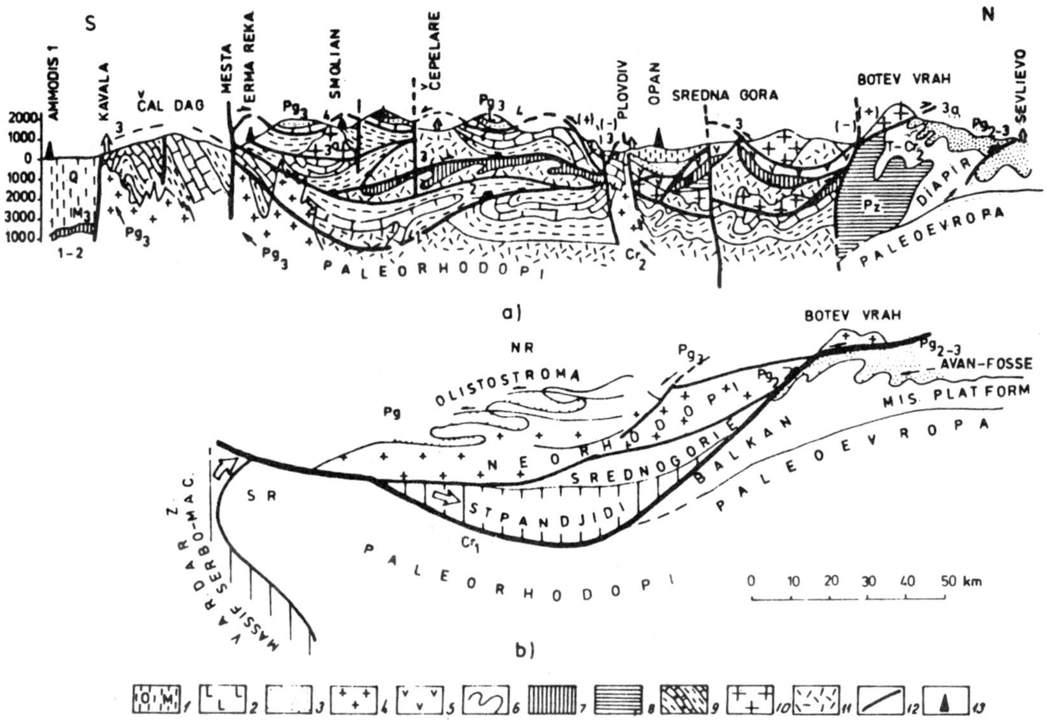

FIG. 9 Cross-sections showing the structure of the Rhodope and Balkanides and the relationship between the cretaceous and late eocene overthrusts (acc. to Gočev, 1980).

1 – Miocene-Quarternary of the north Aegean graben, 2 – Oligocene post-nappe rhyolites, 3 – Upper Eocene and Oligocene within the north Rhodope volcanic graben, 4 – Upper Cretaceous (Srednogorie) and Upper Paleogene (South Rhodope) calko-alkaline intrusions, 5 – Srednogorie volcano-sedimentary Upper Cretaceous, 6 – Mesozoic of the Outer Balkanides, 7 – Mesozoic of the Inner Balkanides, North and South Strandzides, 8 – Variscan basement of the Outer and Inner Balkanides and the Strandzides, 9 – Rhodopian Complex (Proterozoic), 10 – South Bulgarian granitoids (Paleozoic?), 11 – Prarhodopian basement of the Inner Balkanides and Strandzides, 12 – overthrusts, 13 – deep boreholes.

By comparing the Pyrenean overthrusts in the Dinaro-Hellenic area with those of the Balkans and the Alps (and(?) Western Carpathians), a difference should be emphasized: in the former area the deformed units have a continuous trend and an arcuate shape, while in the latter they are discontinuous (no direct connection between the Pyrenean units of the Alps and Balkan) and generally have an east-west direction.

During the Pyrenean tectogenesis the final cicatrization of the Dinaro-Hellenic suture zone took place (Aubouin, 1973). Thus, the continent/continent collision was active, leading to 'hypercollision' in the Eastern Alps (and the Western Carpathians?). In this context the remobilization of the Srednagorie overthrust front is an 'echo' of the collision process which developed along the Vardar zone. The intra-Eomiocene tectogenesis (Old Styrian) was active mostly in the Carpathians, namely in the flysch zone and the Pieniny Klippen belt.

The oldest tectogenetic moment generating nappes in the Carpathian flysch zone has long been considered as Savian (because of the Oligocene age of the youngest formations involved in these nappes). The former 'Oligocene' formations are in fact Oligocene–Lower Miocene in age (Martini and Lebenson, 1971; Stefănescu *et al.*, 1979). Therefore the overthrust of the nappes involved in this tectogenesis is intra-Burdigalian, corresponding more or less to the Old Styrian 'phase' (Dumitrescu and Săndulescu, 1968). To this moment belong: the overthrust of the Pieniny Klippen zone above the Magura nappe (in the Soviet Carpathians, Mahel, 1974) (Fig. 10) as well as their equivalent in the Romanian Carpathians (Botiza nappe, Dumitrescu, 1957), the Magura overthrust, the overthrusts of the Convolute Flysch nappe, the Audia nappe, the Czernahora nappe, the Silezian nappe and, probably, the Subsilezian nappe (Mahel, 1974). Foldings took place also in the most external flysch nappes (namely the Tarcău (Skole), the Marginal folds (Pokutian-Boryslav) and the Subcarpathian (Stebnik) nappes), but without important horizontal displacements. It may be possible that frontal overthrusts of the Alpine flysch zone have the same intra-Eomiocene age.

Intra-Eomiocene (namely intra-Burdigalian) (Cadet *et al.*, 1980) overthrusts were also recorded in the outer Dinaro-Hellenic area (Aubouin, 1973, 1977; Grubič, 1980). Both on the European and on the Apulian continental margin, the intra-Eomiocene tectogenesis developed in a continent/continent collision framework. Here it should be emphasized that this tectogenic moment has produced mainly cover nappes in the Carpathians as well as in

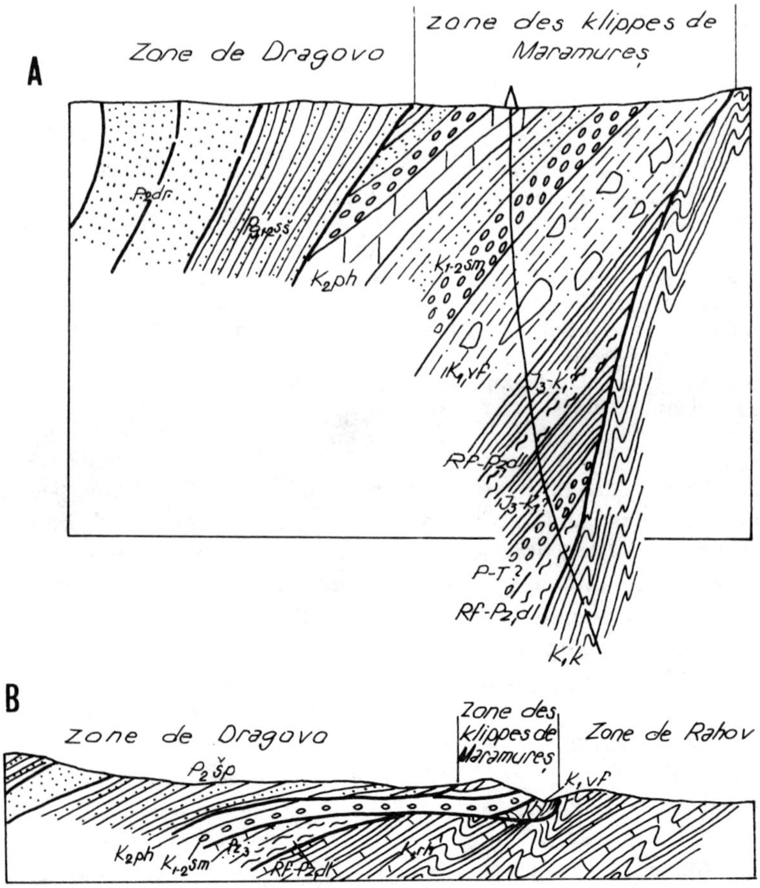

FIG. 10 Geological cross-section showing the Tertiary tectonics of Magura (Dragovo) nappe in the Ukrainian Carpathians (acc. to Bîzova *et al.*, 1972).

the Dinaro-Hellenic area, but the *en bloc* overthrust of the more internal zones such as the Pindus Zone (Cadet *et al.*, 1980) or the Central Western Carpathians (see above) also took place.

Excepting the two marginal zones where the Old Styrian nappes were generated, in the former deformed areas this tectogenic moment is also present with general imbrications or local overthrusting, this being the case in the Southern Carpathians where some parts of the Supragetic frontal line were removed in the post-Aquitanian time (eastern Serbia) or also in the Oligo-Eomiocene molasse of the Petrosani depression which was thrust on its inner limb. Intra-Eomiocene retrothrusts and retrofolding have been recorded in the Central Eastern Carpathians (Săndulescu, 1975); together they deformed the former nappes and their post-nappe cover. These imbrications, retrothrusts and retrofolding in the former deformed zones are reflections of the more external nappe overthrusts.

The intra-Badenian tectogenesis (Young Styrian) developed in the outer Carpathian flysch zone. Cover nappes, similar to those of intra-Eomiocene age (Tarcău or Skole, Marginal folds and equivalents, namely Pocutian and Boryslaw) were generated during this tectogenic moment. The Carpathians are the main area in central and southeastern Europe where Young Styrian nappes are known. Therefore the European margin seems to be the most important host of the Young Styrian tectogenesis (certainly in the area analysed). In a more general framework contemporaneous tectogeneses also occurred on the African margin (in the Maghrebides of the North African Alpine chains). There the Miocene tectogeneses also affected the outer zones of the chains.

It should be stressed that the sedimentary complexes which built the Young Styrian cover nappes were deformed (mainly folded) before the overthrusting, namely during the Old Styrian tectogenic moment (see above). This model (Fig. 11) is common for all cover nappes of the Carpathian flysch zone (Săndulescu, 1980), even Cretaceous or Miocene in age. The overthrusting is preceded by folding and detachment of the sedimentary sequences from their basement (without horizontal displacements); the detachment plane is used as overthrust plane by the later overthrusts.

In the above-mentioned model the underthrusting of the foreland has an important role. Furthermore, the whole nappe structure of the Carpathian flysch is directed by this mechanism. This means that inside the European continental margin the underthrusting of the most rigid part generated the cover nappes issued during several tectogenic moments from the flysch troughs seated in the intracontinental rift zone and outwards.

The intra-Sarmatian tectogenesis (Moldavian) is the youngest tectogenic moment generating nappes—therefore crustal contraction—in central and southeastern Europe.

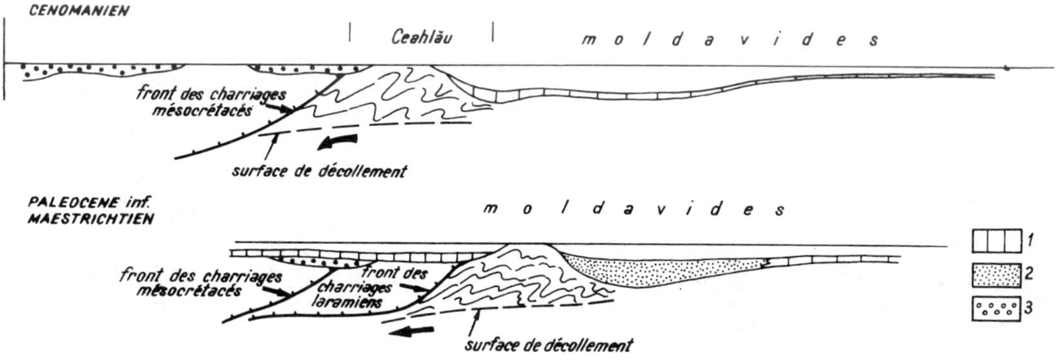

FIG. 11 A model of the successive generation of cover (flysch) nappes (acc. to Săndulescu, 1975).

The Subcarpathian (Stebnik) nappe was generated at that time in the outermost Carpathians.

It is necessary to mention that the timing of this tectogenesis is attributed to the Ponto-Caspian stratigraphic scale; there the Sarmatian represents the timespan between the Badenian and the Meotian. In the inner Carpathians (Western Paratethys) the Moldavian tectogenesis corresponds to the boundary between the 'Sarmatian' and the Pannonian (the latest corresponding to the upper part of the Ponto-Caspian Sarmatian and to the Meotian). The Moldavian tectogenesis is in fact intra-Bessarabian (Dumitrescu and Săndulescu, 1968), namely at the lower boundary of the Pannonian.

Stille (1953) was the first to define the Moldavian 'phase' as the youngest responsible for the outermost overthrusts in the Eastern Carpathians. At that time the age of these overthrusts was placed at the 'Tortonian' (actual Badenian)–Sarmatian boundary. Later it was pointed out that the last overthrusts on the Carpathian border are intra-Sarmatian and consequently the Moldavian 'phase' is of this age (Dumitrescu and Săndulescu, 1968).

The areal extension of the Moldavian tectogenesis is restricted with respect to the oldest ones. Furthermore, the Tertiary tectogeneses are—on the European margin—progressively restricted to smaller and smaller areas. On the eastern Apulian continental margin this tectogenic moment is not important. However, a more or less similar moment occurred in the Maghrebides, Sicily and the Apennines. There too it was restricted to the outermost zone of the Alpine area.

The Moldavian tectogenesis is the last moment when important underthrusting of the cratons took place at the outer border of the eastern and partly the western Carpathians.

The Walachian tectogenesis is restricted to the Romanian Carpathian arc area, namely southern Moldavia and northeastern Muntenia (Walachia). Folding and thrusting, but not overthrusting, were generated during this tectogenesis.

The age of the Walachian tectogenic moment is determined by the age of the Cîindesti Gravels (the youngest deposits involved in the Walachian deformation). If they are considered as of Lower Pleistocene age the tectogenesis is intra-Pleistocene. An upper Romanian age of the Cîindesti Gravels establishes the age of the Walachian tectogeneses at the Pliocene–Quaternary boundary.

Compared with the other Tertiary tectogenetic moments known in the Carpathians, the Walachian has a smaller development area which may be related to the weak intensity of this tectogenesis. It is interesting to note that the area affected by the Walachian tectogenesis is also the zone where the Vrancea earthquakes are active.

The general and specific features of the Tertiary tectogenic period may be summarized as follows:

(1) Except for the Pyrenean tectogenesis, the Old and Young Styrian as well as the Moldavian are condensed into a relatively short time span (4–6 m.y.), with respect to the Mesocretaceous tectogeneses for instance, which last approximately 10 m.y. From this point of view, the whole Tertiary (27–30 m.y.) is shorter than the Cretaceous (45–50 m.y.) tectogenic period.

(2) The Tertiary tectogenic period involved only former continental margins. It developed in the framework of continent/continent collision, leading to hypercollision.

(3) As compared with the Cretaceous tectogenesis the Tertiary involved a more restricted area. In the same way it should be emphasized that the younger Tertiary tectogeneses are areally more limited than the older Tertiary.

(4) While the Pyrenean and the Old Styrian tectogeneses developed—with different intensities—on both the Apulian and the European continental margins, the younger

ones are recorded only on the European margin. This is true only in the case of the Dinaro-Hellenic Apulian margin since in the Apennines, Sicily and North Africa the younger overthrusts may be more or less contemporaneous with the Moldavian tectogenesis.

(5) During the Tertiary tectogenic period no true obduction nappes were generated; in some places *en bloc* overthrusting occurred involving older obduction nappes, but Tertiary obductions are unknown.

(6) Most of the Miocene nappes, at both the Apulian and the European margins, are cover nappes, indicating the importance of foreland underthrusting.

SYNCHRONISM AND HETEROCHRONISM OF THE COMPRESSIVE TECTOGENETIC EVENTS

A thorough analysis of the Cretaceous and Tertiary tectogenic events, recorded in south-eastern and central Alpine Europe, centres on some features of the time/space (temporal/areal) development of these processes.

As a first conclusion it may be considered that the compressive tectogenetic events occur in various different moments. These moments may be correlated with similar events recorded in other parts of the continents (in the Alpine areas). Thus, different 'phases' (or moments) recorded in the analysed area are also recognized in western Alpine Europe, in the Caucasus, Asia Minor, the Himalayas, and so on, emphasizing the synchronism of the tectogenic processes, but not the ubiquity of such synchronous events. In fact the compressive deformations developed within limited areas adjacent to zones where no contemporaneous deformations are recorded. The various examples presented in this paper clearly support the following conclusion: end-Jurassic tectogenesis in one place (Apulian margin) versus continuous sedimentation from the Upper Jurassic up to the Aptian or Albian in other areas; Mesocretaceous tectogenesis (European margin) versus uninterrupted sedimentation in the Dinaro-Hellenic area or in the Moldavian flysch, etc. This picture is valid for each tectogenic moment (Fig. 12).

In a few cases specific sedimentary or tectonosedimentary processes developed in the unaffected areas, simultaneously with the compressive tectogenic events. Breccias, wildflysch or olistostromes are known in the analysed Alpine chains, but they are older or younger with respect to the compressive tectogenic 'phases'. Generally, breccias (and conglomerates) were generated either by active fracturing or by morphogenic uplifts, which may precede, sometimes by a long time, or succeed, generally after a short time, the compressive tectogenesis. The wildflysch or the olistostromes are generally connected with vertical uplift (in the area of provenance of these formations) or with the frontal part of gravity sliding nappes, also generated by uplifts.

As a second conclusion it may be considered that the compressive tectogenic events occur in an orogenic chain as a whole following a heterochronic 'programme'. Thus, different areas are involved in tectogeneses of different ages. This migration of tectogenesis may have a definite polarity ('orogenic polarity', Aubouin, 1965), but some chains—the Carpathians for instance—may show inverse or alternating polarity.

The heterochronism of the compressive tectogenic moments may be followed not only across a chain, but also a long some zones of it. The Alps-Carpathian flysch zone is one of the possible examples.

The heterochronism of the tectogeneses, at least in the area analysed is also emphasized by their alternation in time on both sides of the Tethys ocean, mainly before the continent/continent collision. This is obvious for the Cretaceous tectogeneses. The Tertiary tectogenic episodes are more synchronous on both margins.

Main Compressive Tectogeneses \ Geotectonic Domains	Apulian Margin	Tethyan Ocean (Vardar)	Inner Dacidian (-Austroalp.) Block	Tethyan Ocean (Transylvanides-T) (Pienides-P)	European Margin
End-Jurassic	++	++++++			
Mesocretaceous				(T) (P) +++++++++++	++++++
Pre-Gosau (Mediteranian)			++++++++++++	+++?	
End-Cretaceous ('Laramide')	+++	+++		(P) +++++	++++++
Late-Eocene (Pyreenean)	++++	+++	++?++	++?++	+++
Intra-Eomiocene (Old Styrian)	++++			(P) ++++	+++++
Intra-Badenian (Young Styrian)					+++++
Intra-Sarmatian (Moldavian)					++++++

FIG. 12 Schematic temporal/areal correlation of the main compressive tectogeneses within the central and southeastern Europe.

Following the length of the compressive tectogenic events, it is possible to distinguish: tectogenic periods, tectogenic intervals and tectogenic moments. The tectogenic periods are Cretaceous and Tertiary. Examples of a tectogenic interval are the Mesocretaceous and the Laramidian, covering several tectogenic moments. The tectogenic moment may be compared with the classical tectogenic (or more generally orogenetic) 'phase'. The intervals are more specific for the Cretaceous tectogenic period. During the Tertiary period the tectogenic moments are more obvious. This may be a result of the continent/continent collision.

To summarize it may be emphasized that:

(a) the episodicity of the compressive tectogenic events is important in the geotectonic history of the earth;
(b) these events occur in some correlable timespans which may be intervals or moments; from this point of view they are synchronous;
(c) the compressive tectogenic intervals or moments are not ubiquitous; from this point of view they are heterochronous;
(d) the synchronism and the heterochronism of the compressive tectogenic events are complementary rather than exclusive features.

REFERENCES

(NB. Most of the references include extensive bibliographical notes)

Andjelkovič, M. and Lupu, M. (1977). Die Geologie der Sumadija und Mureş Zone, Carp.-Balk. Geol. Assoc., VIII Congr., *Rep. geotect.*, **I**, Belgrade, pp. 15–28.

Andrusov, D. (1965). Aperçu général sur la géologie des Carpathes occidentales, *Bull. Soc. géol. France*, sér. 7, **7**, 1029–62.

Andrusov, D., Bystricky, J. and Fusan, O. (1973). Outline of the structure of the West Carpathians, 10th Congr. Carp.-Balk. Geol. Assoc., Bratislava, 45 pp.

Agryriadis, I., Gracjansky, P., Marcoux, J. and Ricou, L. E. (1980). The opening of the Mesozoic Tethys between Eurasia and Arabia-Africa in *Geologie des chaines alpines issues de la Tethys*, Coll. C5, 26 Congr. Géol. Intern., Paris, pp. 199–214.

Aubouin, J. (1965). *Geosynclines*, Amsterdam: Elsevier, 355 pp., 67 fig.

Aubouin, J., Blanchet, R., Cadet, J. P., Celet, P., Charvet, J., Chorowicz, J., Cousin, M. and Rampnoux, J. P. (1970). Essai sur la géologie des Dinarides, *Bull. Soc. Géol. France*, sér. 7, **12**, 1060–93.

Aubouin, J. (1973). Des tectoniques superposées et leur signification par rapport aux modèles géophysiques: l'exemple des Dinarides; paléotectonique, tectonique, tarditectonique, néotectonique, *Bull. Soc. Géol. France*, sér. 7, **15**, 5–6, 426–60.

Aubouin, J. (1977). Brève présentation de la géologie de la Grèce, *Bull. Soc. Géol*. France, sér. 7, **19**, 1, 6–10.

Birkenmajer, K. (1970). Pre-Eocene fold structures in the Pieniny Klippen belt (Carpathians) of Poland, *Stud. Geol. Pol.*, **31**, 1–81.

Bleahu, M. (1976). Structure géologique des Apuseni septentrionaux, *Rev. roum.* géol. *géophys. géogr*., ser. géol., **20**, 1, 27–39.

Bleahu, M., Lupu, M., Patrulius, D., Bordea, S., Stefan, A. and Panin S. (1981). *The Structure of the Apuseni Mountains*. Guidebook to Excurs. B3, XII Congr. Carp.-Balk. Geol. Assoc., Bucharest: Publ. Inst. geol. geophys., 116 pp., 24 fig., 2 pl.

Boillot, G. (1977). Modèles actualistiques des Héllénides, *Bull. Soc. Géol. France*, sér. 7, **19**, 1, 82–3.

Cadet, J. P., Bonneau, M., Charvet, J., Durr, S., Elter, P., Ferrière, J., Scandone, P. and Thiébault, F. (1980). Les chaines de la Méditerranée moyenne et orientale, in *Géologie des chaines alpines issues de la Tethys*, Coll. C5, 26 Congr. Géol. Intern. Paris, pp. 98–118.

Chanell, J. E. T., D'Argenio, B. and Horvath, F. (1979). Adria, the African promontory, in the Mesozoic Mediterranean paleogeography, *Earth Sci. Rev*., **15**, 213–92.

Codarcea, Al. (1940). Vues nouvelles sur la tectonique du Banat Méridional, *An. Inst. geol. rom*., **XX**, 1–74.

Debelmas, J., Oberhauser, R., Săndulescu, M. and Trümpy, R. (1980). L'arc alpino-carpathique, in *Géologie des chaines alpines issues de la Tethys*, Coll. C5, 26 Congr. Géol. Intern. Paris, pp. 86–96.

Dumitrescu, I. (1957). Asupra faciesurilor şi orizontării Cretacicului superior şi Paleogenului în bazinul Lăpuşului (nordul Depresiunii Transilvaniei), *Lucr. Inst. Petr. Gaze*, **III**, 19–42.

Dumitrescu, I. and Săndulescu, M. (1968). Problèmes structuraux fondamentaux des Carpathes et de leur avant-pays, *An. Com. Geol.*, **XXXVI**, 195–218.

Durand Delga, M. (1980). Considérations sur les flysches du Crétacé inférieur dans les chaines alpines d'Europe, *Bull. Soc. géol. France*, sér. 7, **22**, 1, 15–30.

Gočev, P. M. (1979). Miastoto na Strandja v alpiskata strucktura na Balkanskya poluostrov., *Rev. Bulg. Geol. Soc.*, **XL**, 1, 27–46.

Grubič, A. (ed.) (1980). *Yugoslavia—An Outline of Geology of Yugoslavia*, Guidebook 15, 26th Int. Geol. Congr., Paris-Belgrade, 97 pp.

Ianovici, V., Giuşcă, D., Ghiţulescu, T., Borcoş, M., Lupu, M., Bleahu, M. and Savu, H. (1969). *Evoluţia geologică a munţilor Metaliferi*, Bucharest: Ed. Acad. R.S.R., 741 pp. 263 fig.

Ianovici, V., Borcoş, M., Bleahu, M., Patrulius, D., Lupu, M., Dimitrescu, R. and Savu, H. (1976). *Geologia munţilor Apuseni*, Bucharest: Ed. Acad. R.S.R., 631 pp., 176 fig.

Lupu, M. (1976). The main tectonic features of the southern Apuseni Mountains, *Rev. roum. géol. géophys. géogr.*, sér. géol., **20**, 1, 21–5.

Mahel, M. (ed.) (1974). *Tectonics of the Carpathian-Balkan Regions* (Explanations to the Tectonic Map of the Carpathian-Balkan Regions and Their Foreland), Bratislava: Edit. Geol. Inst. 453 pp.

Martini, E. and Lebenzon, C. (1971). Nannoplankton—Untersuchungen im oberen Tal des Tarcău (Ostkarpathen, Rumänien) und stratigraphische Ergebnisse, *Neues Jahrb. f. Geol. u. Pal.*, **9**, Stuttgart.

Năstăseanu, S. (1975). Aperçu synthétique sur la géologie des Carpathes Méridionales entre le Danube et le Jiu, *Bull. Soc. géol. France*, sér. 7, **17**, 3, 359–68.

Oberhauser, R. (1968). Beitráge zür Kentniss des Tektonik und der Paläegeographie während der Oberkreide und dem Paläogen in Ostalpenraum, *Jahrb. geol. B.A.*, **111**, Vienna.

Patrulius, D. (1976). Les formations mésozoïques des monts Apuseni septentrionaux: corrélation chronostratigraphique et faciale, *Rev. roum. géol. géophys. géogr.*, sér. géol., **20**, 1, 49–57.

Pop, Gr. (1973). *Depozitele mezozoice din munţii Vîlcan*, Bucharest: Ed. Acad. R.S.R., 155 pp., 36 fig.

Săndulescu, M. (1972). Consideraţii asupra posibilităţilor de corelare a structurii Carpaţilor Orientali şi Occidentali, *D. S. Inst. geol.* **LVIII**, 5, 125–50.

Săndulescu, M. (1975). Essai de synthèse structurale des Carpathes, *Bull. Soc. géol. France*, sér. 7, **17**, 3, 299–358.

Săndulescu, M. (1980). Analyse géotectonique des chaines alpines situées autour de la Mer Noire occidentale, *An. Inst. geol. geofiz.*, **LVI**, 5–54.

Săndulescu, M. (1981). La marge continentale européenne dans l'aréal Carpatho-Balkanique, 12th Congr. CBGA, Abstracts, pp. 143–4, Bucharest: Publ. Inst. geol. geofiz.

Stille, H. (1953). Der geotektonische Wérdegang der Karpathen, *Beih. f. Geol. Jb.*, vol. 8, 239 pp., 33 fig., 1 pl.

Stefănescu, M. (1976). O nouă imagine a structurii flişului intern din regiunea de curbură a Carpaţilor, *D. S. Inst. geol. geofiz.*, **LXII**, 5, 257–9.

Stefănescu, M., Gheţa, N. and Dicea, M. (1979). On the Oligocene-Miocene boundary in the external flysch zone of the Carpathian Bend (between the Teleajen Valley and the Dîmboviţa Valley). A tentative solving by calcareous nannoplankton, *Rev. roum. géol. géophys. géogr.*, sér. géol., **23**, 1, 89–94.

Tollmann, A. (1965). Das Strandscha Fenster, ein neues Fenster des Metamorphiden im alpinen Nordstamm des Balkan, *N. Jb. Geol. Pal. Min.*, **4**, 234–48, Vienna.

Tollmann, A. (1978). Plattentektonische Fragen in den Ostalpen und der Plattentektonische Mechanismus des mediterranen Orogens, *Mitt. öster. geol. Gesel.*, **69**, 291–351.

Trümpy, R. (1973). L'évolution de l'orogenèse dans les Alpes Centrales. Interprétation des données stratigraphiques et tectoniques, *Eclog. geol. Helv.*, **66**, 1, 1–10.

Trümpy, R. (1975). Penninic-Austroalpine boundary in the Swiss Alps: A presumed former Continental Margin and its problems, *Amer. J. Sci.*, **275-A**, 209–38.

Visarion, M. and Săndulescu, M. (1979). Structura subasmentului de presiunii pannonice in România (sectorul central şi nordic), *St. cerc. geol. geofiz. geogr.*, ser. geofiz., **17**, 2, 191–201.

Global Correlation of Tectonic Movements
Edited by Yu. G. Leonov and V. E. Khain

7

Mesozoic Tectonic Movements and Magmatism in Eastern Asia (Correlation Problem of Tectonics and Magmatism)

M. S. NAGIBINA, YU. G. GATINSKY, G. A. GRINBERG, G. S. GUSEV, V. I. KOVALENKO*

During investigations conducted within the framework of the project 'Global correlation of epochs of tectogenesis' the authors carried out research to find dynamic relationships in the development of Mesozoic tectonic structures and magmatism in the eastern and southeastern parts of Asia along three major profiles: the northern—from the Verkhoyansk fold system to the Karyak zone; the central—from Mongolia and the Prebaikalian area along the Mongolo-Okhotsk belt and the Sea of Okhotsk coast; and the southern—through Indochina, southeast China, Indonesia and the Philippines. Detailed correlation of Triassic, Jurassic and Cretaceous stratigraphic key sequences and their lithologic features, angular unconformities and reworking of the structural pattern, as well as of the time of magmatic manifestations has been carried out on the basis of new data so as to ascertain stages in the development of tectonic structures and magmatic activity. The study is based on both geological and isotopic data. This paper deals with some results of the investigations aimed at the correlation of tectonics and magmatism in eastern Asia.

At the beginning of the Mesozoic, the formation of the continental crust was completed over the most of eastern and southeastern Asia (Peive, *et al.*, 1976). The territory of the continent was occupied by separate narrow sea bays and larger epicontinental seas accumulating siliciclastic and clastic-carbonate deposits of different thicknesses. Deeper relict sea basins separating different microcontinents, such as the Sinoburmania, southern and northern Tibet, eastern Asiatic occupied southeastern Asia in the Paleopacific and Palaeotethys junction zones (Fig. 6d).

During the Mesozoic, the eastern part of the Eurasian continent was subject to active tectonism and magmatism and can be subdivided by the type of their manifestations into three distinct regions or, conventionally, megablocks: the northern (the Verkhoyansk-Chukotka), the central (including zones of the Stanovik and Mongolo-Okhotsk belt up to Cathasiya) and the southern (Indochina-Sunda-Philippine) megablocks (Fig. 1). These

* Respectively:
Geological Institute, USSR Academy of Sciences, Moscow.
VNII Zarubezhgeologia, USSR Ministry of Geology, Moscow.
Institute of Geology, Siberian Department of USSR Academy of Sciences, Iakutsk.
Institute of Geology, Siberian Department of USSR Academy of Sciences, Iakutsk.
IGEM, USSR Academy of Sciences, Moscow.

N
C
S
1
2
3
4
5
6

large regions differ in their pre-Mesozoic development and time of formation of granite crust (Nagibina *et al.*, 1981).

The northern region almost throughout its history was the eastern part of the ancient Siberian continent, and its granitic crust was completely formed in the pre-Riphean times (Markov, *et al.*, 1979). Local carbonate troughs (with mobile platform regime, according to Pushcharovskiy, 1960) developed in the Verkhoyansk zone during the Riphean and Early-Middle Palaeozoic. In the Early Carboniferous a large geosynclinal basin was formed on this territory. Between the Late Permian and the Early Triassic, formation of local uplifts took place which were accompanied by breaks in sedimentation. However, no considerable reworking of the structural plan of the territory occurred. A correlation of Mesozoic sequences (Fig. 2) suggests that from the Early Triassic up to and including the Late Jurassic (Callovian) the vast Verkhoyansk-Chukotka region developed as little differentiated major suboceanic troughs inherited from the Palaeozoic (Gusev, 1979). The troughs were separated by synsedimentary fault-bounded uplifts of older (Archaean and Proterozoic) basement. In the course of continuous accumulation of thick (over 10 km in the Verkhoyansk complex) terrigenous deposits formed in the marine shelf environments deep faults developed due to extension in the ancient basement and provided channels for minor intrusions of basaltoid magma such as diabase and gabbro dykes and sills, and sheets of traps.

In the Late Jurassic (at the Callovian and the Oxfordian boundary) nearly the whole territory of the northern megablock was stressed and long zones of linear folding and block fault-folding were developed.

At the end of the Jurassic and Early Cretaceous period, the deformation was manifested by wrench-faults, shifting fold axes and overthrusts. Horizontal displacement along some wrench faults exceeded 20 km. After folding weakened zones of extension developed in some parts of the Verkhoyansk zone. The intrusion of smaller diorite and quartz diorite bodies and gabbro-diabase dykes took place in the Oxfordian–Kimmeridgian.

The graben and rift-like structures originated during the Oxfordian-Neocomian (J_3-K_1) in the extension zones of the northern megablock and were accompanied by accumulations of continental molassoid, partly coal-bearing deposits and volcanogenic complexes of bimodal basalt-liparite association, locally with specific sodic liparites.

Late Mesozoic orogenic granitoid magmatism can be subdivided into two stages: Early Cretaceous and Late Cretaceous (Grinberg, 1973; Grinberg *et al.*, 1979). Lateral series with nearly coeval, although of different composition and origin, magmatic complexes, were formed during each stage. Intrusive rocks of the Okhotsk and Kolyma complexes were areally separated, although they are nearly coeval and represent a good example of lateral consanguinity. For example, intrusion of huge masses of the Kolyma complex granitoids took place within the interval of 150–100 m.y. (predominantly in the Early Cretaceous, 120 ± 5 m.y.) along the continuous Indigirka-Kolyma system of faults within the boundaries of the Inyaly-Debinsk synclinorium. These rocks, varying in composition from granitoids and quartz diorites to biotite granites, form bodies of the main batholithic belt. Smaller bodies of heterogeneous granitoids occur along zones of cryptofaults, surrounding

---

FIG. 1 Areas of Mesozoic continental volcanism and granite formation in eastern Asia.
1 – areas of Mesozoic continental volcanism of different composition (volcanic belts of continental margins and inner-continental areas); 2 – boundaries of Mesozoic granitoid areas (hachures towards areas of granitoid development); 3 – major faults; 4 – wrench faults (arrows show direction of displacement); 5 – fragments of Mesozoic Zavaritsky–Benioff zones; 6 – Early Mesozoic collision zones of microcontinents (after Yu. G. Gatinsky).

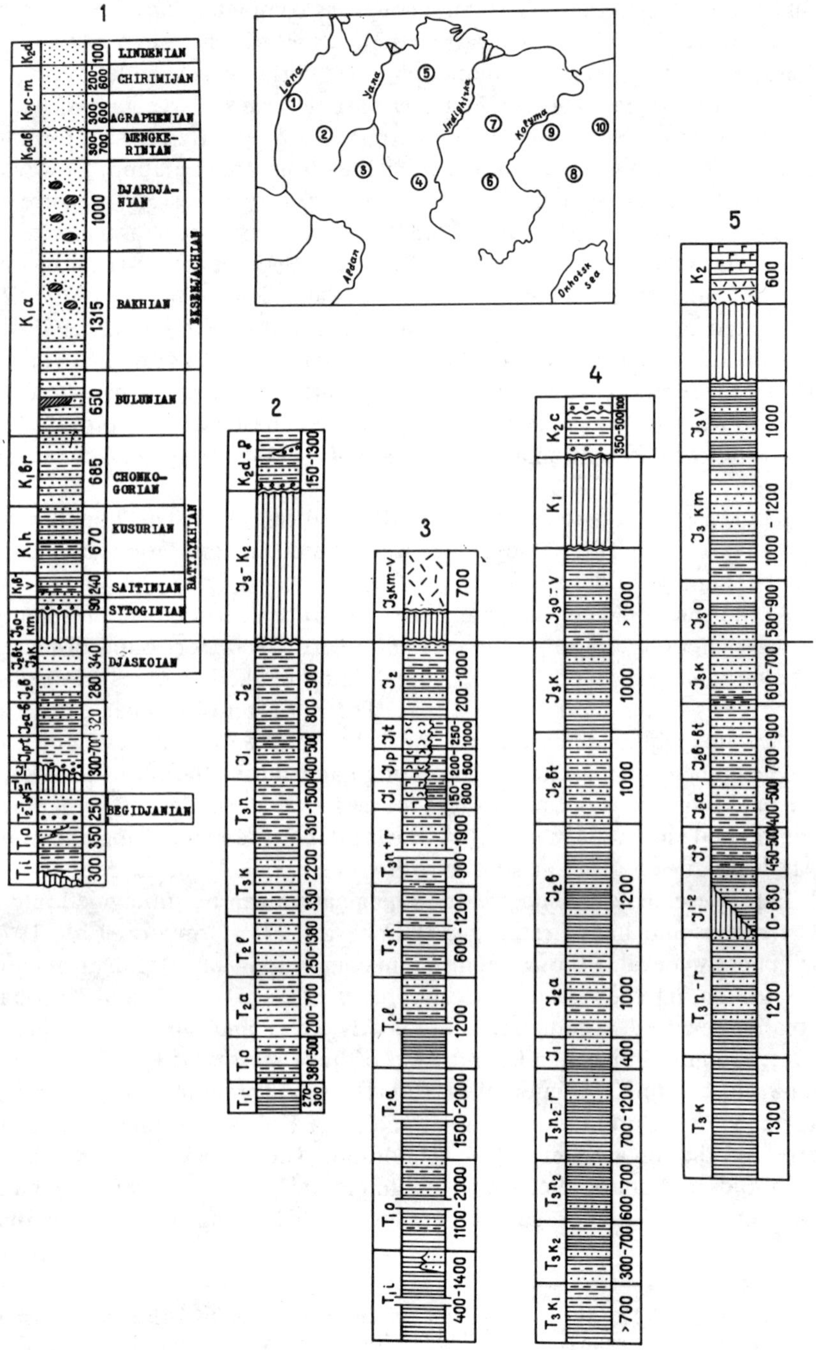
1
2
3
4
5
Lena
Yana
Indighirka
Kolyma
Aldan
Okhotsk sea
LINDENIAN
CHIRIMIJAN
AGRAPHENIAN
MENGKE-RINIAN
DJARDJA-NIAN
BAKHIAN
BULUNIAN
CHONKO-GORIAN
KUSURIAN
SAITINIAN
SYTOGINIAN
DJASKOIAN
BEGIDJANIAN
BATYLYKHIAN

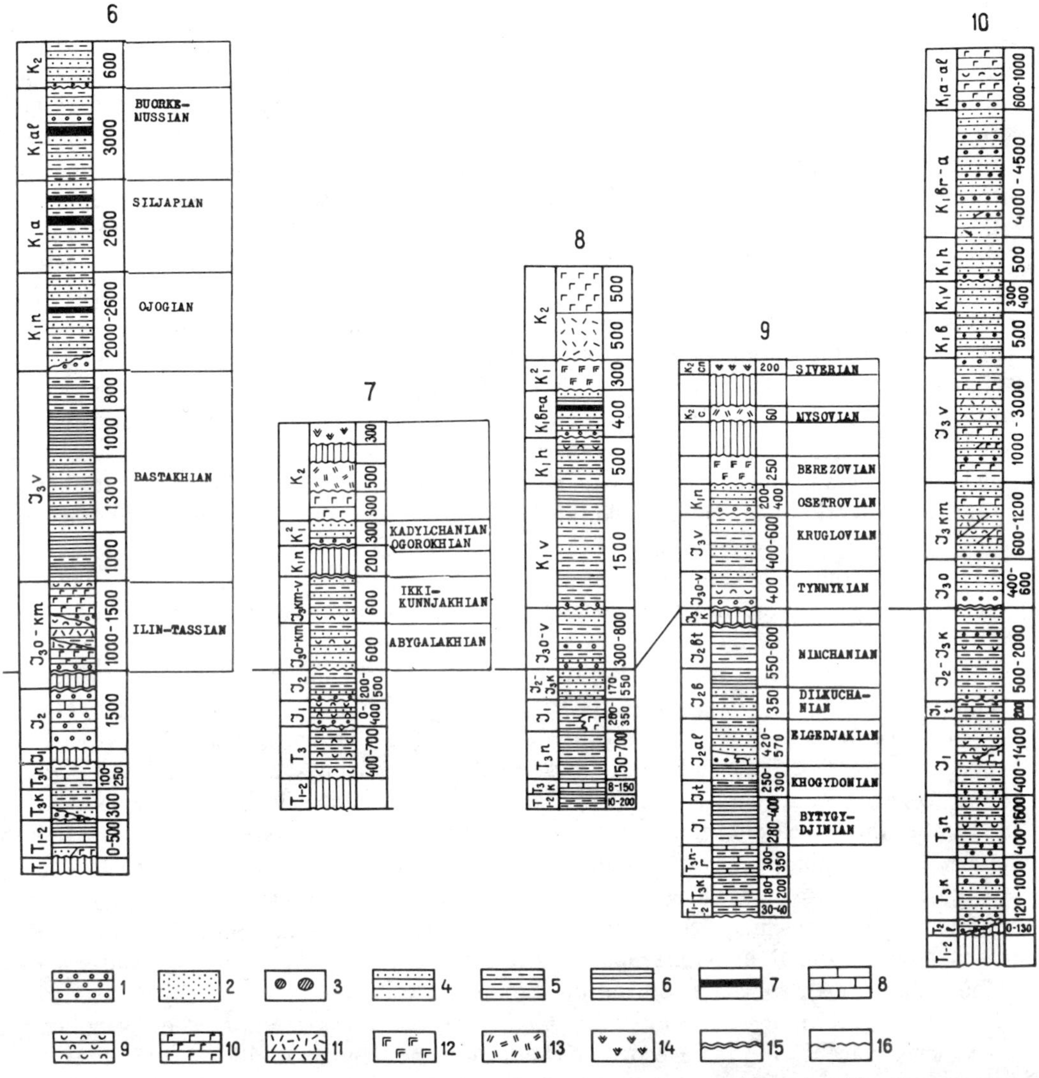

FIG. 2 Scheme of correlation of Mesozoic deposits in the northern cross-section (the Verkhoyansk Province).

1 – conglomerate; 2 – sand; 3 – lenses of current bedding sandstone; 4 – sandstone; 5 – siltstone; 6 – clay shale and clay; 7 – hard coal; 8 – carbonate rock; 9 – tuff; 10 – basalt and andesite-basalt; 11 – liparite; 12 – trachybasalt; 13 – trachyandesite-basalt; 14 – trachyliparite; 15 – angular unconformity; 16 – stratigraphic unconformity.

Position of sequences; 1 – Verkhoyansk marginal trough; 2 – eastern limb of the West Verkhoyansk meganticlinorium and the Sartangan synclinorium; 3 – Yane zone; 4 – Inyaly-Debin synclinorium; 5 – Polousnens synclinorium; 6 – Momsk horst-anticlinorium and Momo-Zyryan superposed basin; 7 – Alazei uplift; 8 – interior part of the Omolon massif; 9 – southern part of the Oloi trough; 10 – eastern part of the Oloi trough.

the main batholithic belt and cross-cutting different structural elements of the Verkhoyansk miogeosynclinal fold system nearly across the strike.

Magmatic activity was intense along a continuous (more than 3000 km) zone of the Okhotsk-Chukotka volcanic belt (Belyi, 1968) within the continental margin of the northern megablock. The eruption of andesitic basalts and andesites of calc-alkaline series and the intrusion of the complex Okhotsk granitoids (granodiorites, quartz diorites and biotite granites with K/Ar ages of 126–111 m.y.) were related to the beginning of the Late Jurassic–Lower Cretaceous stage of the generation of the external zone of the belt. Intrusive bodies of gabbro, gabbro-diorites, quartz diorites and plagiogranites were formed within the interior zone of the belt during the above-mentioned period. Thus, the Lower Cretaceous granitoids of the Okhotsk and Kolyma complexes belong to a single lateral series of magmatic rocks, close in age, but different in composition and origin. The transition zone from the continental margin to the interior of the continent shows the replacement of the gabbro-granodiorite-plagiogranite complex of the inner zone of the Okhotsk-Chukotka belt, associated with mantle sources and formed within an ocean-continent suture zone, by the volcano-plutonic association of the Okhotsk granodiorites and calc-alkaline andesites in the interior of the belt. The latter, in their turn, give place to typical crustal granitoids of the Kolyma complex, distributed within the inner continental regions of the Verkhoyansk and Chukotka fold belts and showing various compositions and facies.

Late Cretaceous magmatic rocks also form a coeval lateral series. For example, the terrestrial eruption of diorites, trachydacites, liparites and, more rarely, andesitic basalts, as well as the intrusion of biotitic and rare-metal lithium-fluoric and agpaitic granites took place in the interior (inner continental) zone of the Okhotsk-Chukotka belt during the Late Cretaceous time (110–70 m.y.) (Grinberg, 1976). Granitoid intrusions together with andesites occurred in the interior zone of the belt at the same time. Thus the Late Cretaceous lateral series of magmatic rocks of the Verkhoyansk-Kolyma fold belt represents a change of probably mantle granodiorite-granite intrusions and the associated andesites of the transitional zones to the continent by typical crustal leucogranites including the rare-metal lithium-fluorine and alkaline types within the inner continental zone.

Pre-Mesozoic history of the central megablock was rather complicated. The granitic crust is here heterogeneous in age (Nagibina, 1963; *Tectonics of Eurasia*, 1966). There are some crustal blocks formed in Archaean and Proterozoic times (the Stanovoy Range zone, Chinese Platforms), as well as during the Baikalian, Caledonian and Hercynian stages of tectogenesis (the Mongolo-Okhotsk and Cathasiya zones). Differentiated deformation of blocks took place during the Mesozoic. These movements produced local synsedimentary troughs which separated vast dome-like uplifts of the pre-Mesozoic basement. Marine and continental (coal-bearing) molasse and terrestrial volcanogenic deposits accumulated in such zones (Fig. 3). Mesozoic block structures of the Mongolo-Okhotsk zone are morphologically rather long linear graben-synclinal structures, related to deep faults, isometric and linear dome-like uplifts and volcano-tectonic, ring and semi-ring structures. During the Lower Cretaceous, a considerable part of the region was subject to imbricate thrusting and oblique deformation with the amplitude of horizontal displacement ranging from hundreds of metres up to 15–40 km.

The development of these structures and the magmatism associated with them shows that these features—closing of troughs, block fault-fold deformation, terrestrial volcanism of different composition and intrusions of the calc-alkaline series (gabbre, diorites and granitoids) with rare-metal lithium-fluorite granites of subalkaline and scarce alkaline series—took place in the western part of the Mongolo-Okhotsk zone in the interior parts of the continent during the Late Triassic–Early Jurassic and, partly, in the Middle Jurassic.

In the Late Jurassic–Early Cretaceous and Late Cretaceous–Palaeogene times, manifestations of uniform block fault-fold deformations, terrestrial volcanism and granite formation of the same composition gradually shifted to the east towards the continental margins (Fig. 4). Late Cretaceous–Palaeogene volcano-tectonic structures developed within the eastern margin of the continent (the Uda branch of the Okhotsk-Chukotka belt, as well as some superposed volcano-tectonic structures and granite massifs in the Okhotsk part of the Mongolo-Okhotsk belt).

Mesozoic magmatism within the central (Mongolo-Okhotsk) megablock penetrated inside the continent for 2500–3000 km along the Mongolo-Okhotsk fault. Abundant plutonic volcanism is observed in large block fault-fold uplifts (e.g. the Daur'khentei uplift) and volcanism in troughs surrounding such uplifts (Zonnenshein *et al.*, 1976; *Mesozoic and Cenozoic Tectonics*, 1975). The megablock generally shows a rather wide range of composition of magmatic rocks varying from mafic and even ultramafic up to ultrafelsic and from low-alkaline (tholeiitic basalts, plagiogranites) up to high alkaline (melaphonolites, trachytes with carbonatites, alkaline granites) (Kovalenko *et al.*, 1977; Kovalenko, 1979). It also comprises abundant crustal rare-metal lithium-fluorite granites (initial ratio Sr^{87}/Sr^{86} = 0.714) and their effusive analogues, the ongonites (Kovalenko, 1976). Magmatic rocks of the calc-alkaline series predominate here.

Ultramafic rocks, as a rule, are represented by xenoliths in granitoids and volcanites (the suture of the Okhotsk-Chukotka and Sikhote-Alin belts according to Nagibina and Potapov) or occur in a tectonic breccia (melange) in the Preshtilkino zone of the Mongolo-Okhotsk fault (Misnik and Shevchuk, 1977). The ultramific rocks of the breccia also occur as xenoliths in the gabbroids and plagiogranites. Extremely rare plagiogranites and volcanites from the tholeiitic series are also associated with the Preshilkino zone (Rutstein, 1970).

As was mentioned above, the central megablock was subject to two stages of magmatism: Early Mesozoic and Late Mesozoic. The Early Mesozoic stage of magmatism was strongest in the west of the megablock (Nagibina, 1963; Zonnenshein *et al.*, 1976). This stage can be subdivided into two substages: Late Triassic–Early Jurassic (predominantly granodiorite-granite magmatism) and Middle–Late Jurassic (predominantly granodiorites of normal geochemical type). To the east, towards the margin of the Asian continent, magmatic rocks become younger, Late Mesozoic, Upper Jurassic–Early Cretaceous and Late Cretaceous–Palaeogene types predominating.

The magmatic rocks are regularly distributed over the central megablock, forming coeval lateral series (*Mesozoic and Cenozoic Tectonics*, 1975). Thus, volcanic zones of the Early Mesozoic stage are symmetric relative to the central part of the Mongolo-Transbaikalian area: its central part is predominantly composed of volcanites of calc-alkaline series with rare low-potassium tholeiitic basalts, while in its marginal parts (Western-Transbaikalia and Orkhon-Selenga in the northwest and Dashibalar in the southwest) there are volcanites of higher alkalinity. Similar zonation was found in plutonic rocks of the Triassic–Early Jurassic stage with potassium-argon age of 230–175 m.y. The central part ('core') of the magmatic belt (the Daur-Khentei and Ingodin-Shilkin zones) is composed of plutonic complexes belonging essentially to the calc-alkaline series forming dimensionally 'batholithic' bodies of granodiorite-granites and a zone of smaller bodies of leucogranites, including the rare-metal lithium-fluorine granitoids (the zone of 'dispersed' magmatism). At the same time, the higher alkalic rocks tend to occur in the northern and southern marginal zones of the magmatic belt.

The K/Ar 175–130 m.y. geochronological interval (Middle–Late Jurassic stage) is represented by a lateral series of plutonic rocks, very similar to that described above. In the central part of the area there are 'batholithic' granodiorite-granite massifs (the Uda

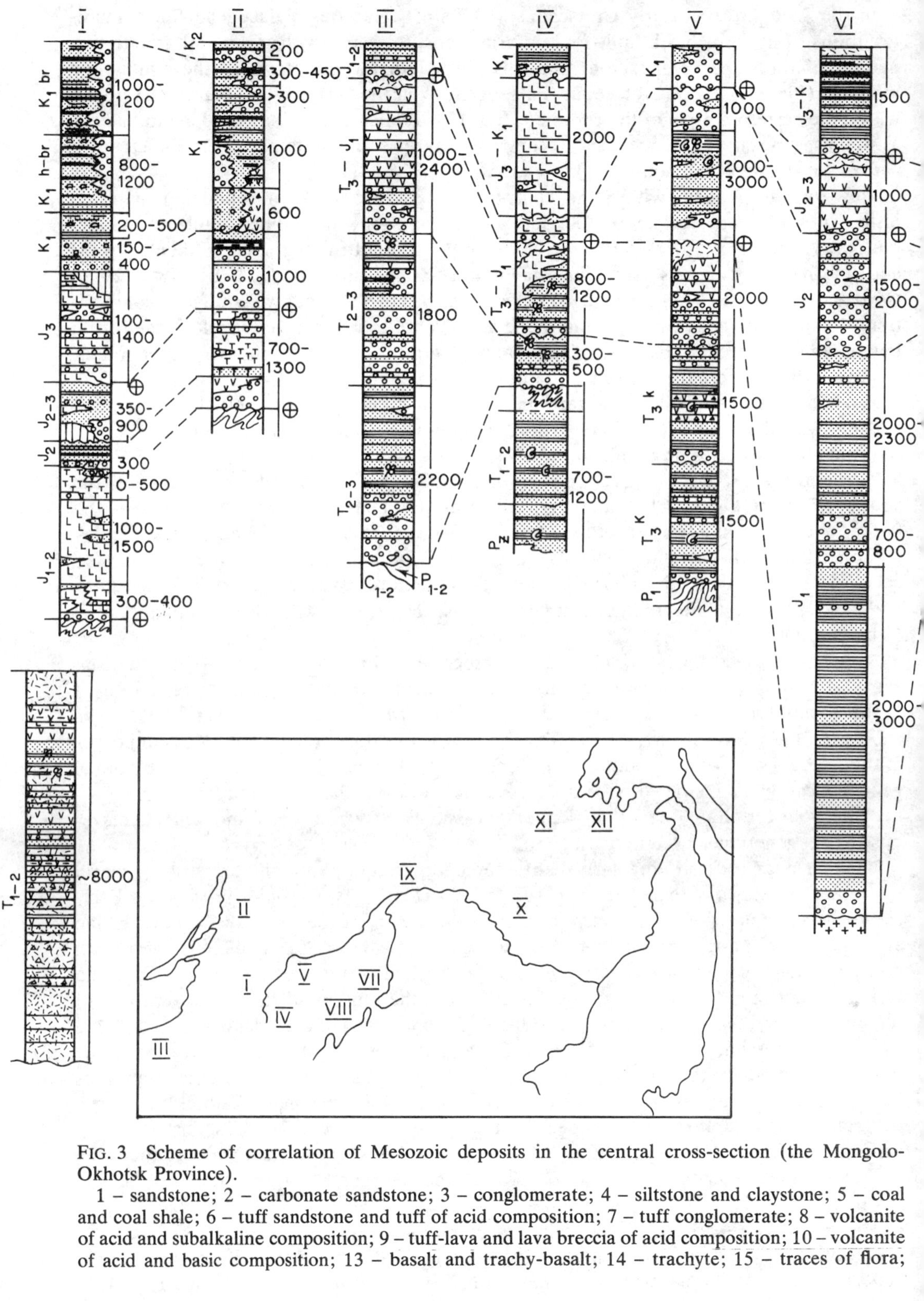

FIG. 3 Scheme of correlation of Mesozoic deposits in the central cross-section (the Mongolo-Okhotsk Province).

1 – sandstone; 2 – carbonate sandstone; 3 – conglomerate; 4 – siltstone and claystone; 5 – coal and coal shale; 6 – tuff sandstone and tuff of acid composition; 7 – tuff conglomerate; 8 – volcanite of acid and subalkaline composition; 9 – tuff-lava and lava breccia of acid composition; 10 – volcanite of acid and basic composition; 13 – basalt and trachy-basalt; 14 – trachyte; 15 – traces of flora;

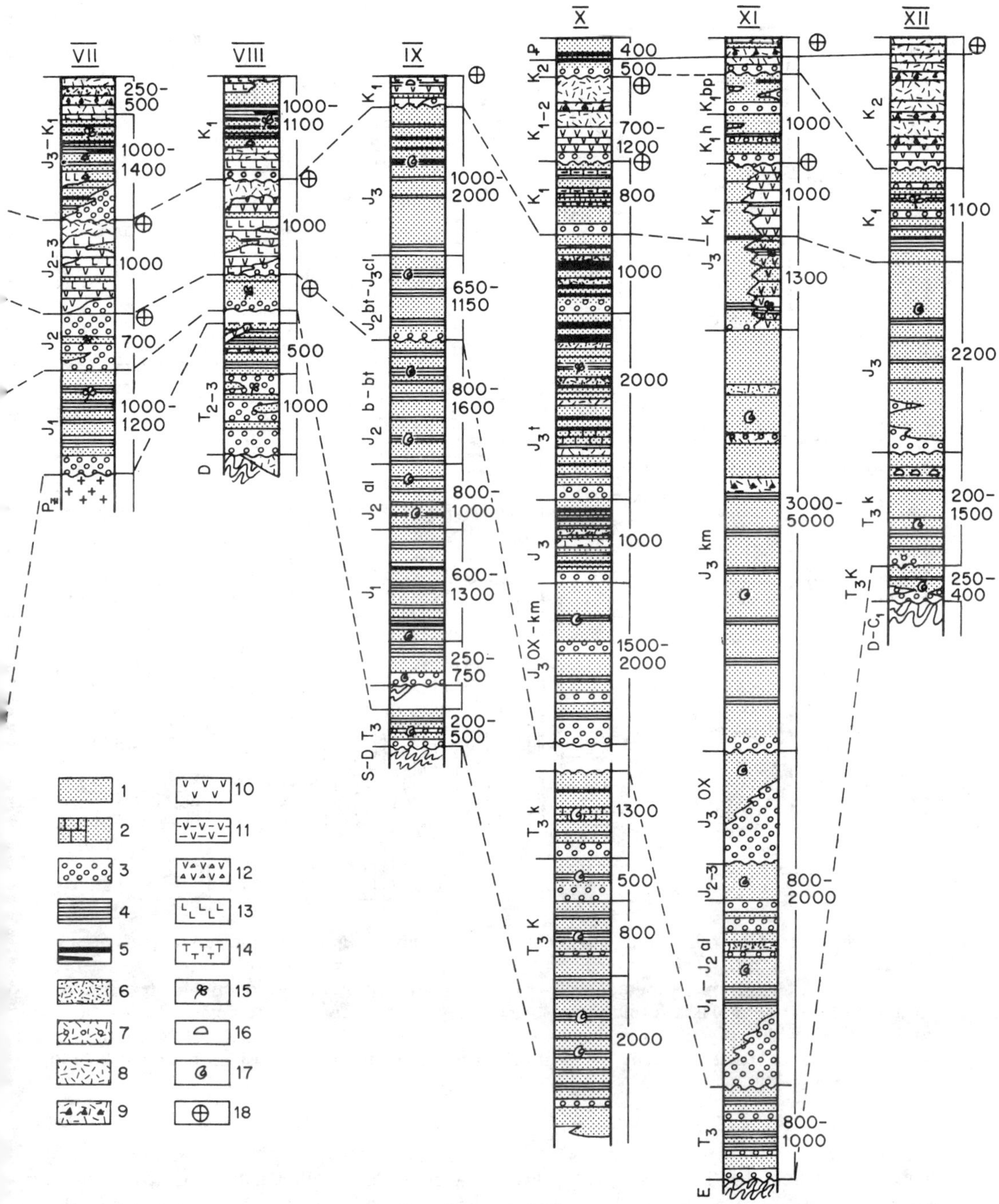

of acid and basic composition; 13 – trachyte; 14 – basalt and trachy-basalt; 15 – traces of flora; 16 – fresh-water fauna remains; 17 – sea fauna remains; 18 – granitoid intrusion.

Position of sequences: I – Dzhadin-Tugnui zone; II – Undin-Vitim zone; III – Otkhon-Selenga zone; IV – Dzhargalantyin zone; V – Ingodin-Shilkin zone; VI – central eastern Transbaikal zone; VII – eastern Mongolian zone; VIII – Ol'doi zone; IX – Amur-Zeya zone (Zeya part); X – Uda zone; XI – Toroma zone.

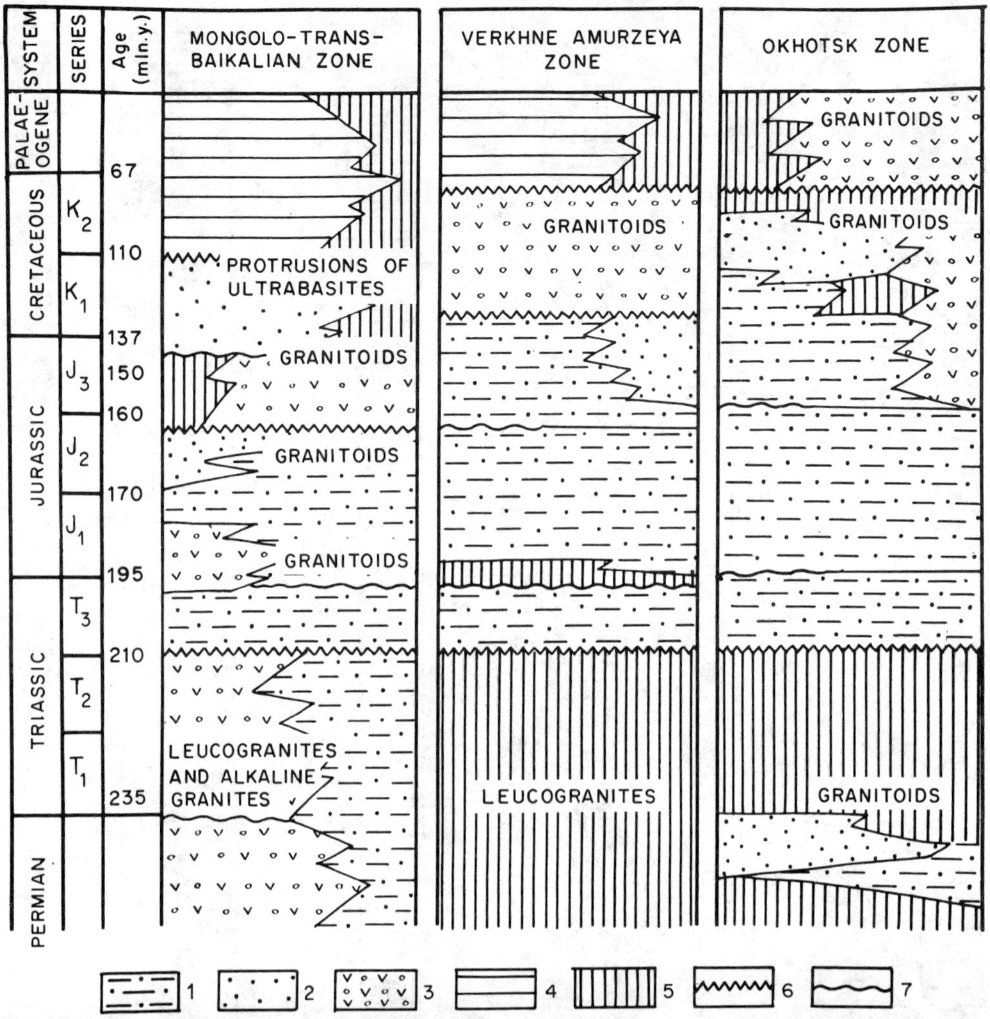

FIG. 4 Scheme of correlation of sedimentation, magmatism and tectonic movements in the Mongolo-Okhotsk province.

1 – marine, terrigenous, predominantly shallow-water deposits; 2 – continental terrigenous, locally coal-bearing molassa formations; 3 – continental sedimentary-volcanogene formation of variable composition; 4 – platform cover sediments; 5 – hiatus; 6 – block fault-fold deformations with reworking of structural plan; 7 – local unconformity without any considerable reworking of structural plan.

complex), which, to the west, give place to a zone of smaller leucogranites, including rare-metal lithium-fluorine granites (zone of 'dispersed' magmatism). All these rocks belong to the calc-alkaline series.

The alkalinity of plutonic rocks increases northwest of the Mongolo-Okhotsk fault, intrusive potassium-alkaline rocks appear in the Aldan shield and southwest of it, and potassium-alkaline intrusives occur in central Gobi (Mongolia). Rare-metal carbonatites and apatite-bearing rocks with 'mantle' values of the intial ratio of Sr^{87}/Sr^{86} close to 0.705 also occur in this area.

Late Cretaceous–Palaeogene magmatism was most intense in the west Okhotsk area of the central megablock. Its eastern part is composed predominantly of calc-alkaline rocks (andesite-basalts, andesites, liparites, dacites, diorites, granodiorites), whereas its western parts are of higher alkalinity rocks (trachybasalts, trachyandesites, trachyliparites). The Yan-Shan and Cathasiya zones within the megablock are of very similar Mesozoic tectonic development, magmatism and metallogeny.

The southern (Indochina-Sunda-Philippines) megablock was united into a single large structure of continental earth's crust only at the end of the Triassic. Correlations of stratigraphic sequences (Fig. 5) and data on morphology of the structures show that in the Late Permian, Early and Middle Triassic rather complex and varied geodynamic environments existed within the megablock at the conjunction of the Tethys and Pacific tectonic belts. There were several blocks of continental crust (microcontinents), separated by basins with oceanic crust (Fig. 6d). This is confirmed by a regular change at the margins of the blocks from deep-sea to shelf and terrestrial sedimentary facies and by Permian and Triassic calc-alkaline volcanic and granite belts with distinct lateral zonation from granodiorites and normal granites to leucogranites in the more inner parts of the blocks (Mitchel, 1976; Gatinsky, 1978). Tensional conditions existed in the margins of some microcontinents during the end of the Permian and Triassic, causing rift structures (the Black River zone in Vietnam, for example), preceded by eruptions of continental tholeiites and alkaline olivine basalts (Hawaiian type). Bimodal volcanic series were formed during the mature stage of the rifts (Gatinsky and Dao Dinh Thuk, 1982).

Intense alpinotype deformation accompanied by intense intrusive activity took place within the southern megablock during the second half of the Late Triassic. These dislocations are concentrated along relatively narrow belts in the interior parts of Indochina, Malaya and Kalimantan (Fig. 6c). The belts were, probably, collision zones of Palaeozoic microcontinents and were formed under compression. Permian and Triassic rocks within the belts were deformed into linear folds frequently overturned and complicated by thrusts. The zones locally comprise metamorphic rocks of high-pressure facies, including galucophane schists, tectonic melange and protrusions of 'alpinotype' (differentiated) ultramafites (Hutchison, 1975; Thanasuthipitak, 1978).

At the end of the Triassic to the beginning of the Jurassic, the collision zones were transformed into uplifts surrounded by continuous troughs with accumulation of littoral and terrestrial Late Norian, Rhaetic and Jurassic molasse. The uplift was accompanied by the intrusion of large bodies predominantly composed of leucocratic granites with a high Sr^{87}/Sr^{86} ratio ($\geqslant 0.0710$), which proves their crustal origin (Bignell *et al.*, 1978). Calc-alkaline and subalkaline volcanism took place locally in the adjacent molasse troughs. Simultaneously, the territory away from the zones of intense deformation suffered block faulting-folding with the development of brachyfolds, fault-proximate basins and displacement along transcurrent faults (the Dien-Bien-Lai-Tau fault in northwest Vietnam with horizontal displacement about 60 km (Chan Van Chi, 1977).

Beginning with the Middle Jurassic, manifestations of tectonomagmatic activity were related exclusively to the margins of the continental southern megablock. They are displayed as block faulting-folding, intrusion of numerous granites and leucogranites (including rare-metal) and eruptions of calc-alkaline volcanite. These processes took place in the Burmese-Malayan and Cathasiyan marginal continental belts which possess zonal structure (Fig. 6b). Magmatic rocks show a regular increase in the K_2O content across the strike of the belts towards the inside of the continent, granite-granodiorite association being replaced by leucogranites (Gatinsky, 1979). In the course of time, magmatic activity died out towards the Pacific and Indian oceans. In the Middle Mesozoic the eastern part of the

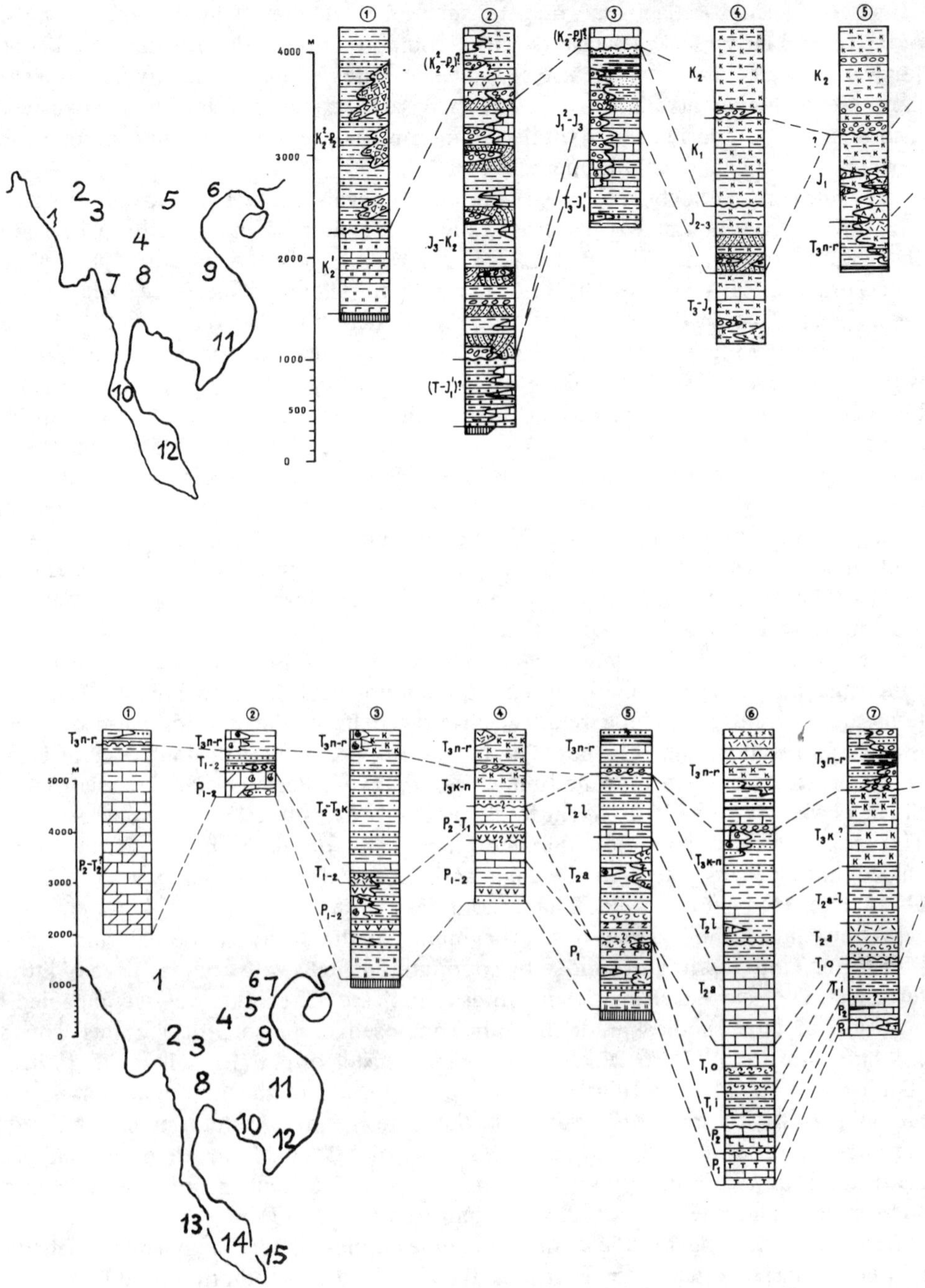

FIG. 5 Scheme of correlation of Permian and Mesozoic in the southern cross-section.

At the bottom – Permian and Triassic deposits: 1 – Shan plateau in northeast Burma; 2 – western Thailand; 3 – central Thailand, the Lampang zone; 4 – northern Laos; 5 – the Dien-Bien-Samnya zone at the boundary of Vietnam and Laos; 6 – the Shongda (Black River) zone in northern Vietnam; 7 – northeast Vietnam; 8 – the western margin of the Khorat plateau; 9 – Trungson Range; 10 – western Campuchea; 11 – southern Laos and northern Campuchea; 12 – southern Vietnam; 13 – Malaya Peninsula, Kedah-Perak; 14 – Malaya Peninsula, Kelatan-Pahang; 15 – Malaya Peninsula, Dzhokhor-Singapore.

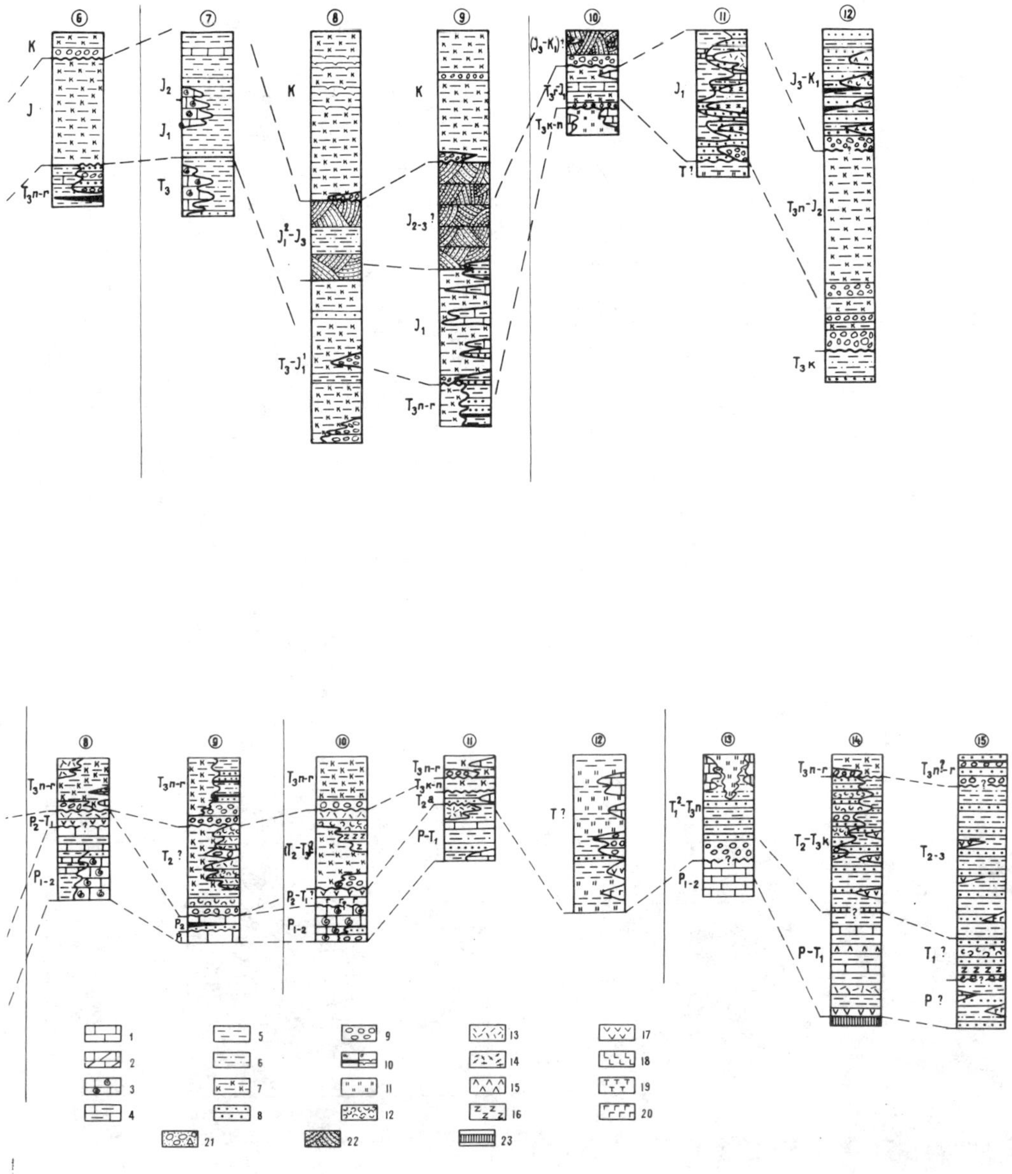

Above – Jurassic and Cretaceous deposits: 1 – Arakan-Yoma and the Yanbe Island in western Burma; 2 – the western margin of the Shan Plateau and the upper Iravadi River in northern Burma; 3 – Shan Plateau; 4 – northern Thailand, northwestern Laos and southern China; 5 – northwestern Vietnam; 6 – northeastern Vietnam; 7 – the Maesot-Thongphaphum region in western Thailand; 8 – Khorat Plateau; 9 – southern Laos and central Vietnam; 10 – Kra Isthmus; 11 – southern part of Vietnam; 12 – Malaya Peninsula.

1 – limestone; 2 – dolomitized limestone; 3 – reef limestone; 4 – clay limestone, marl; 5 – claystone, clay shale; 6 – siltstone, 7 – red terrigenous deposits; 8 – sandstone; 9 – conglomerate, gravelite; 10 – (a) coal, (b) salt; 11 – siliceous rocks; 12 – tuff, tuffite; 13 – rhyolite; 14 – alkaline rhyolite, comendite; 15 – orthophyre, trachyte; 16 – dacite; 17 – andesite, 18 – alkaline olivine basalt; 19 – trap (continental tholeiite and differentiated gabbroids); 20 – oceanic tholeiites; 21 – olistostromes; 22 – cross-bedded sandstone; 23 – rocks of the lower part of the ophiolite complex ('alpinotype' ultramafic, gabbroids, locally metabasites).

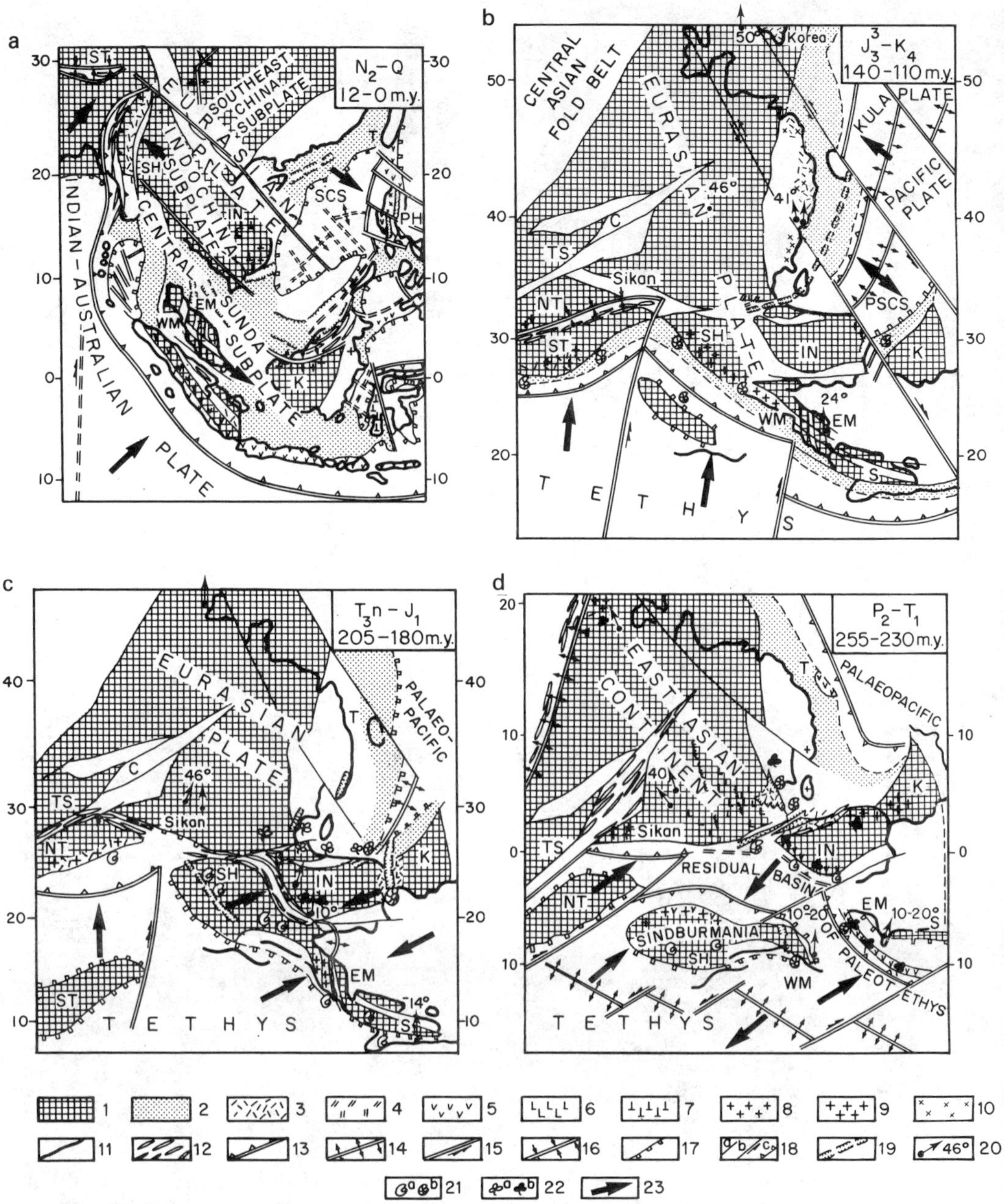

FIG. 6 Schematic palinspatic reconstructions of southeast Asia (the end of the Palaeozoic and the Mesozoic). Compiled by Gatinsky (1981).

1 – blocks of the ancient continental crust development (Precambrian platforms and Palaeozoic–Mesozoic shelves); 2 – Cenozoic shelves and older shelves on the Palaeozoic and Mesozoic basement; 3 – calc-alkaline volcanites of the felsic composition; 4 – subalkaline volcanites of the felsic and intermediate composition; 5 – calc-alkaline volcanites of the intermediate composition; 6 – continental tholeiite; 7 – subalkaline olivine and alkaline basalt; 8 – leucocratic essentially potassium granite; 9 – granite-granodiorite; 10 – gabbro-plagiogranite; 11 – alpinotype ultramafics; 12 – folding; 13–15 – plate boundaries: 13 – convergent, 14 – divergent, 15 – transform; 16 – collision zone; 17 – innerplate ('passive') continental margin; 18 – faults (a – non-differentiated, b – wrench fault, c – thrust); 19 – rift; 20 – palaeomagnetic data with indexes of palaeolatitudes (after Stauffer,

megablock (eastern China) suffered horizontal displacements of up to 200 km along sinistral strike-slip faults (Kuo, 1978).

As mentioned above, the Late Cretaceous–Palaeogene tectonic movements and magmatism were intense along the whole eastern margin of the Asian continent and within the zones transitional to the Pacific and the Tethys. The territory from southwest Japan to southeast Kalimantan was invaded by volcanites of the calc-alkaline series and by granites. Similar processes took place along the boundary of the Asian continent with the Tethys in central Burma, Sumatra and Java (Mishina, 1979). Simultaneously, in the interior parts of the transitional zone in the Shimanto belt (Japan), on the Philippine Islands, in western Burma and other regions, deep-sea environments existed, eruption of oceanic tholeiites of ophiolite association (chert-spilitic formation) taking place there together with the accumulation of clastic turbidite-flysch deposits. The earliest recent marginal seas in the Pacific Ocean—the South China Sea—seemed to originate in the Middle–Late Cretaceous. Rifting within its boundaries was accompanied by dolerite intrusions in the eastern part of Malacca Peninsula (Haile, 1974) and eruption of alkaline volcanites in Cathasiya (*Regional Stratigraphy*, 1963). The development of that deep-sea basin was accompanied by an accumulation of Upper Cretaceous and Palaeogene siliciclastic, chert-spilitic and flysch deposits in North Kalimantan.

DISCUSSION OF THE RESULTS

The suggested correlations permit us to recognize close temporal and areal connections between tectonic movements and magmatism displayed differently in different structural zones.

Tectonic movements in the earth crust are continuous and manifested irregularly in time and place. Relatively longer stages of quiet synsedimentary development gave place to short pulses of intense movements—phases of folding, formation of nappes, and so on. These movements are displayed differently in different areas. Phases of compression, folding and piling up are synchronous with tension in some zones and with subsidence in others.

Mesozoic tectonic movements and magmatism in the northern, central and southern megablocks point to certain similarities and differences among them. In general they are as follows (Fig. 7). First, in the east of the vast Asian continent intense tectonic movements took place in the Mesozoic, accompanied by extremely powerful, predominantly granitoid magmatism. The following main stages of activity can be recognized: (1) Triassic (usually late)–Early Jurassic; (2) Early–Middle Jurassic; (3) Late Jurassic–Early Cretaceous and (4) Late Cretaceous–beginning of Palaeogene. At the boundaries of these stages, phases of intense tectonism are pronounced, as at the boundary of Middle and Upper Triassic, at the boundary of Middle and Late Jurassic (or at the beginning of Late Jurassic–Callovian–Oxfordian) and at the boundary of Early and Late Cretaceous.

1974; Gorodnitsky *et al.*, 1978); 21–22 – places of fossil findings: 21a – faunda of the Tethys province, 21b – mixed fauna of the Tethys and Pacific provinces, 22a – Mesozoic Hongai flora and that related to it, 22b – Permian Cathasyian flora and that related to it; 23 – direction of movement of the lithosphere plates and subplates.

BR – Black River zone; C – Cinlin; EM – eastern part of the Malaya Peninsula; IN – Indochina Peninsula; K – Kalimantan; NT – northern Tibet; PH – the Philippines; PSCS – the Palaeosouth China Sea; S – Sumatra; SCS – the South China Sea; SH – the Shan Plateau; ST – southern Tibet; T – Taiwan; TS – Tsaidam; WM – the western part of the Malaya Peninsula. a – the present picture; b – the late Mesozoic; c – the early Mesozoic; d – the end of the Permian–the beginning of the Triassic.

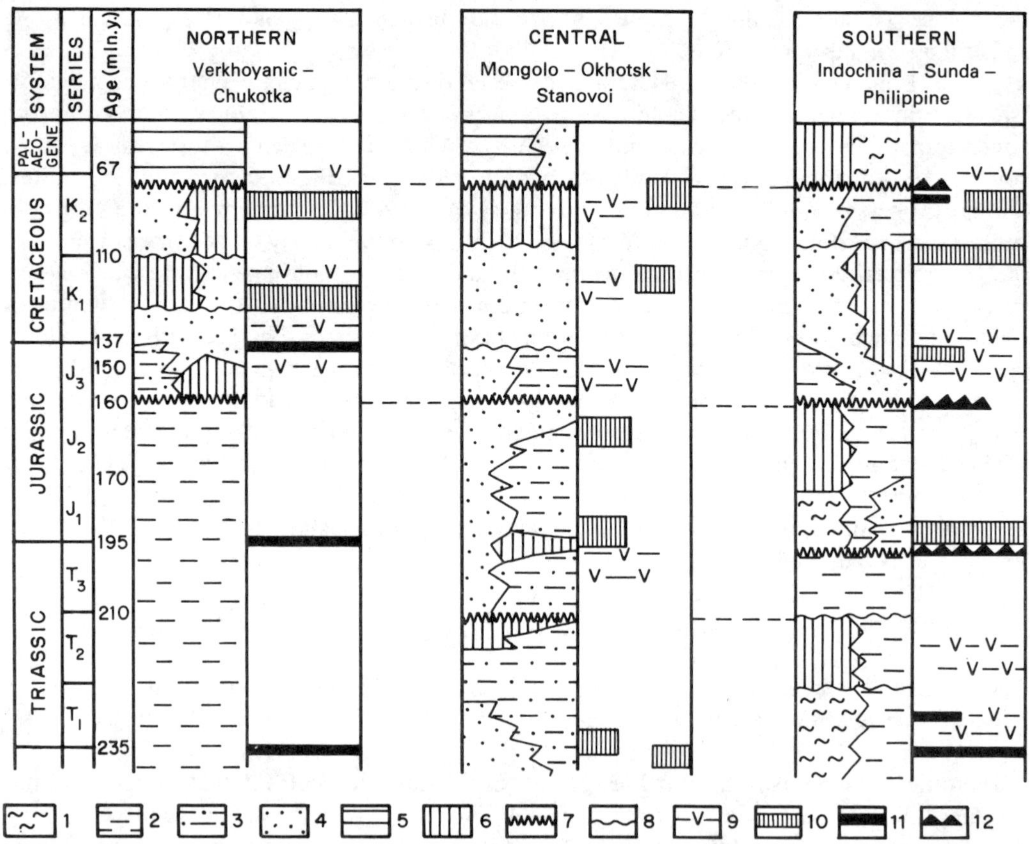

FIG. 7 Scheme of correlation of sedimentation, magmatism and tectonic movements in different segments of southeastern Asia.

1 – marine deep-water deposits; 2 – marine carbonate-terrigenous deposits of various depths; 3 – marine terrigenous, predominantly shallow-water deposits; 4 – continental molassoid deposits; 5 – continental deposits of the platform cover; 6 – hiatus; 7 – fold and block fault-fold deformation with reworking of the structural plan; 8 – local unconformity without considerable reworking of the structural plan; 9–12 – magmatic manifestations: 9 – subaerial volcanism of different composition; 10 – granitoid intrusions; 11 – magmatic manifestations of basic composition, 12 – ultramafic protrusions.

The first two stages in the central megablock show a peculiar shift of areas of tectonic structures and magmatism deep into the continent, as far as 2500–3000 km (Fig. 1) along the major lineaments (e.g. the Mongolo-Okhotsk and the Central Mongolian lineaments). Thus the distribution of such tectonomagmatic areas was controlled by two structural elements: the continental margin and the major faults transverse to it. As a rule, all younger tectonomagmatic areas shifted to the east towards the Asian continental margin. During the Late Cretaceous–Palaeogene stage the tectonomagmatic areas were controlled by the continental margin.

Second, the most distinct qualitative differences in tectonic development of megablocks occurred in the Early Mesozoic. During the Triassic and the beginning of the Late Jurassic, in the northern megablock few differential tectonic movements prevailed; troughs with the thickest terrigenous sedimentation developed on a wide shelf sea. Magmatism occurred here only as small bodies of tholeiitic mafics (trap formation) appearing along deep faults in the granite basement. The central megablock during this time developed under predominantly continental conditions, whereas the southern megablock formed in both continental and marine regimes with different depths of basins. Both of these megablocks were subject to intense differential tectonic movements and widespread magmatism of different composition. Tectonic movements here were reflected by the formation of zones of extension and compression, development of linear and block structures and continental volcanism and granite intrusion.

The magmatism of the central megablock gradually changed laterally from calc-alkaline in the 'core' of the largest Daur-Khentei dome uplift (Mongolo-Transbaikal area) to highly alkaline at its northwestern and southeastern margins. Immediately within the calc-alkaline magmatic zone, large granodiorite-granite intrusions, apparently related to subcrustal sources, were replaced from east to west by small bodies of typically crustal leucogranites, including those containing rare metals. It is noteworthy that the area of granitoid magmatism was far away (up to 3000 km) from the zone of the coeval eugeosynclinal Sikhote-Alin system towards the west along the Mongolo-Okhotsk fault (Fig. 1).

Most of the marine troughs in the southern megablock closed up at the end of the Triassic as a result of collision of the microcontinents (Gatinsky, 1981). These collisions were accompanied by the formation of narrow zones with intense alpinotype dislocations and ophiolite belts (Fig. 6c). Powerful crustal (sialic) magmatism (including formation of rare metal-bearing granites) developed along the same zone after the collision and folding. Thus during the Early Mesozoic stage, the intensity of tectonic movements and magmatism increased from the northern to the central and then to the southern megablocks.

Third, in the Late Mesozoic, rather active tectonic movements and magmatism took place within all the megablocks. As a result dislocations of different morphology such as zones of linear folding, block fault-fold (brachy-shaped, box-like, mosaic) and volcano-tectonic (linear, ring and semi-ring) structures were formed.

During the Early Cretaceous, sinistral and dextral strike-slip faulting of different horizontal displacements took place throughout almost all the megablocks. In the northern and central megablocks this amplitude is up to 15–40 km. At the southern megablock within the suture of the Pacific and Tethian belts, the horizontal displacement of up to 200 km (Fig. 6b) along the faults was completed over a longer time interval (the end of Middle Jurassic–Early Cretaceous).

After Late Jurassic folding, the northern megablock was subjected during the Early Cretaceous to a powerful intrusion of granitoids and volcanic eruptions. During the formation of zones of volcano-plutonic belts, calc-alkaline granodiorite-granites and the related andesites gave place, from the active margin towards the internal parts of the continent, to crustal granites and leucogranites (including rare-metal lithium-fluorine and alkaline) and to volcanites of higher alkalinity.

In the central megablock, tectonic movements continued and magmatism similar to that of the Early Mesozoic was initiated. However, the tectonomagmatic area of this age was shifted to the east. Contrary to the Early Mesozoic, some alkaline-carbonatite complexes occur in the marginal parts of the area, among the Late Mesozoic magmatic rocks.

In the southern megablock, all the tectonomagmatic activity shifted towards the boundaries of the continent with the Pacific and Tethys oceans and was manifested as block faulting-folding and formation of long marginal-continental belts composed of calc-alkaline rocks with tin-bearing granitoids.

Fourth, Late Cretaceous and Palaeogene tectonic movements occurred throughout the eastern and southeastern margins of the Asian continent. In many parts of the transition from the continent to oceans, tectonic dislocations within all the megablocks were accompanied by major horizontal displacements of oceanic crust over the continental margin (Ruzhentsev and Byalobzheskiy, 1978). At the same time, the interior parts of the transitional zone (in the Philippines, western Burma and other regions) suffered eruptions of oceanic tholeiites in deep-sea environments. Intense continental volcanism and intrusion of granites continued within active continental margins. In the central and southern megablocks, these processes took place only in a relatively narrow (a few hundred kilometres) marginal zone of the continent. In the northern megablock, manifestation of granitoid magmatism and continental volcanism occupied, during the Late Cretaceous, a wide intracontinental zone trending off the Okhotsk-Chukotka marginal volcanic belt for more than 1500 km. At the same time, the interior parts of the central and southern megablocks were areas of quiet accumulation of the platformal cover deposits with local eruption of trap basalts.

Fifth, in conclusion it should be emphasized that the main stages of tectonic and magmatic activity in eastern Asia during the Mesozoic were relatively synchronous. This suggests a dynamic interconnection of different dislocations and manifestations of magmatism including emplacement of granites not only within the elongated suture zone of continental and oceanic structures of the Palaeopacific and Palaeotethys, but also over a wide territory within the continent. This circumstance may help us to discover the general global causes of tectonic and magmatic activity.

These causes may be due to such patterns in interaction of continental and oceanic structures in eastern Asia, which were not just phenomena within active continental margins, but also involved the interaction between lithosphere and astenosphere far in the interior of the continent. The following hypothetic models are advanced as possible explanations.

(1) Interaction between the Asian continent and oceanic structures of the Pacific, Tethian and Indian oceans is related to the development of Zavaritskiy–Benioff zones plunging under the continent with the formation of lateral series of magmatic zones above them. This model clearly explains the formation of the calc-alkaline magmatism of the Okhotsk-Chukotka-Cathasiya, Burma-Malaya belts, and so on; but it presents problems in explaining why tectonomagmatic areas along the Mongolo-Okhotsk and other faults are situated in the far interior of the continent (up to 2000–3000 km in the central block and up to 1500 km in the northern block along the Indigirka-Kolyma fault system) from a supposed Mesozoic Zavaritskiy–Benioff palaeozone.

(2) Collision of microcontinents and continents causing Alpine deformation zones, occasionally accompanied by ophiolite belts (only in the southern megablock) and with extensive development of crustal granitoid magmatism (the southern megablock—in the Early Mesozoic; the northern megablock—in the Late Mesozoic).

(3) Manifestations of marginal-continental rifting with its specific magmatism (the zone of the Black River, South China Sea, and so on). A probable cause of these processes was the subduction of the Mesozoic midoceanic Palaeotethys and Palaeopacific ridges beneath the margins of the eastern Asian continent (Gatinsky, 1981; Hilde *et al.*, 1977). In conformity with this model, the tectonomagmatic activity in eastern Asia also occurred deep in the interior of the continent over zones of spreading (Zonnenshein *et al.*, 1976; Kovalenko *et al.*, 1977a). For the central megablock, a shift of the spreading zone intensity from west to east should also be assumed.

Tectonic and magmatic processes under the conditions of earlier continental crust were called reactivation (revival) processes (Nagibina, 1963). In the northern and central megablocks, these processes were going on in continental environments, far from the oceanic margin. Therefore a model accepting the existence of specific continental processes in reactivation of intracontinental faults should not be altogether excluded.

Development of fairly active tectonic and magmatic reactivation within the interior parts of the continent in the regions under review may be attributed to major deep horizontal thrusts in the Mesozoic, affecting the upper mantle. These dislocations can be regarded as continuations of the seismofocal surface (Zavaritskiy–Benioff zone) essentially flattening towards the interior of the continent and passing into a surface of the lithospheric as a deep horizontal thrust. Such dislocations generated powerful tectonic processes of reactivation (Fig. 1), involving vast amounts of territory and resulting in the development of a mosaic of block fault-fold deformations, accompanied by the initiation of new and the 'revival' of more ancient deep faults breaking the granite crust of the continents around the Pacific Ocean.

Penetration of deep sources of energy into the interior of the continent from the deep lithospheric thrust was most intense along the zones of deep faults cutting the earth's crust and, probably, the lithosphere. Numerous deep faults are known within the northern, central and southern megablocks. These faults separated heterogeneous (by age) pre-Mesozoic granite crusts, for example within the Prebaikal region, Stanovoi Range and Mongolo-Okhotsk belt, and determined the most active Mesozoic intracontinental differential tectonic movements, as well as the compositionally most diverse Mesozoic magmatic rocks.

The development of the intracontinental magmatism or granitoid reactivation resulted in important deep processes of transformation of the granite crust formed earlier. Deep heat flow and flux of mantle matter caused partial melting and remobilization of the earth's crust and, hence, its secondary enrichment by lithophytic elements. It is these processes that can explain the increased concentration of these elements in the intrusive and effusive Mesozoic rocks. Distinctly leucocratic, frequently lithium-fluorine granites perspective for tin, tungsten, molybdenum and other rare metals are particularly abundant in the granitoids of the intracontinental zones. Similar ore deposits are widely developed in the continental margin zone and in more interior continental regions of the inner zone of the Pacific tectonic belt.

The authors therefore think that the peculiarities of Mesozoic tectonic movements and magmatism in eastern Asia may be most reasonably treated from the point of view of harmonious combination of horizontal and vertical movements. The models above which interpret the geological phenomena under review can be explained partly from a plate-tectonic standpoint close to its classical variant and the hypothesis of major horizontal deep thrusts in the earth's crust and upper mantle. According to Gatinsky, the development of the Alpinotype dislocations and Early Mesozoic magmatism now occupying the interior of the southern megablock (Fig. 6a), as well as the formation of marginal-continental volcano-plutonic Late Mesozoic belts regularly distributed along the margins of the Asian continent, can be well explained from the viewpoint of plate tectonics.

The suggested presence of large horizontal deep thrusts in the earth's crust and upper mantle (probably accompanied by frontal Zavaritskiy–Benioff zones), can explain the synchroneity of active tectonic and magmatic Mesozoic processes throughout the intracontinental zones of eastern Asia, particularly typical of the northern and central megablocks and demonstrated by correlation.

REFERENCES

Belyy, V. F. (1968). Formations and tectonics of the Okhotsk-Chukotka volcanogene belt, Moscow: Nauka, 212 pp. (Russian).

Bignell, J. D., Shelling, N. J. and Teggin, D. E. (1978). Geochronology of the tin granites of Southeast Asia, Geol. Open-File Rept., 701, p. 37–8.

Chan Van Chi (ed.) (1977). Geology of Vietnam (northern part). Explanatory notes to the geological map of the scale 1:1,000,000. Hanoi, 354 pp. (Vietnamese).

Gatinsky, Yu. G. (1978). Permian and Triassic of mainland Southeast asia, *Izvest. vysshykh. ucheb. zaveden., geolog. i razvedka*, no. 11, 47–64 (Russian).

Gatinsky, Yu. G. (1979). Jurassic and Cretaceous stage of evolution of continental Southeast Asia, *Izvest. vysshykh. ucheb. zaveden., geolog. i razvedka*, no. 4, 31–45 (Russian).

Gatinsky, Yu. G. (1981). Geokinematic reconstructions of continental Southeast Asia in the Late Phanerozoic, *Izvest. vysshykh. ucheb. zaveden., geolog. i razvedka*, no. 4, 61–9; no. 5, 10–19 (Russian).

Gatinsky, Yu. G. and Dao Dinh Thuk. (1982). An outline of geology and evolution of the Songda (Black River) palaeorift zone in Vietnam, *Bull. Moscow obshch. isp. prirody*, otd. geolog., **57**, no. 3, 12–25 (Russian).

Gorodnitskiy, A. M., Zonnenshein, L. P. and Mirlin, E. C. (1978). Reconstructions of Phanerozoic continent's position, Moscow: Nauka, 122 pp. (Russian).

Grinberg, G. A. (1973). Late Mesozoic formations of granitoids in the USSR northeast, in *Magmatism of Northeastern USSR*, Moscow; Nauka, pp. 6–26 (Russian).

Grinberg, G. A. (1976). Volcanic formations of the southwest Okhotsk-Chukotka belt, in *Volcanic and Intrusive Formations of the Okhotsk Region*, Novosibirsk: Nauka, pp. 4–14 (Russian).

Grinberg, G. A., Gusev, G. S. and Mokshantsev, K. B. (1979). Tectonics, formation of the continental crust and ore deposits in the Verkhoyansk-Chukotka region, in *Tectonics of the USSR*, Moscow: Nauka, pp. 197–207 (Russian).

Gusev, G. S. (1979). *Fold Structures and Faults of the Verkhoyansk-Chukotka Mesozoic System*, Moscow: Nauka, 243 pp. (Russian).

Haile, N. S. (1974). Are the Kuantan dolerite dykes and extrusive basalt related to each other or to the Segamat basalt? Palaeomagnetic evidence, *Geol. Soc. of Malaysia N.L.*, **52**, 25–6.

Hilde, T. W. C., Uyeda, S. and Kroenke, L. (1977). Evolution of the western Pacific and its margins, *Tectonophysics*, **38**, 145–65.

Hutchison, C. S. (1975). Ophiolites in southeast Asia, *Bull. Geol. Soc. Amer.*, **86**, no. 6, 797–806.

Kovalenko, V. I. (1979). Magmatism and problems of metallogeny in Mongolia, in *Geology and Magmatism of Mongolia*, Moscow: Nauka, pp. 32–46 (Russian).

Kovalenko, V. I. and Kovalenko, N. I. (1976). *Ongonites as Subvolcanic Analogues of Rare-metal Lithium-Fluoritic Granites*, Moscow, Nauka, 127 pp. (Russian).

Kovalenko, V. I., Kuzmin, M. I., Antipin, V. S., Kovalh, P. V. and Tsypukov, Yu. P. (1977a). Mesozoic intrusions of Mongolia. Space distribution, geochemistry and mineralization, in *The Main Problems of Mongolian Geology*. Moscow: Nauka, pp. 133–44 (Russian).

Kovalenko, V. I., Samoilov, V. S., Vladykin, N. V., Goregliad, A. V. and Makagon, L. D. (1977b). Geochemical characteristics of subsurface carbonatite complex's rocks from the Gobi desert (MPR), *Geokhimiya*, no. 9, 1312–27 (Russian).

Kuo, Wen-kuei, (1978). General aspects of the geological evolution in China, *Proceed. of 3rd Reg. Conf. on Geol. and Miner. Res. of SE Asia*, Bangkok, pp. 867–73.

Markov, M. S., Pushcharovskiy, Yu. M. *et al.* (1979). Tectonics of East Asia and Far East seas. Geotektonika, no. 1 (Russian).

Mesozoic and Cenozoic Tectonics and Magmatism of the Mongolian People's Republic, Moscow: Nauka, 308 pp. (Russian).

Mishina, A. V. (1979). Late Palaeozoic and Mesozoic of the Indonesian-Philippine Region, *Izvest. vysshykh. ucheb. zaveden., geolog. i razvedka*, no. 4, 46–57 (Russian).

Misnik, Yu. F. and Shevchuk, V. V. (1975). Prishilkinskaya zone of the Mongolo-Okhotsk deep fault, Vyssha shkola i izd. pri Lvovskom universitete (Russian).

Mitchel, A. H. G. (1976). Southeast Asian tin granites: magmatism and mineralization in subduction and collision related setting, *CCOP Newsletter*, **3**, nos. 1–2, 10–14.

Nagibina, M. S. (1963). Tectonics and magmatism of the Mongolo-Okhotsk belt, Trudy Geolog. Instituta; izd. AN SSSR (Russian).

Nagibina, M. S., Antipin, V. S., Gatinsky, Yu. G., Grinberg, G. A., Gusev, G. S., Kovalenko, V. I.,

Kovalh, P. V., Komarov, Yu. V., Kuzmin, M. I. and Mishina, A. V. (1981). Correlation of Mesozoic tectonic movements and magamatism in Southeast Asia, *Geotektonika*, no. 6, 77–91 (Russian).
Peive, A. V., Yanshin, A. L., Zonnenshein, L. P., Knipper, A. L. and Markov, M. S. (1976). Main principles in compilation of the tectonic map of North Eurasia. Moscow: Nauka (Russian).
Pushcharovskiy, Yu. M. (1960). Verkhoyansk marginal trough and mesozoids of northeast Asia, Vyp Tektonika, V. V.: izd. AN SSSR (Russian).
Regional Stratigraphy of China (1963). vol. II, Moscow: Inostrannaya Literatura, 273 pp. (Russian).
Rutstein, I. G. (1970). Structure of Early Mesozoic troughs and structural conditions during formation of Mesozoic and Jurassic intrusions in Central East Transbaikal province, Candidate thesis (Russian).
Stauffer, P. H. (1974). Malaya and Southeast Asia in the patterns of Continental Drift, *Bull. Geol. Soc. Malaysia*, **7**, 89–138.
Tectonics of Eurasia (1966). Explanatory note to the tectonic map of Eurasia, Scale 1 : 5 000 000, Moscow: Nauka, 488 pp. (Russian).
Thanasuthipitak, T. (1978). Geology of Uttaradit area, its implication on tectonic history of Thailand, *Proc. of 3rd Reg. Conf. on Geol. and Miner. Res. of SE Asia*, Bangkok, pp. 187–97.
Zonnenshein, L. P., Kuzmin, M. I. and Moralev, V. M. (1976). *Global Tectonics, Magmatism and Metallogeny*, Moscow: Nedra, 232 pp. (Russian).

Global Correlation of Tectonic Movements
Edited by Yu. G. Leonov and V. E. Khain

8

A Chronological Synopsis of Events of Tectogenic Significance in Italy

ROBERTO MALARODA*

INTRODUCTION

This paper was first prepared in the summer of 1977, and distributed as the 'First Report on the Italian Data' at the Izmir (Turkey) meeting of the International Geological Correlation Programme, Project 107, in September of the same year. In this second version, the data have been updated to 1981 and the beginning of 1982, though the original arrangement of the synoptic table, in which events are distinguished by tectonic units and geographical areas, has been retained.

What has been done is to make a first attempt to construct a model from all the available data, even though it has not always been possible to check them carefully. In the absence of more precise evidence, even vague references or single reports have been included. The indications of stratigraphic and sedimentological facts may be found particularly unsatisfactory in view of the enormous literature on the subject, the often discordant dates, and the difficulty of determining the regional, or only local, extent of phenomena that nevertheless are not devoid of significance, even if they are no more than local. The following works provide a general overview.

General References

Accademia Nazionale dei Lincei (1973). Atti del Convegno sul tema: Moderne vedute sulla Geologia dell'Appennino. Quaderno 183.
Desio, A. (ed.) (1973). *Geologia dell'Italia*, Torino: Utet.
Ogniben, L., Parotto, M. and Praturlon, A. (eds) (1975). *Structural Model of Italy*, Rome: CNR.

METHODOLOGY

In addition to the works cited in the introduction, and an examination of all the recent literature, the paper is based on data collected from an inquiry conducted in 1977, and in some cases renewed in 1982, in which all Italian research institutions took part. I should like to express my thanks to all those who supplied information, in particular: Professors G. Ferrara and A. Del Moro, Istituto di Ricerca per la Geocronologia e la Geochimica

*Istituto di Geologia, Torino University and CNR, Centro di Studio sui Problemi dell'Orogeno delle Alpi Occidentali, Torino.

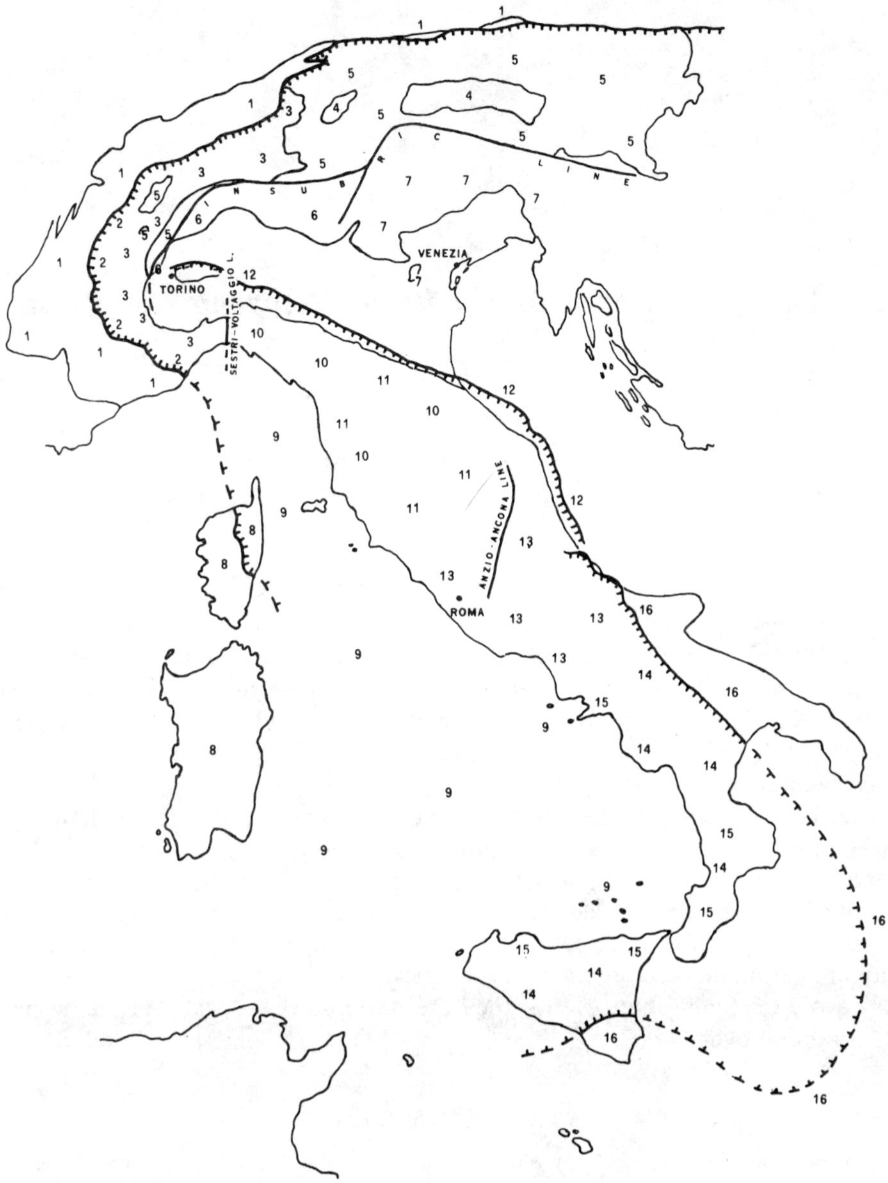

FIG. 1 Geological sketch of the Italian region. Numbers refer to columns of the table and sections of the text.

Isotopica, Consiglio Nazionale delle Ricerche, Pisa; Professor M. Fornaseri, Centro di Studio per la Geocronologia e Geochimica delle Formazioni Recenti, Consiglio Nazionale delle Ricerche, Rome; and Professors B. Zanettin and F. P. Sassi, Istituto di Mineralogia e Petrografia, University of Padua.

The data in the table are arranged by tectonic complexes and large geographical areas (Fig. 1).* This method has certain disadvantages and, apart from problems of delimitation,

* In the case of the Alps, it has been found clearer to give greater prominence to an arrangement by tectonic units, whereas the columns for the rest of Italy correspond predominantly to large geographical zones.

the most significant of these will be dealt in the respective sections of the paper. It is clearly suitable for the correlation of sedimentological and stratigraphic data and, up to a point, even for tectonic data. On the other hand, it is manifestly unsatisfactory for such phenomena as magmatic intrusion, metamorphism and thermal events; all of which, when they are later than orogenic movement phases, would be more appropriately referred to in a purely geographical setting; a good example is that provided by the Periadriatic Plutons, or the Eastern Alpine metamorphism. Moreover, many tectonic events, especially those of distensional and gravitational nature, and their associated sedimentological events, are often linked to geographical zones running across major tectonic units.

As in all summaries of this kind, the data are variably reliable in function of the phenomenon analysed and the sources employed. Uncertainties particularly arise with regard to both the interpretation and the dating of certain tectonic events, such as nappes, and above all conditions of extension or compression. Indications of extension or compression are examples of situations where the greatest chronological, and often geographical, uncertainty attends the expression of an opinion. When the symbols for both phenomena are shown, this means, in general, that two sources disagree, though compression and tension may occur at the same time in neighbouring or adjacent areas of the same tectonic unit.

To prevent the chart from being overburdened with detail, the localities or tectonic subunits of individual events are shown in only a few cases (and solely where radiometric ages are reported). For the same reason, individual data sources are not indicated, though they can in all cases be traced through the references given after the section of the paper dealing with the column in question augmented by the three works listed in the introduction.

1 – HELVETIC

The Helvetic crystalline massifs display traces of Precambrian and Alpine metamorphism, though their main pattern involves Caledonian and Hercynian metamorphic and plutonic changes. Anatexis between the Middle Carboniferous and the Upper Permian is particularly evident. The non-metamorphic or only weakly metamorphic cover goes back to the Middle–Upper Carboniferous, and there are marked transgressions at the base of the Middle Permian, Lower Trias, Cenomanian and Palaeogene (Upper Palaeocene, Lower, Middle or Upper Eocene, depending on the area). The appearance of fault-bound troughs and rises and hence the genesis of breccias, are observed at the Lower Triassic–Middle Triassic boundary, and in the Middle–Upper Triassic. The Upper Cretaceous of the Argentera autochthon is marked by extensive debris flows.

Flysch is solely Upper Eocene–Lower Oligocene, contemporary with the arrival of the Alpine nappes in the Helvetic domain, while more recent fault rejuvenation, further nappes and especially the conspicuous backfolding and backthrusting phenomena are attributable to frequent pulses of uplift.

References

Baggio, P., Ferrara, G. and Malaroda, R. (1967). Results of some Rb/Sr age determinations of the rocks of the Mont Blanc Tunnel, *Boll. Soc. Geol. It.*, **86**.

Biancotti, A. (1974). La Tettonica della regione circostante il Lac des Mesches (estremo sud-orientale del Massiccio dell'Argentera), *Rend. Acc. Naz. Lincei*, ser. 8, **56**.

Ferrara, G. and Malaroda, R. (1969). Radiometric age of granitic rocks from the Argentera Massif (Maritime Alps), *Boll. Soc. Geol. It.*, **88**.

Malaroda, R. (1963). Les faciès à composante détritique dans le Crétacé autochtone des Alpes-Maritimes italiennes, *Geol. Rund.*, **53**.
Malaroda, R. (1974). Prime osservazioni sulla Tettonica ed il metamorfismo in corrispondenza al prolungamento sud-orientale della sinclinale intracristallina Lago del Vei del Bouc-Colle del Sabbione (Argentera meridionale), *Mem. Soc. Geol. It.*, **13**.
Malaroda, R. (1981). Permian granitization in the Roja basin (southern Argentera, Maritime Alps), *Boll. Soc. Geol. It.*, **100**.
Romain, J. (1978). Etude pétrographique et structurale de la bordure sud-occidentale du Massif de l'Argentera. Thèse, Université de Nice.

2 – SUBBRIANÇONNAIS AND BRIANÇONNAIS

The main feature of these megaunits is their extreme fractionation into subunits, each of which rapidly dies out both across and longitudinally. The submerged, or even partly emerged, morphology acquired by these palaeogeographic domains during the Triassic and Jurassic is demonstrated, *inter alia*, by repeated layers of conglomerates and breccias, while the sharp delimitation of the sedimentary basin of each subunit is indicative of environments consisting of basins confined and bounded by repeated active fault escarpments. The age of the breccias probably extends from the Upper Cretaceous (Nappe de la Brèche) to the Eocene (Colle di Tenda Unit). The two domains differ from the Helvetic in their abundance of Upper Palaeozoic volcanic manifestations, mainly of a rhyolitic nature, and the greater prominence of their syntectonic flysch, which runs from the Turonian to the Upper Oligocene.

Mention may be made of the probable presence of an Eocene backfolding translative stage in the Ligurian Alps and the middle Valle Stura di Demonte.

References

Malaroda *et al.* (1970). Carta Geologica del Massiccio dell'Argentera alla scala 1 : 50.000 e Note Illustrative, *Mem. Soc. Geol. It.*, **9**.
Vanossi, M. (1980). Les unités géologiques des Alpes Maritimes entre l'Ellero et la Mer Ligure: un aperçu schématique, *Mem. Sc. Geol. Padova*, **34**.

3 – PIEDMONTESE OF THE WESTERN ALPS

Here the older intrusion bodies, metamorphism and anatexis are Hercynian; the metamorphism and anatexis seem to be essentially Lower Carboniferous. A characteristic high-pressure metamorphism is a distinguishing feature of the Middle Cretaceous (Eoalpine) phase. The so-called Lepontine metamorphism marks the Oligocene–Miocene, mostly after the emplacement of nappes. Its complete chronology is shown in the chart and embraces marked differences in cooling rates. Its climax was between 40 and 35 m.y. The variable phengite-biotite K/Ar age differences (from 1.1 to 4.8 m.y.) observed on passing from one zone to another are generally ascribed to zonal differences in uplift rates, it being taken for granted that the geothermal gradient in the region remained unchanged.

Fault tectonics certainly characterise the embryonic part of the history of the ocean basin. Except at the edges of the basin, ophiolitic manifestations and ore deposits abound from the Dogger to the Cenomanian.

Immediately prior to these phenomena, and perhaps associated with the Eoalpine metamorphism, there occurred the first nappe movements of the Piedmontese overriding the Briançonnais, and those of the nappes derived from the adjacent external Austroalpine basin overriding the Piedmontese (Second Diorite-kinzigitic Zone). The end of the main nappe emplacements, sealed by the transgression of the Middle Oligocene, is very clearly chronologically defined.

References

Dal Piaz, G. V., von Raumer, J., Sassi, F. P., Zanettin, B. and Zanferrari, A. (1975). Geological outline of the Italian Alps, in *Geology of Italy*, Tripoli, 1975.

Hunziker, J. Ch. (1974). Rb-Sr and K-Ar age determinations and the Alpine tectonic history of the Western Alps, *Mem. Ist. Geol. Min. Univ. Padova*, **31**.

Pasquare', G. (1968). La 'Serie di Montenotte': un elemento alloctono sovrapposto al bacino oligocenico di Santa Giustina (Alpi Liguri), *Riv. It. Pal.*, **74**.

Purdy, J. W. and Jäger, E. (1976). K-Ar ages on rock-forming minerals from the Central Alps, *Mem. Ist. Geol. Min. Univ. Padova*, **30**.

Wagner, G. A., Reimer, G. M. and Jäger, E. (1977). Cooling ages derived by apatite fission-track, mica Rb-Sr and K-Ar dating the uplift and cooling history of the Central Alps, *Mem. Ist. Geol. Min. Univ. Padova*, **30**.

4 – PIEDMONTESE OF THE EASTERN ALPS

Recent radiometric data have indicated Hercynian ages for parts of the Periadriatic Plutons, previously regarded as being entirely Tertiary. High-pressure Cretaceous metamorphism, confined to the southern edge of the tectonic window, and in common with the adjacent Austroalpine zone, has also been observed in the Piedmontese Pennines of the Alti Tauri.

References

Borsi, S., Dal Moro, A. and Ferrara, G. (1972). Età radiometriche delle rocce intrusive del Massiccio di Bressanone-Ivigna-Monte Croce (Alto Adige), *Boll. Soc. Geol. It.*, **91**.

Sassi, F. P., Zanettin, B. and Zirpoli, G. (1980). Quadro della storia termica alpina nelle Alpi Orientali, *Rend. Soc. It. Min. Petr.*, **36**.

Satir, M. (1975). Die Entwicklungsgeschichte der westlichen Hohen Tauern und der südlichen Oetztalmasse auf Grund radiometrischer Alterbestimmungen, *Mem. Ist. Geol. Min. Univ. Padova*, **30**.

5 – AUSTROALPINE

The new data mainly refer to the Sesia-Lanzo Zone, the Dent Blanche Klippe in the Western Alps and the Alto Adige (Südtirol). Five metamorphic cycles have now been clearly demonstrated for the Eastern Alps, though their tectonic significance is still uncertain: Caledonian, Hercynian, Eoalpine, Palaeogene, and Oligo-Miocene. The radiometric age boundaries can be regarded as sufficiently precise for each of these cycles.

Caledonian metamorphism (500–450 m.y.) was immediately followed by anatexis and intrusive-volcanic activities (450–420 m.y.). In the same way, Hercynian metamorphism (350–300 m.y.) gave way to silicic plutonism and volcanism (300–240 m.y.). In the case of the plutons, the distinct separation between the two events is very clear in the histogram of Fig. 2.

The three Alpine metamorphic phases took place in the Pennine and Austroalpine regions at the same time. Intrusions of the junction area (Periadriatic Massifs) appear to be virtually contemporaneous with the Palaeogene phase though some of the small andesitic dykes are also Cretaceous (120–80 m.y.).

The Palaeogene event, a weak greenschist facies metamorphism, has long been known as 'Tauern crystallization'. Recent chronological data fall into narrow time limits

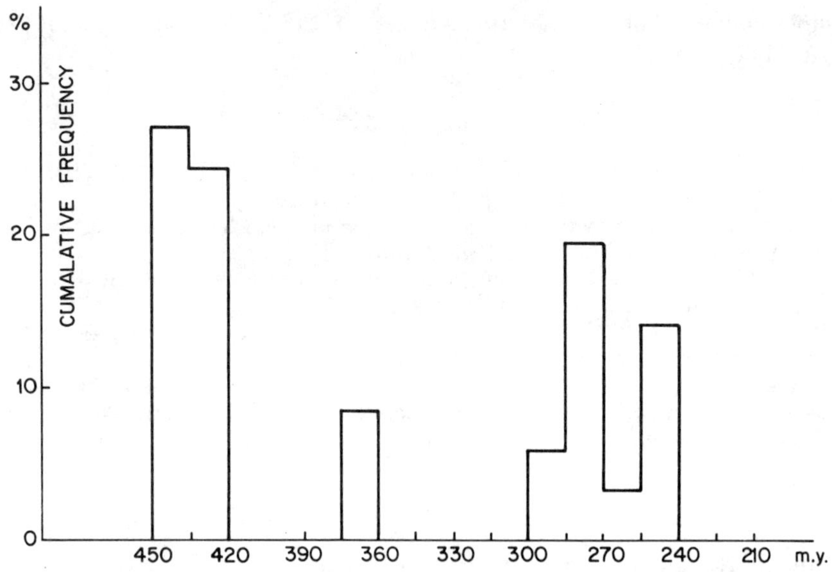

FIG. 2 Distribution of Rb/Sr age frequencies (whole rock isochron) for single granitoid bodies in the Austroalpine of the Eastern Alps (from Sassi and Zirpoli, 1979).

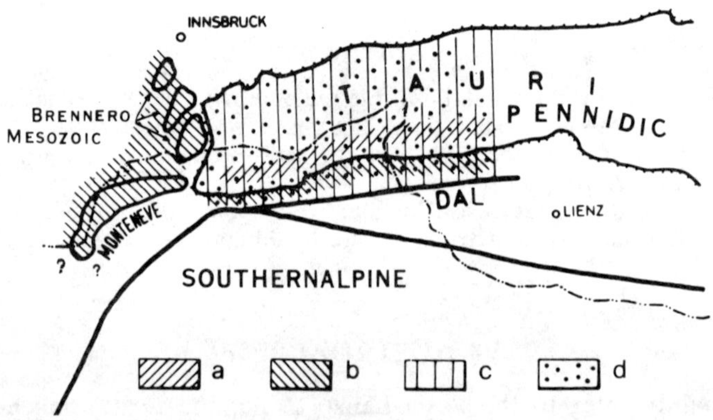

FIG. 3 Distribution of four Alpine metamorphisms over part of the Eastern Alps: (a) Cretaceous metamorphism hp/lt; (b) Barrovian Cretaceous metamorphism; (c) Barrovian Oligocene metamorphism ('Tauern crystallization'); (d) Low-grade Oligo–Miocene metamorphism; DAL, Deferegger-Anterselva Line (from Sassi, Zanettin and Zirpoli, 1980).

(40–30 m.y.). To the south it is sharply bounded by the Deferegger-Anterselva Line (Fig. 3), which means that either this line underwent a marked displacement after the metamorphism, or else the fault set up a discontinuity that effectively blocked the metamorphism.

References

Borsi, S., Del Moro, A., Sassi, F. P. and Zirpoli, G. (1973). Metamorphic evolution of the Austridic rocks to the south of the Tauern Window, *Mem. Soc. Geol. It.*, **12**.

Borsi, S., Del Moro, A., Sassi, F. P., Zanferrari, A. and Zirpoli, G. (1978). New geopetrological and radiometric data on the Alpine history of the Austridic continental margin south of the Tauern Window, *Mem. Sc. Geol. Padova*, **32**.

Borsi, S., Del Moro, A., Sassi, F. P., Visona', D. and Zirpoli, G. (1980). On the existence of Hercynian aplites and pegmatites in the lower Aurina Valley (Ahrntal, Austrides, Eastern Alps), *Neues Jahrb. Min. Monatsch*.

Del Moro, A., Sassi, F. P. and Zirpoli, G. (1980). First results on the radiometric age of the Hercynian metamorphism in the South Alpine basement of the Eastern Alps, *Neues Jahrb. Geol. Pal. Monatsch*.

Sassi, F. P. and Zirpoli, G. (1979). Stato delle conoscenze sulla storia pre-ercinica delle Alpi Orientali, *Mem. Soc. Geol. It.*, **20**.

Sassi, F. P. and Zirpoli, G. (1980). Quadro delle conoscenze sulla storia pre-alpina delle Alpi Orientali, *Rend. Soc. It. Min. Petr.*, **36**.

Sassi, F. P. and Zanettin, B. (1980). Schema degli eventi metamorfici e magmatici nelle Alpi Orientali, *Rend. Soc. It. Min. Petr.*, **36**.

Zanettin, B. and Justin Visentin, E. (1971). Considerazioni geologico-petrografiche sul 'Tratto di Monteneve' (Alto Adige), *Mem. Ist. Geol. Min. Univ. Padova,* **29**.

6 – SOUTHERNALPINE AREA OF PIEDMONT-LOMBARDY ALPS

The following subunits are included here in the Southern Alps, though some workers attribute them to the Austroalpine: Ivrea-Verbano Zone, Strona-Ceneri Zone (or Serie dei Laghi) and the Canavese Zone. For the purposes of radiometric dating, a distinction must be drawn between the Ivrea-Verbano Zone, the associated Valpelline Series and Second Diorite-kinzigitic Zone, and the Strona Ceneri Zone. The first metamorphic event, in fact, is reported for the Ivrea-Verbano Zone only. This is the oldest metamorphic event recorded for Italy. It is assigned to the Lower and Middle Cambrian (580–520 m.y.).

The event that has left the dominant impression in both crystalline zones, however, is the Hercynian, which mainly lies in the Carboniferous (350–270 m.y., and particularly 300–285 m.y.).

A third group of data is clustered around 240–140 m.y. (essentially Triassic and Jurassic). These can hardly be attributed to the Eoalpine event of the Middle–Upper Cretaceous. For this reason, they are regarded as cooling ages, though it may be objected that the hiatus (virtually all of the Permian) is against their being thought of as a consequence of the Hercynian event. On the chart, the Triassic–Jurassic event is shown as thermal. It left a deeper overprint on the Ivrea-Verbano Zone, and the Strona-Ceneri Zone was no more than marginally affected.

Unpublished K/Ar radiometric data obtained by the CNR Centre of Rome point to the occurrence of two metamorphic events, one corresponding to an undisputed Hercynian age between 351–327 m.y. (mainly Lower Carboniferous) and the other between 258 and 179 m.y. (Upper Permian–Lower Jurassic). Considerations similar to those used for the Ivrea-Verbano and Strona-Ceneri zones suggest that the latter was a separate thermal event. The ages recorded for certain pegmatites in the Lombardy Alps fall within a much narrower range (230–220 m.y.), but are equally referable to a thermal event.

The magmatites display well-known acidic intrusions, and more particularly extrusions of the Upper Carboniferous–Lower Permian, and Triassic mafic volcanics. Small mafic or intermediate bodies, intruded into Triassic formations, are assigned to the Upper Jurassic–Lower Cretaceous (148–117 m.y.: K/Ar dates from the same CNR Centre in Rome). A small tonalite stock, intruding the Ivrea-Verbano gneisses near the Canavese Line at Miagliano, Biella, has been dated as 31 m.y. It must be regarded as aligned with the Periadriatic Massifs.

The main stratigraphic events are the two transgressions in Middle and Upper Permian times, respectively overlain by the Collio Series and Verrucano Lombardo. The former was preceded by intense horst and graben movements, and either preceded or accompanied by initially andesitic and then rhyolitic volcanism.

Mesozoic faulting and folding are indicated by: (a) numerous local transgressions; (b) megabreccia horizons, especially in the Jurassic; (c) the Upper Cretaceous flysch in the Bergamo district.

References

Bocchio, R., De Capitani, L., Liborio, G., Mottana, A., Nicoletti, M. and Petrucciani, C. (1981). K-Ar radiometric age determinations of the South Alpine metamorphic complex western Orobic Alps, *Neues Jarhb. Geol. Pal.*

Carraro, F. and Ferrara, G. (1968). Alpine 'tonalite' at Miagliano, Biella (Zona Dioritico-Kinzigitica). A preliminary note, *Schweiz. Min. Petr. Mitt.*, **48**.

Casnedi, R. (1981). Segnalazione di megabrecce carbonatiche gravitative nel Giurassico bresciano, *Boll. Soc. Geol. It.*, **100**.

Ferrara, G. and Innocenti, F. (1974). Radiometric age evidences of a Triassic thermal event in the Southern Alps, *Geol. Rund.*, **63**.

Hunziker, J. C. (1979). The Ivrea-Verbano Zone: a geochronological overview of the data until 1975, *Mem. Sc. Geol. Padova*, **33**.

Martina, E. (1966). La granodiorite di Val Torgola-Val Navazze e la Linea della Val Trompia (Prealpi Bresciane), *Rend. Ist. Lomb., Cl. Sc.*, ser. A, **100**.

7 – SOUTHERNALPINE OF VENETIANS ALPS

Stratigraphic data for Carnia reveal Caledonian plutonism and Middle Carboniferous volcanism (lava flows intercalated in the flysch deposits). Abundant radiometric dates indicate a concentration of magmatic activity in the Upper Carboniferous–Lower Permian (Adige rhyodacitic-rhyolitic platform; granitic-granodioritic plutons in the Periadriatic and Cima d'Asta batholiths and stocks). A horizon of the Adige upper ignimbrites can be exactly dated as 270 m.y., while a hydrothermal event associated with the deposition of barite has been set at 225 m.y.

In contrast, there is little evidence of Triassic plutonic and volcanic activity in the Ladinian Middle Triassic (Predazzo-Monzoni and Val Posina, Recoaro: 230 m.y.). There are numerous data for the Palaeocene–Lower Miocene Lessinian-Beric-Euganean volcanism. This was essentially basaltic, with acidic and intermediate types in the Oligocene of the Euganean Hills only.

Metamorphism affects only the quartz-phyllites of the basement and their intercalations and is low-grade in greenschist-amphibolite facies. Rb/Sr dates of porphyroid intercalations in the Bressanone Phyllites show that it can be ascribed to a Visean and Westphalian phase (350 m.y. and 317 m.y. respectively). The Hercynian tectonic event is also indicated by the Namurian flysch and olistoliths.

A first tectogenic phase in the Triassic is observed at the Werfenian-Anisian boundary. Its presence is shown by the Richthofen Conglomerate in Trentino-Cadore and the Ugovizza Breccia in Carnia. At the Ladinian-Carnian boundary, the presence of a new tectonic phase is indicated by a distinct regression, the appearance of synsedimentary faults and the volcanic activity that gave rise to the Riofreddo rhyolites (Julian Alps) and to the andesites of the Dolomites.

Discordant brecciated limestones of the Middle Lias, and similar occurrences in other areas of the Julian and Carnian Alps point to the existence of a series of Jurassic tectonic phases (Lower Lias, Middle Lias, Bajocian). Subsequent movements occurred in the Malm, or at the Jurassic–Cretaceous boundary. Starting from the Lower Cretaceous, the several episodes of the Alpine orogeny are locally illustrated by flysch deposits, sedimentary breccias, and by back-thrust in the Upper Miocene–Lower Pleistocene. These are particularly obvious in the Friuli Prealps and are also supported by subsurface geological data of the Friuli Plain.

References

Baldi, P., Ferrara, G. C., Massari, F., Medizza, F. and Sedea, R. (1975). Ricerche stratigrafiche e calcimetriche sulla Scaglia Rossa euganeo-berica, *Mem. Ist. Geol. Min. Univ. Padova*, **30**.

Brigo, L., Dal Piaz, G. V., Ferrario, A., Omenetto, P. and Zuffardi, P. (1978). Correlazioni tra Paleogeografia e mineralizzazioni. Alcuni esempi delle Alpi, *Mem. Soc. Geol. It.*, **19**.

D'Amico, C., Del Moro, A., Freddo, A. and Pardini, G. (1980). Studio radiometrico delle ignimbriti riolitiche atesine, gruppo superiore, *Rend. Soc. Min. Petr. It.*, **36**.

Del Moro, A., Sassi, F. P. and Zirpoli, G. (1980). Preliminary results on the radiometric age of the Hercynian metamorphism in the South Alpine basement of the Eastern Alps, *Neues Jahrb. Geol. Pal. Monatschr*.

Massari, F. and Medizza, F. (1973). Stratigrafia e Paleogeografia del Campaniano-Maastrichtiano nelle Alpi Meridionali, *Mem. Ist. Geol. Min. Univ. Padova*, **28**.

Massari, F., Medizza, F. and Sedea, R. (1976). L'evoluzione geologica dell'area euganea tra il Giurese superiore e l'Oligocene inferiore, *Mem. Ist. Geol. Min. Univ. Padova*, **30**.

Pisa, G., Farabegoli, E. and Ott, E. (1978). Stratigrafia e Paleogeografia dei terreni anisici della conca di Agordo e dell'alta Val di Zoldo (Dolomiti sudorientali), *Mem. Soc. Geol. It.*, **18**.

Sassi, F. P., Zanferrari, A. and Zirpoli, G. (1979). Nuovi dati sulla Stratigrafia e i caratteri chimico-petrografici dei porfiroidi del Comelico (Alpi Orientali), *Mem. Soc. Geol. It.*, **20**.

Spalletta, C., Vai, G. B. and Venturini, C. (1979). Il flysch ercinico nella Geologia dei monti Paularo e Dimon (Alpi Carniche), *Mem. Soc. Geol. It.*, **20**.

8 – CORSICA AND SARDINIA

Like the Eastern Alps, the Sardo-Corsican Massif has a history of Caledonian magmatic and metamorphic events, and a marked correlation between its tectogenetic and metallogenetic epochs. The Hercynian and Caledonian tectogeneses tend to fit together owing to the presence of metamorphic and sedimentary events in, or near, the lower and upper limits of the Devonian. A new feature is the discovery of Middle Carboniferous overthrusts in central and southeastern Sardinia.

The chronological linkage is of particular interest. For the Hercynian events, this runs as follows: surface sedimentary events → metamorphism → plutonism → volcanism. Something of the kind is also true for the Hercynian and Caledonian events of the Austroalpine and the Hercynian events of the eastern Southernalpine region. It is in the Corsica and Sardinia column of the table, however, that this relation is most clearly decipherable.

Recent volcanics are to be found within the timespan of the Oligocene to the Middle Miocene, and then from the Pliocene to the Pleistocene. Felsic volcanism shifts from west to east between the Upper Oligocene and the Middle Miocene. Three well-separated calc-alkaline volcanic episodes (29–26 m.y.; 23–13 m.y.; 4–1.5 m.y.) occur in western Sardinia. The basic alkaline volcanism began with Late Messinian movements that preceded the Pliocene transgression and are linked to horst and graben tectonics similar to those that gave rise to the Campidano. Such tectonics and volcanism continue together until the Lower Pleistocene.

In the Alpine Corsica, transgressions, flysch and nappes are mere episodes in a history of predominant extension. As in the Tyrrhenian basin, the faults, horst and graben extension pattern continued until the Recent time.

References

Alvarez, W. and Cocozza, T. (1973). The tectonics of central-eastern Sardinia and the possible continuation of the Alpine chain to the south of Corsica, *Rend. Sem. Fac. Sc. Univ. Cagliari*, suppl. **43**.

Ardanese, L. R. and Nicoletti, M. (1978). Datazione K-Ar di un filone microgranitico dell'Isola di Mal di Ventre (Sardegna occidentale), *Rend. Soc. It. Min. Petr.*, **34**.

Barca, S. and Marini, C. (1979). Lineamenti litostratigrafici e sedimentologici del 'Postgotlandiano' del Fluminese (Sardegna SW), *Mem. Soc. Geol. It.*, **20**.

Beccaluva, L., Deriu, M., Macciotta, G., Savelli, C. and Venturelli, G. (1977). Geochronology and magmatic character of the Pliocene–Pleistocene volcanism in Sardinia, *Bull. Volcan.*, **40**.

Bellon, H. (1980). Chronologie 'potassium-argon' des manifestations magmatiques autour de la Méditerranée occidentale entre 33 et 1 m.a., Abstr. Int. Conf. 'Sedimentary basins of the Mediterranean margins, Urbino, 20–22 Ottobre 1980'.

Carmignani, L., Cocozza, T., Minzoni, N. and Pertusati, P. C. (1978). Falde di ricoprimento erciniche nella Sardegna a Nord-Est del Campidano, *Mem. Soc. Geol. It.*, **19**.

Cherchi, A., Marini, A., Murru, M. and Ulzega, A. (1978). Movimenti neotettonici nella Sardegna meridionale. *Mem. Soc. Geol. It.*, **19**.

Cocozza, T. and Scäfer, K. (1973). Cenozoic graben tectonics in Sardinia, *Rend. Sem. Fac. Sc. Univ. Cagliari*, suppl. **43**.

Del Moro, A., Di Simplicio, P., Ghezzo, C., Guasparri, G., Rita, F. and Sabatini, G. (1975). Radiometric data and intrusive sequence in the Sardinian batholith, *Neues Jahrb. Min. Abh.*, **126**.

Di Simplicio, P., Ferrara, G., Guasparri, G., Pellizzer, R., Ricci, C. A., Rita, F. and Sabatini, G. (1974). Il metamorfismo e il magmatismo paleozoico nella Sardegna, *Rend. Soc. It. Min. Petr.*, **30**.

Ferrara, G., Ricci, C. A. and Rita, F. (1978). Isotopic ages and tectono-metamorphic history of the metamorphic basement of northeastern Sardinia, *Contr. Min. Petrol.*, **68**.

Nicoletti, M., Ardanese, L. R. and Colasanti, S. (1983). La granodiorite di Capo Carbonara (Sardegna). Età K-Ar di fasi minerali in paragenesi, *Rend. Soc. It. Min. Petr.* **38**.

Padalino, G., Tocco, S. and Violo, M. (1978). Present hypothesis on the correlation between paleogeographic environments and ore deposits. Examples from Sardinian mining district, *Mem. Soc. Geol. It.*, **19**.

9 – TYRRHENIAN SEA

The Tyrrhenian Sea is of recent origin. The subsidence that led to its creation is associated with a general expansion that began before the Messinian and, according to the latest data, in the Lower Miocene. This subsidence did not cancel the marked transgression of the Pliocene sea reflected on distinct erosional surfaces. It has continued until today, generating horst and graben topographies, volcanic events and gravitational displacement of sediments. It was preceded by a compressive phase that proceeded from the same area to give rise to the Piedmontese and Liguride nappes of Corsica and the Apennines.

Under the Miocene deposits, the bottom of the Tyrrhenian is mainly composed of basalts. The non-basaltic pre-Miocene outcrops in its southern part, especially at the borders of the basin, are primarily related to the crystallines of the Sardo-Corsican Massif. Their ages range from Hercynian (295 m.y.) for a granite near the southeastern coast of Sardinia to Eoalpine (97 m.y.) for a micaschist from the southwestern Tyrrhenian.

The plutonism of Elba and the other Tuscan islands is, on the basis of radiometric and sedimentological data, clearly assignable to the Lower Messinian. The volcanoes of Capraia and the commencement of the more prolonged volcanism of the Pontine Islands are more or less of the same age. Extension along a north-south fault system accompanied the passage from andesitic to rhyolitic magmatism of the Aeolian Islands.

References

Barberi, F., Ferrara, G., Innocenti, F., Keller, J. and Villari, L. (1974). Evolution of Eolian Arc volcanism (southern Tyrrhenian Sea), *Earth Plan. Sc. Lett.*, **21**.

Bartole, R. (1980). Seismic evidence of an earlier Pliocene erosional surface in the deep part of the Tyrrhenian Sea, Abstr. Int. Conf. 'Sedimentary basins of Mediterranean margins, Urbino, 20–22 Ottobre 1980'.

Bossio, A., El-Bied Rakich, K., Giannelli, L., Massei, R., Russo, A. and Salvatorini, G. (1976). Corrélation de quelques sections stratigraphiques du Mio-Pliocène de la zone atlantique du Maroc avec les stratotypes du Bassin Méditerranéen sur la base des Foraminifères planctoniques, nannoplancton calcaire et Ostracodes, *Atti Soc. Tosc. Sc. Nat.*, ser. A, **83**.

Fabbri, A., Gallignani, P., Curzi, P. and Zitellini, N. (1980). Geologic evolution of the peri-tyrrhenian sedimentary basins, Abstr. Int. Conf. 'Sedimentary basins of Mediterranean margins, Urbino, 20–22 Ottobre 1980'.

Locardi, E. and Nappi, G. (1979). Tettonica e vulcanismo recente nell'Isola di Lipari, *Boll. Soc. Geol. It.*, **98**.

Savelli, C. (1979). K-Ar age of micaschist rock from the SW Tyrrhenian Sea, *Rend. Soc. Geol. It.*, **2**.

Selli, R. (1975). Laboratorio di Geologia Marina. Attività svolta nel 1974, *Giorn. Geol.*, ser. 2, **40**, 1972.

Selli, R., Lucchini, F., Rossi, P. L., Savelli, C. and Del Monte, M. (1977). Dati geologici, petrochimici e radiometrici sui vulcani centro-tirrenici, *Giorn. Geol.*, ser. 2, **42**.

10 – NORTHERN APENNINES: LIGURIDS

These Ligurids were originally the most internal regions, in which the allochthony is especially pronounced, particularly because of the resulting chaotic structure. Their history includes an initial distensional period, marked by the occurrence and cooling of ophiolitic rocks, followed by the appearance of breccias, slumps and olisthostromes. Starting from the Upper Cretaceous, there are flysch deposits of various composition and thickness. These form the boundary of widely differing and not very extensive palaeogeographic environments that underwent displacement towards the Adriatic while sedimentation continued in their interior.

The total distances covered by these displacements were increasingly greater in more recent times and their subsequent history is best described in the column of the unit they reached then, namely the Tuscanian and External Trough. On the Tyrrhenian margin there is clear evidence of a very recent distensional tectonic event.

References

Bettelli, G., Bonazzi, V. and Fazzini, P. (1980). Il complesso alloctono ligure nella Toscana meridionale, *Mem. Soc. Geol. It.*, **21**.

Fanucci, F. and Nosengo, S. (1977). Rapporti tra la Neotettonica e fenomeni morfogenetici del versante marittimo dell'Appennino Ligure e del margine continentale, *Boll. Soc. Geol. It.*, **36**.

11 AND 12 – NORTHERN APENNINES: TUSCANIDS AND THE NORTHERN PART OF THE EXTERNAL TROUGH

The crystalline Tuscanid area has recently been explored by numerous drillings in the Grosseto and Siena thermal fields. These have established its tectonic, metamorphic and volcanic history prior to the Middle Carboniferous transgression, of which only hints existed previously.

After the beginning of the Mesozoic, significant features include the Middle Triassic transgression, the thick clastic formations of the Triassic–Lower Jurassic and of Lower Cretaceous, and the numerous successions of displacement tectonics separated (or accompanied?) by fault distension from the Oligocene to the Lower Pliocene.

There are abundant olistostromes in the stratigraphic series of the Macigno and 'Marnoso-arenacea' (base of the Upper Oligocene-Tortonian; Fig. 4). The higher one is in the series, the more it is composed of elements from older terrains and more internal units.

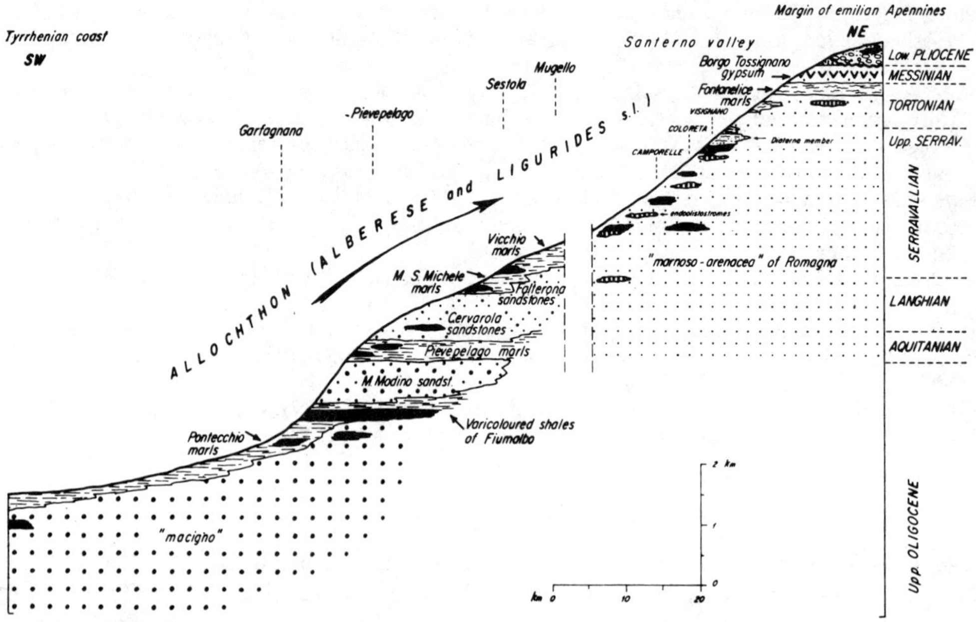

FIG. 4 Stratigraphic occurrence of olistostromes (black) in the 'Macigno', in the 'Mt. Modino Sandstones', in the 'Mt. Cervarola–Mt. Falterona Sandstones' and in the Romagna 'Marnoso-arenacea'. The ages of the olistostromes date the progressive advancement of the allochthon from the Upper Oligocene to the Lower Pliocene, proceeding from the Tyrrhenian coast toward the Emilian Apennines (from Bruni, 1973, in Ogniben, Parotto and Praturlon, 1975).

In Umbria, the Marches, Lazio and the Abruzzi, the Tortonian compressive tectonic phase is particularly important. This has generated a large number of small, narrow, north-south trending basins bounded by folds and faults and filled with turbidites. Oil prospecting in the Romagna and Marches areas of the Preapennines and off the Adriatic coast has made it possible to detect the arrival, during Upper Miocene–Lower Pliocene times, of Ligurid overthrusts into the parautochthonous or autochthonous sedimentary basins of the External Trough.

Moreover, the Messinian–Lower Pliocene tectonics have also resulted in structural highs and depressions. Accumulation of breccias and olistostromes fill such minor troughs, sometimes only in the Middle Pleistocene.

References (see also section 13)

Carmignani, L., Dessau, G. and Duchi, G. (1976). Una mineralizzazione sin-tettonica; il giacimento di Valdicastello (Alpi Apuane). Rapporti tra Tettonica e minerogenesi in Toscana, *Boll. Soc. Geol. It.*, **94**.

Cerrina Feroni, A. and Patacca, E. (1975). Considerazioni preliminari sulla Paleogeografia del dominio toscano interno tra il Trias superiore ed il Miocene medio, *Atti Soc. Tosc. Sc. Nat.*, ser. A, **82**.

Cremonini, G. and Farabegoli, E. (1978). Tettonica sinsedimentaria messiniana nell'Appennino romagnolo, *Rend. Soc. Geol. It.*, **1**.

Decandia, F. A. and Giannini, E. (1977). Studi geologici nell'Appennino Umbro-marchigiano, *Boll. Soc. Geol. It.*, **96**.

Elter, P., Giglia, G., Tongiorgi, M. and Trevisan, L. (1975). Tensional and compressional areas in the recent (Tortonian to present) evolution of the Northern Apennines, *Boll. Geof. Teor. Appl.*, **17**.

Gianelli, G. and Puxeddu, M. (1979). An attempt at classifying the Tuscan Palaeozoic geochemical data, *Mem. Soc. Geol. It.*, **20**.

Gianelli, L., Mazzanti, R., Mazzei, R. and Salvatorini, G. (1981). I sedimenti del Pliocene e del Pleistocene inferiore della zona compresa fra Riparbella e Bibbona (province di Pisa e Livorno), *Boll. Soc. Geol. It.*, **100**.

Pertusati, P. C., Plesi, G. and Cerrina Feroni, A. (1980). Un episodio di raccorciamento interposto tra fasi di distensione nel Calcare di Risignano (Neoautoctono) del bacino della Fine (Toscana meridionale), *Boll. Soc. Geol. It.*, **99**.

Puxeddu, M., Raggi, G. and Tongiorgi, M. (1979). Descrizione di alcuni sondaggi e osservazioni geologiche nel Paleozoico della zona di Monticano (Siena), *Mem. Soc. Geol. It.*, **20**.

13 – APENNINES OF CENTRAL ITALY

The geological boundary between the Northern and Central Apennines is not the geographical border. It more or less corresponds with the Ancona-Anzio Line, which was active from the Lias onwards. Even so, the difference between the two sectors is primarily determined by the development of the Jurassic–Cretaceous carbonate platform facies in the central sector, often with horizons of breccia, and by the minor importance of the Ligurid-Sicilid units. As far as the external trough in the Adriatic offshore of Abruzzi is concerned, the data are virtually the same as those for the more external land units.

In substance, this column shows, with undefined limits to the north and south, data for Lazio, the Abruzzi, and a part of Campania. These regions also have in common the extension on the Tyrrhenian side of intense silicic and mafic, alkaline Plio-Pleistocenic volcanism. The flysch of northern Lazio and southern Tuscany, on the other hand, is shown in column 11 of the table.

References (see also sections 11 and 12)

Ambrosetti, P., Azzaroli, A., Bonadonna, F. P. and Follieri, M. (1972). A scheme of Pleistocene chronology for the Tyrrhenian side of Central Italy, *Boll. Soc. Geol. It.*, **91**.

Barberi, F., Innocenti, F. and Ricci, C. A. (1971). Il magmatismo nell'Appennino centro-settentrionale, *Rend. Soc. It. Min. Petr.*, **27**, fasc. spec.

Cantelli, C., Castellarin, A. and Praturlon, A. (1978). Tettonica giurassica lungo l' 'Ancona-Anzio' nel settore Monte Terminillo-Antrodoco, *Geol. Romana*, **17**.

Centamore, E., Chiocchini, U., Cipriani, N., Deiana, G. and Micarelli, A. (1978). Analisi dell'evoluzione tettonico-sedimentaria dei 'bacini minori' torbiditici del Miocene medio-superiore nell'Appennino umbro-marchigiano e laziale-abruzzese, Parte 5, *Mem. Soc. Geol. It.*, **18**.

Lombardi, G., Nicoletti, M. and Petrucciani, C. (1974). Età delle vulcaniti acide dei complessi Tolfetano, Ceritee Manziate (Lazio nord-occidentale), *Per. Min.*, **43**.

Praturlon, A. and Sirna, G. (1975). The Stratigraphy and Palaeontology of the Cretaceous in Italy, in *Geology of Italy, Tripoli, 1975*.

SOUTHERN APENNINES: GENERAL REMARKS

The tectono-palaeogeographic units of the Southern Apennines are extremely fractionated. Their terminology, too, is very varied, so that a precise arrangement is difficult. The scheme used here was first proposed by Ogniben L. and can be found in the *Structural Model of Italy* (see *General References* of the introduction).

A foreland can easily be distinguished. This consists of the Apulian Plateau and the Ragusa Plateau, with the Ionian Bathial Plain between them. The Panormid units lie above and more in the interior of this foreland. They have a carbonate platform facies belonging to a miogeosynclinal environment (Appennino Calcareo *auct.*). Ligurid and Sicilid nappes, from more internal origin and a eugeosynclinal environment, separate the Panormids from the Calabrids which are also of internal provenance and are divided into many allochthonous overthrusts. Unlike the Ligurids and Sicilids, however, they consist of a crystalline basement covered by different series of Mesozoic–Eocene rocks of mainly shelf facies. For this reason, the Calabrid complex is regarded as more internal than the eugeosynclinal Ligurid and Sicilid units.

Lastly, overthrusts composed of eugeosynclinal rocks and known as the Antisicilids are tectonically superposed on the Calabrids. The main feature that distinguishes them from the other units is that they verge inwards (backthrusts).

By comparison with the Alps the orogenesis of the Apennines, especially its southern sector, lasted longer. Repeated compression episodes, sometimes generating backthrusts, alternated with phases of distension, especially during the Miocene.

14 – SOUTHERN APENNINES: LIGURID AND SICILID

The lithology of the Ligurid units consists of ophiolites, radiolarites, pelagic limestones and, in the end phase of the history of the basin, of flysch. During the last phase, the basin had already began its nappe movements. The Sicilids have an external eugeosynclinal character. This is proved by the lack of ophiolites and presence of andesite tuffites. Both the Ligurid and Sicilid rocks were once called, in general terms, 'Argille Scagliose' or 'Argille Varicolori'.

Ophiolites and basic volcanics appear in the Jurassic, and the final calcareous flysch sedimentation continued as late as the Cenomanian. Overthrusts began in the Upper Oligocene and were periodically rejuvenated until the Pliocene; only the main ones could be indicated on the table. Small parts of the Ligurids and Sicilids, that were very external at that time, continued to advance until the Pleistocene.

As in the Northern Apennines, displacements occur in steps and, for example in the Agri Valley, many Sicilid units appear to be interlayered at different levels in the parautochthonous flysch. Chronologically, the span runs from the Lower Miocene to the Upper Pliocene, and namely in the Burdigalian, Upper Helvetian–Lower Tortonian, Messinian, Lower Pliocene and Upper Pliocene.

References

Lentini, F. (1979). Le unità sicilidi della Val d'Agri (Appennino Lucano), *Geol. Romana*, **18**.

Sgrosso, J. (1981). Fasi distensive durante la tettogenesi miocenica nell'Appennino Meridionale: considerazioni preliminari, *Rend. Soc. Geol. It.*, **4**.

15 – SOUTHERN APENNINES: PANORMID AND CALABRID

This group of units is extremely heterogeneous. In the basement, the crystalline plutonics are dated radiometrically as Lower Permian and are probably more or less directly associated with a weak anchimetamorphism to epimetamorphism. Two distinct series of volcanics have been affected by this metamorphism, and must therefore predate it. One of these is certainly Middle–Upper Devonian (it is intercalated between rocks bearing *Tentaculites*), while the other passes from subalkaline or calcalkaline metabasites into metarhyolites and is definitely younger.

Intrusive and volcanic events occurred in the Triassic–Jurassic (Mt. Judica Unit, allochthonous on the Hyblean Plateau). These were accompanied by thermal events during the Jurassic. The distensional tectonics involved are better illustrated by flysch deposits (one, of

Carnian age, near the foredeep, and a second, an Upper Jurassic–Lower Cretaceous deposit), and the syntectonic breccias.

The Alpine metamorphism is less uniform. There is evidence of an Upper Palaeocenic event (56 m.y.), sometimes referred to as Upper Cretaceous–Palaeogene, and in the Sila Crystallines, an Oligocene event (32 m.y.), and an Aquitanian date (18 m.y.) exist, all of them epizonal.

The existence of several orogenic phases and the differences between the subunits mean that numerous transgressions are shown in the diagram. Those of a more regional character are the Tortonian, which marks the end of the more important displacements, and that of the Middle Pliocene, which brings to an end a series of mostly intra-Messinian fault movements since the Upper Messinian–Lower Pliocene sedimentary cycle, with intercalated flows of Ligurid and Sicilid argillaceous materials, lies unconformably on the Lower Messinian.

The flysch deposits are of different ages. In northeastern Calabria an Eocene transgressive turbidite complex particularly rich in slumps and olistoliths (Paludi Formation) is coaeval with andesitic volcanics; in the Sorrento Peninsula a Langhian flysch is clearly related to a paroxysmal tectonic phase, and contains imposing megabreccias. We cannot overlook the most ancient flysch deposits, such as those of the Longobucco unit (northeast of the Sila), which are of Middle Jurassic–Lower Cretaceous age, and the Carnian flysch of some of the outer Panormid units, transitional into the Hyblean Plateau.

The basic volcanism of the Neoautochthon is remarkable and can be readily separated chronologically into Upper Cretaceous–Lower Eocene in the Hyblei, Pleisto-Holocene in Etna, and post-Calabrian in the Vulture.

True compressional events and the first overthrusts appear at the Eocene–Oligocene boundary, and are repeated in the subsequent displacements (partly illustrated in column 14), some of which have been seen as backthrusts. In the Middle Miocene, shortly before the great Tortonian transgression, the strongest horizontal displacements took place, and the Ligurid and Sicilid nappes for the first time reached the foredeep areas in which, however, even after the Mio-Pliocene molasse sedimentation, folds, slices and also true overthrusts once again occurred.

During the Middle Pliocene, compression along the inner margin of the Calabrian arch, contemporary with extension along the transverse longitudinal trenches, is described.

References

Abate, B., Catalano, R. and Renda, P. (1978). Schema geologico dei Monti di Palermo, *Boll. Soc. Geol. It.*, **97**.

Amodio Morelli, L. *et al.* (1976). L'arco calabro-peloritano nell'orogene appenninico-magrebide, *Mem. Soc. Geol. It.*, **17**.

Atzori, P. and Ferla, P. (1979). Caratteristiche del metamorfismo ercinico sulle successioni sedimentarie e magmatiche del basamento paleozoico delle unità inferiori dei M. Peloritani, *Mem. Soc. Geol. It.*, **20**.

AA.VV. (1979). L'arco calabro-peloritano nell'orogene appennino-maghrebide. Atti 68° Congresso S.G.I., *Mem. Soc. Geol. It.*, 1976.

Bonardi, G., Giunta, G., Perrone, V., Russo, M., Zuppetta, A. and Ciampo, G. (1980). Osservazioni sull'evoluzione dell'arco calabro-peloritano nel Miocene inferiore: la Formazione di Stilo-Capo d'Orlando, *Boll. Soc. Geol. It.*, **99**.

Bonfiglio, L. (1970). Facies biodetritica tardopliocenica nei Peloritani a 1250 metri d'altitudine, *Boll. Soc. Geol. It.*, **89**.

Borsi, S. and Dubois, R. (1968). Données géochronologiques sur l'histoire hercynienne et alpine de la Calabre centrale, *C.R. Acad. Sc. Paris*, **266**.

Borsi, S., Hieke Merlin, O., Lorenzoni, S., Paglianico, A. and Zanettin Lorenzoni, E. (1976). Stilo Unit and 'Diorito-kinzigitic' Unit in Le Serre (Calabria), *Boll. Soc. Geol. It.*, **95**.
De Blasio, J., Lima, A., Perrone, V. and Russo, M. (1981). Nuove vedute sui depositi miocenici della Penisola Sorrentina, *Boll. Soc. Geol. It.*, **100**.
Ferla, P. and Azzaro, E. (1978). Natura del vulcanismo precedente al metamorfismo ercinico nelle unità inferiori dei M. Peloritani orientali, *Boll. Soc. Geol. It.*, **97**.
Ghisetti, F. (1979). Evoluzione neotettonica dei principali sistemi di faglie della Calabria centrale, *Boll. Soc. Geol. It.*, **98**.
Lentini, F. and Vezzani, L. (1978). Tentativo di elaborazione di uno schema strutturale della Sicilia orientale, *Mem. Soc. Geol. It.*, **19**.
Ogniben, L. (1973). Schema geologico della Calabria in base ai dati odierni, *Geol. Romana*, **12**.
Ortolani, F. (1978). Alcune considerazioni sulle fasi tettoniche mioceniche e plioceniche dell'Appennino Meridionale, *Boll. Soc. Geol. It.*, **97**.
Sgrosso, J. (1981). Fasi distensive durante la tettogenesi miocenica dell'Appennino Meridionale: considerazioni preliminari, *Rend. Soc. Geol. It.*, **4**.
Zuffa, G. G. (1978). Rapporti fra petrofacies ed evoluzione tettonica delle successioni mesozoiche della Sila nord-orientale, *Mem. Soc. Geol. It.*, **19**.
Zuffa, G. G. and De Rosa, R. (1978). Petrologia delle successioni torbiditiche eoceniche della Sila nord-orientale, *Mem. Soc. Geol. It.*, **18**.

16 – SOUTHERN PARTS OF THE EXTERNAL TROUGH

Carnian flysch and extension breccias at the edge of the Ragusa Plateau are worthy of mention. In the same marginal trough a synsedimentary and synvolcanic faulting phase continued from the Lower Hettangian until the Titonian. It started again in the Senonian. Extension tectonics dominates the history of all the external basins, where sporadic basic volcanic eruptions and, starting from the Palaeocene, olistostromes also occurred followed by true nappes that intercalate with normal sediments. Breccias are very common between the Upper Miocene and the Pleistocene.

Discordances and flows of Ligurid and Sicilid materials have been observed above the Middle Pliocene clays in the Bradanic Trench.

References

Crescenti, V. (1975). Sul substrato pre-pliocenico dell'avanfossa appenninica dalle Marche allo Jonio, *Boll. Soc. Geol. It.*, **94**.
Di Grande, A. and Grasso, M. (1977). Lineamenti geologici del margine occidentale ibleo nei dintorni di Comiso-Ragusa, *Boll. Soc. Geol. It.*, **96**.
Patacca, E., Scandone, P., Giunta, G. and Liguori, V. (1979). Mesozoic paleotectonic evolution of the Ragusa Zone, *Geol. Romana*, **18**.

CONCLUSIONS

Clear signs of tectogenic phenomena corresponding to the Precambrian do not exist in Italy. The only exception, reported because of its interest, can be seen in the dates of the Aar Helvetic massif in Switzerland.

Caledonian magmatic, anatectic and metamorphic events have been demonstrated in the Austroalpine, the Southern Alps, the Helvetic, the Sardo-Corsican craton and Tuscany. The Pennidic units also provide sure evidence of them, and Devonian events are also reported for the crystalline of Southern Italy (Peloritani).

Evidence for the mobility of the crust in the Hercynian age, on the other hand, is much more common. This is everywhere recognizable when the upper Palaeozoic formations can be reached, either in outcrops or in boreholes. It is particularly well documented for

Sardinia and the Carnic Alps in the form of typically syntectonic sedimentary products, such as breccias, flysch and olistoliths, and also plutonic and volcanic magmatism, anatexis and metamorphism extending from the base of the Carboniferous to the Middle Permian. There are numerous stratigraphic discordances, and occasional reports of true overthrust structures also exist. All these phenomena are divided into several phases, and our knowledge is as yet too fragmentary to enable correlations among them.

The Triassic and Jurassic were once regarded as periods of orogenic quiescence in the Italian geosynclinal basins. That this was too simple a view has been made evident by the analysis of discordance, and sedimentary data, especially the conspicuous deposits of breccias, and on some occasions the flysch and the olistostrome deposits. Overthrusts did not occur during the Lower Mesozoic. Even so, there were generalized uplifts and depressions, mostly bounded by faults. Thermal events, particularly extensive examples in the Ivrea-Verbano and Strona-Ceneri zones, have also been described, as well as ore deposits. The latter do not extend very far geographically, but are present in various tectonic units, from the Argentera Helvetic to Calabria.

The Cretaceous saw the commencement of the compressional phase of the Alpine orogenesis. This is indicated, especially in the Apennines, by breccias, flysch, slumps, olistostromes and their chronological and palaeogeographical position in the different units and subunits.

If we leave aside the precursory movements during the Cretaceous (in the Alps) and Cretaceous–Eocene (in the Apennines) and neglect the final overthrust movements of the Upper Oligocene–Lower Miocene in the Alps, and the Upper Miocene–Lower Pleistocene in the Apennines, we obtain the chronological span of the main tectogenesis. This lasts from the Upper Eocene to the Lower Oligocene in the Alps, and from the Lower Oligocene to the Middle Miocene in the Apennines. Thus, the length of the paroxysmal compressional history of the Alpine orogenesis was shorter in the Alps than in the Apennines.

Global Correlation of Tectonic Movements
Edited by Yu. G. Leonov and V. E. Khain

9

Tectonomagmatic Activity in the Western Carpathians (Czechoslovak Part) from Pyrenean to Rhodanian Epochs

D. VASS, Z. STRÁNÍK AND I. KRYSTEK*

The Western Carpathians in Czechoslovak territory may be divided into the following zones (cf. Mahel *et al.*, 1974): Foredeep, Outer Flysch Belt, Klippen Belt, Inner Carpathians. We may also add the backdeep, formed in the course of the development of the Carpathians, partly in the southernmost zones of the inner Carpathians and mostly on the Pannonian block. The backdeep lies mainly in Hungary (Fig. 1).

Tectonomagmatic epochs (Pyrenean to Rhodanian) follow the main tectogeneses of the inner western Carpathians (post-Subhercynian and/or post-Mediterranean, i.e. post-Upper Cretaceous) and during those epochs the molasse was formed. Alpine molasse of the Western Carpathians is divided into the following time segments: early (Uppermost Cretaceous–Egerian), main (Eggenburgian–Sarmatian) and late (Pannonian–Pliocene). Molasse division into areal units is as follows: outer (in the Foredeep), inner (in the intramontane basins and depressions) and back (in the backdeep).

It should be mentioned that the intramontane basins and depressions of the Western Carpathians, which will be discussed later, are dominantly in the Inner Carpathian region but the Vienna basin is partly, and the Oravská Kotlina depression mostly, in the Outer Flysch Belt.

EVIDENCE OF THE PYRENEAN TECTONOMAGMATIC EPOCHS IN THE OUTER CARPATHIANS

The Upper Eocene Illyrian folding is attributed to the Pyrenean orogeny occurring in the Alps between the Upper Eocene and the Oligocene (Fig. 2). Later orogenic movements between the Upper and Lower Oligocene in the Alps are referred to as Helvetian. It is difficult to date exactly the manifestations of folding in the Flysch Belt of the Western Carpathians because of the difficulties in stratigraphical correlation and detailed division of the Oligocene into stages. The Lower Oligocene age of the Menilite Beds (Cicha, Hagn and Martini, 1971) is indicative of a lesser significance of the Pyrenean movements in the Carpathian Flysch Belt. Manifestations of the Pyrenean movements in the Magura Group

* Respectively:
Geologický ústav D. Štúra 809 40 Bratislava, Mlynská dolina 1.
Ústřední ústav geologický, 602 00 Brno, Leitnerova 22.
Přírodověd. fakulta University J. E. Purkyně, 600 00 Brno, Kotlářska 2.

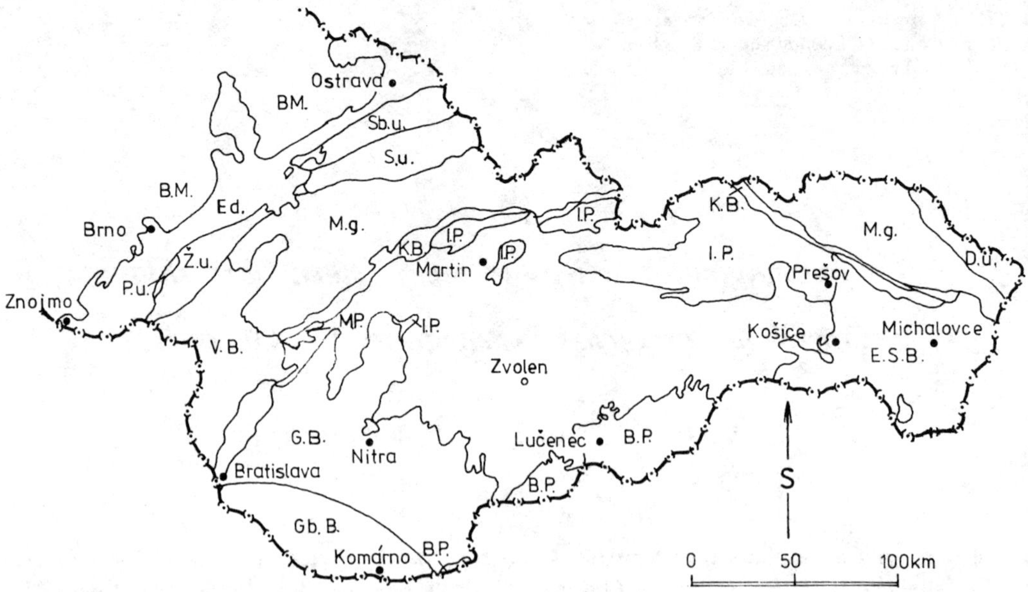

FIG. 1 Space distribution of the units and/or molasse basins mentioned in the paper.

B.M.	Bohemian massif (foreland)	
Fd	Foredeep	
P.u.	Pouzdřany unit	flysch zone
Ž.u.	Ždánice unit	
Sb.u.	Subsilesian unit	
S.u.	Silesian unit	
D.u.	Dukla unit	
M.g.	Magura group	
K.B.	Klippen Belt	
M.P.	Myjavska pahorkatina	
I.P.	Inner Palaeogene	
V.B.	Vienna basin	large intramontane basins
G.B.	Galanta basin	
E.S.B.	Eastern Slovakia basin	
B.P.	Buda Palaeogene basin partly covered by episodic basins	basins of backdeep
Gb.B.	Gabčíkovo basin	

were proved in the Čerhov unit (Stráník, 1967) and in the inner part of the Rača unit (Nemčok, 1961) in eastern Slovakia. Pebbles and concretions of carbonates at the base of *Globigerina* marls in Dlhá Lúka are indicative of a new cycle of sedimentation. In the Magura nappe the Lower Oligocene sediments are only known in the marginal part of the frontal anticlinorium in Chřiby (Chmelík *et al*., 1967) and in the Hostýnské vrchy hills (Pesl, oral communication). The relationship between the Lower Oligocene sediments and the underlying beds is not quite clear. Oligocene sediments were also found in the Jarošov 1 borehole (Špička *et al*., 1976). These occurrences, although sporadic, call for a critical review of the formerly generally accepted opinion about the termination of sedimentation in the Magura Group is in the Upper Eocene (Matějka and Roth, 1956; cf. Andrusov, 1968, p. 134).

Pyrenean movements are associated with a gradual Upper Eocene transgression from the Carpathian Flysch Belt onto the mobile margin of the North European platform. It is indicated by the autochthonous Palaeogene sediments (Nesvačilka Beds—Pícha, 1979) in the Vranovice and Nesvačilka depressions where the Upper Eocene sea advanced into the Bohemian Massif—Brno borehole 41, Malešovice 102 (Pícha *et al.*, 1978).

The origin of the Menilite Beds, characteristic of the external units of the Carpathian Flysch Belt is not exactly known. There are two different opinions about the depth and character of its sedimentary envionent: one suggests lagoonal and the other bathyal.

Bathyal fauna present in the Menilite Beds is indicative of sedimentation in calm, reducing deep-sea environment (Andrusov, 1965). The deposition of non-calcareous Menilite Beds on the underlying *Globigerina* ooze indicates rapid change of sedimentation at the Eocene–Oligocene boundary beneath the compensation level of carbonate. The decrease in CCD may be due to deepening of the basin, indicating Pyrenean movements or climatic changes caused by extensive worldwide cooling (Bartow, 1978) at the original dept (Stráník in Samuel, Gašpariková, 1983).

Slump bodies and deformational structures in the Menilite Beds (Roth *et al.*, 1962a,b, 1967), and/or at the base of the Šitbořice Beds (Stráník *et al.*, 1964), in the Silesian and Ždanice units are indicative of increased Lower Oligocene tectonic activity in the area of sedimentation. The tectonic activity is considered to be among the latest manifestations of the Pyrenean orogeny.

The appearance of the Krosno lithofacies which follows the Lower Oligocene in external units of the Flysch Belt indicates a substantial change in the evolution of the Outer Carpathians. The change of lithofacies and transport of detrital material from the uplifted orogenic belt are associated with a profound rearrangement of the Tethian area caused by the collision of tectonic blocks (Stráník in Samuel, Gašpariková, 1983). These changes are a part of the Helvetian orogeny.

Klippen Belt

The Klippen Belt has a palaeo-Alpine structure of the Laramid epoch; later on it was repeatedly refolded (Mahel', 1974), and might also have been in the Pyrenean epoch. The Illyrian (Pyrenean) folding is indicated by the Manín nappe,* the Klippen Belt of the Palaeogene Magura Group—east, southeast of Spišská Stará Ves (Mahel', 1974) has the Pyrenean–Savian structures.

Inner Carpathians. Myjavská pahorkatina (Gosau)

In this region sedimentation continued from the Cretaceous through the Palaeogene to the Middle Eocene with a break due to Illyrian movements. The sediments were folded in the Pyrenean epoch (Mahel' 1974, p. 153).

Inner Carpathian Palaeogene

The shallowing of sedimentation in the Upper Eocene, resulting in lithofacial change of the mainly pelitic strata into sandy beds and/or into a sequence of predominant sandstone. Fault tectonics affect the Levočské pohorie mountains and the Šarišská vrchovina mountains (Mahel', 1974, p. 135).

* In the area of the Súl'ov anticline the Krížna nappe rests on the Manín unit and they are both overlain by transgressive, partly folded, Palaeogene rocks (Andrusov, 1974, p. 154). The folding of the Palaeogene rocks might also, according to Andrusov, have been post-Palaeogene.

DIAGRAM OF TECTONOMAGMATIC ACTIVITY FROM PYRENEAN TO RHODANIAN EPOCHS

(WEST CARPATHIANS, CZECHOSLOVAK PART.)

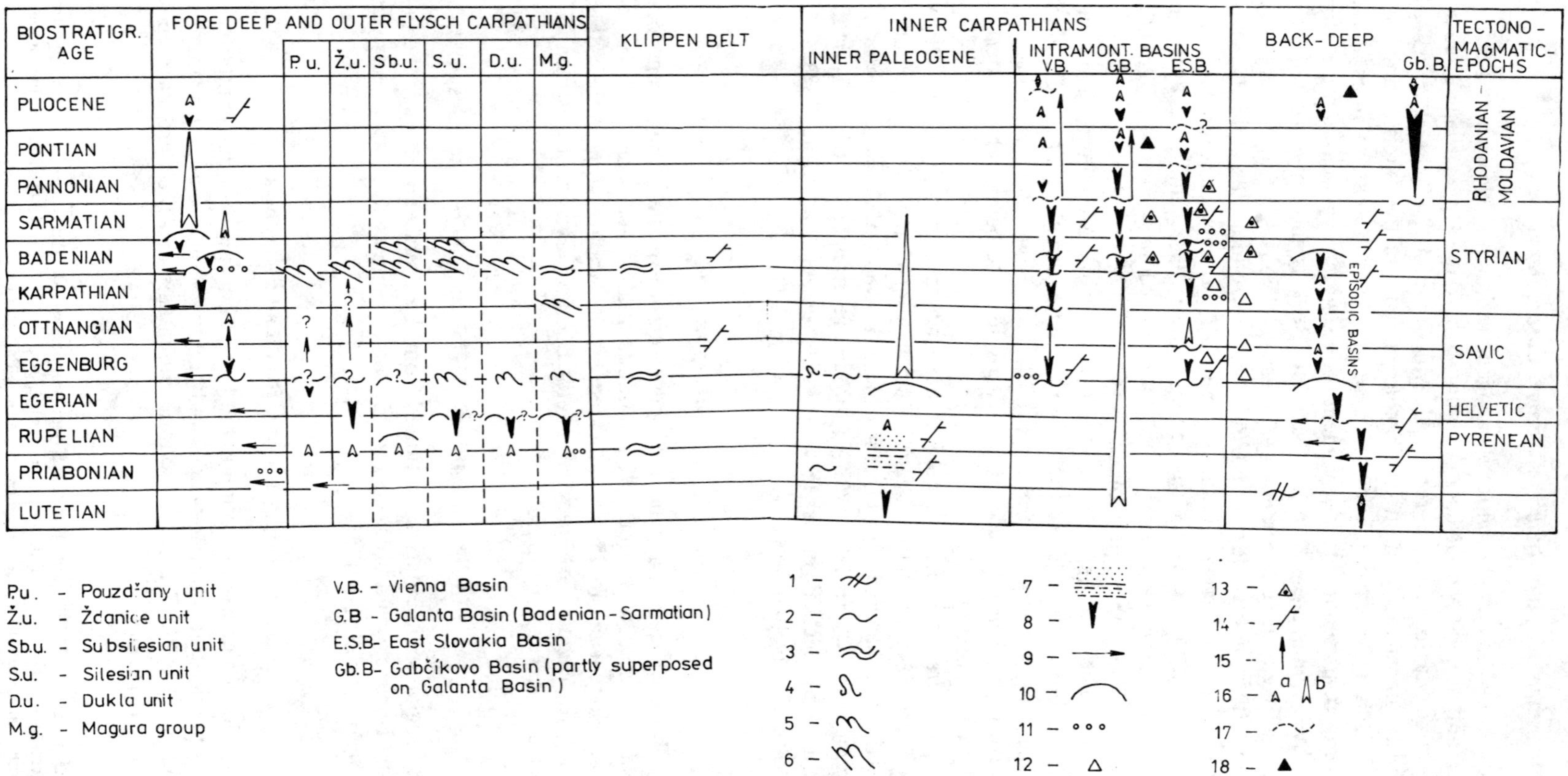

Backdeep (Buda) Palaeogene Basin

Gentle germanotype folding in the area of Štúrovo-Nová Vieska (Samuel, 1973, p. 69) occurred between the Eocene and the Oligocene. Synsedimentary movements along the faults proceeded in the Oligocene.

Formation of megafolds

This type of structure is well represented by the so called Šahy elevation. The first indications of the uplift were observed in the Rupelian (Kiscelian) and it also controlled sedimentation in the Egerian and the Lower Miocene (Vass, 1979, pp. 160, 168–70).

The spread of subsidence from the Pannonian region to the southern margin of the inner Carpathians (the Ipel'ská, Lučenecká, Rimavská kotlina depressions) in the Rupelian also continued in the Egerian (Vass, 1979, pp. 171–3).

The area of continuous uplift in the Pyrenean tectonomagmatic epoch was the Vepor-Spiš region which supplied detrital material transported laterally into the flysch trough of the intramontane Palaeogene, mostly during the Priabonian and Lower Oligocene (Andrusov and Köhler, 1963, pp. 179, 182; Andrusov, 1965, pp. 282–6).

Evidence of the Savian Tectonomagmatic Epoch in the Neogene Foredeep and in the Outer Flysch Carpathians

In the period between the Egerian and the Eggenburgian or during the Eggenburgian to Ottnangian folding took place in the outer Carpathians, but its results are not equally apparent in all areas. Lower Miocene sediments, that facilitate reliable determination of the role of Savian movements, are preserved in the Flysch Belt of the Czechoslovak Western Carpathians only in southern Moravia. The Lower Miocene of the Vienna Basin, with local basal clastic rocks, rests transgressively and unconformably on different lithistratigraphic members of the Magura Group. So in the Magura Group, prior to the Lower Miocene transgression (Eggenburgian), the regional uplift associated with the termination of the

FIG. 2

1. Germanotype folding (folds and faults)
2. Folding without nappes
3. Refolding with overthrust, backthrust and/or restricted tectonic transport
4. Isoclinal folding
5. Beginning of nappe formation
6. Folding with tectonic transport of nappes
7. Change of main lithologic complex from flysch to sandstone
8. Subsidence (movements of sinking)
9. Lateral spread of subsidence and/or transgression
10. Cessation of sedimentation and uplift
11. Conglomerates
12. Outset of ensialic volcanism (silicic) and its manifestation
13. Outset of ensialic volcanism (intermediate and silicic) and its manifestation
14. Active faults or generation of new faults
15. Continuous sedimentation
16. (a) Uplift where the rate of sedimentation fell down; uplift and break in sedimentation (at least partly i.e. regression)
    (b) Prolonged uplift without sedimentation or only basal continental sedimentation
17. Unconformity
18. Onset of basalt volcanism

flysch sedimentation, intensive folding and advanced denudation of the folds took place during the Helvetian and/or Savian orogenies. In the marginal (Pouzdřany and Ždánice) units of the Flysch Belt, the sedimentation represented by the Krosno flysch lithofacies lasted to the end of the Egerian. In the Eggenburgian, deep-sea pelagic Šakvice Marls were deposited over the flysch. The relationship between the Krosno lithofacies and the Šakvice Marls is not thoroughly explained either in the Pouzdřany or in the Ždánice unit (Stráník and Molčíková, 1980). Since the Šakvice Marls are entirely above the Krosno lithofacies, which used to be regarded as the youngest lithostratigraphic members of the Ždánice and Pouzdřany units, and since there is no evidence for basal conglomerates or an angular unconformity, the continuous sedimentation between these units of Egerian and the Eggenburgian ages might have occurred. One of us (Stráník) thinks that in such a case the marginal units of the Outer Flysch Carpathians represented in the Eggenburgian the axial approximately east-west area of sedimentation of the main molasse, and that from this position the sea transgressed northwards onto the craton and southwards into Carpathian intramontane basins.

Savian folding in the Western Carpathian Flysch Belt resulted in the incorporation of the southwestern and/or southern parts of the North European platform into the Carpathians. The folding and the regional uplift of the Inner Carpathians at the Oligocene–Miocene boundary caused the Lower Miocene (Eggenburgian) sea to transgress gradually from the Carpathian area onto the foreland. The Lower Miocene sediments rest mainly on the pre-Tertiary basement of the platform. Only in lower Austria and southern Moravia, especially in the Vranovice and the Nesvačilka grabens, do the Eggenburgian sediments rest on Palaeogene strata (the Melker Series, the Nesvačilka Beds), representing marginal facies of flysch troughs of the Outer Carpathians.

Uplifts and vertical dissection of relief of the pre-Miocene basement in the Vienna Basin prior to and during the Eggenburgian sedimentation are demonstrated by basal conglomerates (debris conglomerates at the base of the Lužice Beds, the Chropov conglomerates) and by marked differences in the primary thicknesses of the Eggenburgian in the areas of elevations and depressions (Jiříček, oral communication).

The Lower Miocene distant volcanism is represented by acid tuffs within the Eggenburgian-Ottnangian sediments. Products of basaltic volcanism from the Eggenburgian near Ostrava are reported by Cicha and Krystek, 1979).

Evidence of the Savian Tectonomagmatic Epoch in the Klippen Belt and Inner Carpathians

In the Eggenburgian to Badenian and/or the Lower Sarmatian the Klippen Belt was refolded. Tectonic phenomena and deformations of the Klippen Belt, controlled by the Savian and later movements were presented by Andrusov (1974, pp. 154–6) as follows:

(a) formation of south-verging fold system;
(b) formation of anticlinal structure embracing the whole of the Klippen Belt;
(c) blocks of Laramide nappes were turned into verticals and a system of south-verging lenticular bodies was formed;
(d) formation of the present-day configuration and delimitation of the Klippen Belt during Helvetian-Savian movements;
(e) particularly intense post-Palaeogene movements occurred at the outer margin of the Klippen Belt;
(f) deformation by faults.

Refolding of the cover of the Klippen Belt and back overthrust of the Outer Carpathians on the belt also took place. Before the Eggenburgian the thrust of the Inner Western Carpathians or at least of their outer units over the Magura Group was accomplished.

Refolding of the Inner Western Carpathians before the Eggenburgian, more accurately in the period between the Oligocene and the Eggenburgian, is shown by the following features.

(1) The termination of subsidence and sedimentation in depressions filled with the Inner Carpathian Palaeogene. The youngest members are found in the Banskobystrická kotlina depression (Rupelian, cf. Bystrická, 1979), and around Prešov—Prešov 1 borehole, the Solivar Beds (Čverčko *et al.*, 1979; Lehotayová, personal communication).
(2) The Palaeogene folding of the inner Carpathians. The intensity of folding increased from the south (monoclinal structures) to the north from broad flat to isoclinal folds, in the periklippen zone with backthrusts (i.e. southwards); these can also be younger—Styrian (Mahel', 1974, p. 135).
(3) The transgressive and unconformable position of the Eggenburgian on Inner Carpathian Palaeogene strata.

Evidence of the Savian Tectonomagmatic Epoch in Intramontane Basins

The Intramontane Basins* of the Western Carpathians main molasse also resulted from movements placed between the Egerian and Eggenburgian times. The basins have a new structural pattern reflecting the structure of the basement. The large basins, especially, have a character of intramontane depressions (Buday *et al.*, 1967) but do not coincide with Palaeogene depressions.

Transgressions in the Intramontane Basins are frequently associated with basal clastic facies indicative of a steep relief of source areas. There are, for example, coarse detrital beds in the eastern part of the Vienna Basin (the Dobrá Voda conglomerates).

Volcanic activity of local or distant centres during the Eggenburgian time resulted in tuff and tuffite layers (e.g. the eastern Slovakian basin, the Bánovská kotlina depression). Fault tectonics were active in the Eggenburgian time.

Folding proceeded to the end of the Eggenburgian and caused uplift, regression and perhaps partial refolding of the Eggenburgian in the eastern Slovakian basin. It looks as if the basin was uplifted throughout the Ottnangian, since deposits of this age are missing.

There are no indications of movements between the Eggenburgian and the Ottnangian in the Vienna Basin. The tectonic regime of the small intramontane depressions cannot be described accurately because of the lack of data. In the Bánovská kotlina (depression) there are indications of gradual shallowing of the depositional environment during the Ottnangian (Brestenská, 1979).

Evidence of the Savian Tectonomagmatic Epoch in the Backdeep

In the Backdeep the Savian movements produced uplift and regression and the *extinction* of the Buda Palaeogene Basin (Early Molasse).

In the Eggenburgian a new transgression occurred according to a new structural plan. It was of a smaller lateral extent and underwent lesser subsidence. In the Backdeep ephemeral

* Intramontane basins in the western Carpathians are of two types: large longitudinal basins (Vienna, Galanta and eastern Slovakia) and small interior intermontane depressions.

basins were formed (Vass, 1976, 1979). Eggenburgian basal sediments in these basins consist of coarse detrital rocks (the Ipel'ská kotlina, the Lučenská kotlina depressions).

In the Savian epoch the ensialic silicic volcanism commenced in the Backdeep. The first products of the volcanism are tuffite layers among the Eggenburgian marine sediments.

At the end of the Eggenburgian time the area of the Backdeep was uplifted and sedimentation proceeded in a continental environment. The sediments comprised coarse clastic material. Silicic volcanism of increased intensity produced thick layers of tuffs (welded tuffs) with radiometric ages proving the Eggenburgian age of the continental sequence.

During the Ottnangian a new transgression took place in the Backdeep. It commenced with a slow subsidence associated with the formation of coal seams, and was followed by increasing subsidence rate. In Hungary the sea transgressed into the Backdeep; in Slovakia sedimentation proceeded in a lacustrine environment and was sporadically associated with silicic tuffites which were a product of distant volcanism.

Evidence of the Styrian Tectonomagmatic Epoch in the Foredeep and the Flysch Belt of the Western Carpathians

The Styrian folding between the Karpathian and the Sarmatian had a decisive role in the formation of the present nappe structures of the outer western Carpathians. Sediments of the Flysch Belt deformed by the Savian and/or Helvetian folding were thrust over the foreland. The folds were compressed and ultimately squeezed into tectonic slices and partial nappes. The flysch nappes formed chronologically at the periphery of the Carpathian arc from northwest to northeast and/or from west to east (Buday *et al*., 1961). In southern and central Moravia the overthrust movements ceased before the Badenian time. The Pouzdřany, Ždánice units and the Maguta Group were thrust over the Lower Miocene (including the Karpathian). The Lower Badenian clastic rocks rest unconformably on the completed nappe structure. In northern Moravia the sub-Silesian and Silesian nappes rest upon Karpathian and the Lower Badenian beds. The Middle Badenian there is post-orogenic. Further eastward, in the flysch Carpathians of eastern Poland Middle Badenian strata lie below the nappes.

Styrian folding in the Flysch Belt caused significant changes in the evolution of the main Molasse. At the Ottnangian–Karpathian boundary the tectonic movements were followed by a regression and by locally interrupted sedimentation. Schlieren lithofacies of the transgressive marine Karpathian time have a character most resembling the terminal flysch of geosynclinal areas. The origin and different extension of the lithofacies in comparison with older Lower Miocene sediments (the Carpathian foredeep in the Ostrava region) reflect the tectonic reconstruction of the orogenic zone, of its foreland, and the lateral mobility of the sedimentary basin.

The folding movements between the Karpathian and Badenian times produced the earlier sedimentation cycle of the Main Molasse in the Foredeep. The Lower Miocene sediments were in the most part folded, imbricated during the movement of the nappes, and extensively covered by them. The degree of tectonic deformation of these sediments decreased from the marginal units of the Flysch Belt and the frontal anticlinorium of the Magura nappe into the Carpathians and their foreland.

In the Lower Badenian time the sea transgressed from the Carpathians to the new sedimentary area of the Carpathian foreland. In the initial stage the contact between the Lower Badenian foredeep and the Carpathian orogen was very limited. This is indicated by the Slavkov-Těšín ridge terminating the Lower Badenian foredeep in the southeast, and by the pebbles in the Lower Badenian clastic rocks. Among these pebbles lithologies from the Bohemian Massif are prevalent and those from the Carpathians are practically absent. The first mass presence of the Carpathian boulders appeared during the sedimentation of the Lower Badenian in marginal Terešov conglomerates. Data indicate the completion of thrusting of the flysch nappes in central Moravia at this time. The nappe fronts reached beyond the top of the Slavkov-Tešín ridge.

By the end of the Lower Badenian an extensive regression (Lithothamnian limestones and sands), as well as the decay of the foredeep, took place. Middle and Upper Badenian sediments between Hlučín and Opava in Silesia represent a spur of the Badenian foredeep of the Polish outer-Carpathians (Buday *et al.*, 1967).

During the Styrian folding the axis of the main Molasse sedimentation migrated southwards, and/or northwestwards. The Miocene sea transgressed further over the foreland. The Upper Badenian in Moravia has the most external position. The areas of subsidence were eliminated by the moving flysch nappes (Buday, *et al.*, 1967). Layers of silicic tuffites in this area are indicative of distant volcanic activity. In the northern part of the Moravian foredeep there are also local basalts.

Evidence of the Styrian Tectonomagmatic Epoch in the Klippen Belt and the Inner Carpathians

The deformation mentioned in connection with the Savian tectonomagmatic epoch continued in the Klippen Belt during the Styrian tectonomagmatic epoch. Then in the periklippen area the Upper Cretaceous and Palaeogene rocks of the Myjavská pahorkatina highland were imbricated. This was prior to the Upper Badenian which rests with an angular unconformity upon the deformed beds (Leško *et al.*, 1978, p. 47). Leško *et al.* suggest that the imbrication was caused by mobile internal units of the western Carpathians which by then were near to or at their present position.

Evidence of the Styrian Tectonomagmatic Epoch in Intramontane Basins

The end of the Ottnangian saw the uplift, regression and refolding (Vienna basin). The beginning of the Karpathian witnessed a new transgression associated with intensive subsidence.

In the Eastern Slovakian basin, subsidence and transgression took place in the early Karpathian and coarse clastic rocks were formed. During Karpathian times the eastern Slovakian area lost its connection with the open sea and changed into a saline lagoon with evaporitic sedimentation. Subsidence was revived in the Bánovská kotlina depression. The sedimentation environment was marine. The area of the Galanta basin, except for its northern spurs in the Bánovská kotlina, emerged during the Karpathian times.

At the end of the Karpathian the slow folding, emersion and regression took place. The sea completely withdrew from the Bánovská kotlina.

At the beginning of the Badenian subsidence occurred and a new transgression took place; the first subsidence followed a long Palaeogene and Lower Neogene immersion in the central parts and at the southern periphery of the Galanta basin. The transgression particularly in eastern Slovakia is associated with a strong subsidence.

During the Badenian uplift, regression and a break in sedimentation occurred. All these were formed in the Vienna basin and partly in the Galanta basin, at the end of the Lower Badenian. At the beginning of the Middle Badenian renewed subsidence and transgression took place. In the northern parts of the Galanta basin the subsidence is indicative of the first transgression during the Badenian (the Lower Badenian is missing, cf. Adam and Dlabač, 1969, Fig. 2, p. 163). In the eastern Slovakian basin no conspicuous manifestations of tectonic activity between the Lower and Middle Badenian were found. The uplift of the region began during the Middle Badenian and had a coastline against the open sea; thus saline lagoons with evaporites were formed. At the end of the Middle Badenian the movements in the Vienna and Galanta basins were less intense. In the Galanta basin a flyschoid formation was deposited in the Semerovce depression (Vass *et al.*, in press) whose origin was primarily associated with the volcanism of adjacent regions.

At the beginning of the Upper Badenian the new transgression was related to a strong subsidence in the eastern Slovakian basin. The beginning of the peak period of subsidence is contemporaneous with subsidence in small interior depressions (Turčianska, Žiarska kotlina and the Bánovská and the Hornonitrianska kotlina). Sedimentation in these depressions proceeded under continental conditions.

During the Badenian active volcanism continued. This is indicated by tuffites transported to the Vienna basin from distant centres. In the Galanta basin there are andesitic lava flows and volcanoclastic deposits from local centres. In the inner depressions there are also products of silicic volcanism. During the Lower Badenian the silicic (rhyodacite) volcanism was active, while in the Upper Badenian andesite volcanism was also significant. At the end of the Badenian the uplift, regression and/or weak folding took place in the Vienna, Galanta and eastern Slovakian basins. No definite manifestations of tectonics can be detected in small interior depressions.

At the beginning of the Sarmatian there was a new subsidence, transgression of brackish sea, a strong downwarping in the eastern Slovakian basin and, in contrast to the Badenian, in the inner depressions as well. In the Vienna and Galanta basins the intensity of the subsidence decreased.

During the Sarmatian the andesitic and, in part, silicic volcanism continued in the eastern Slovakian basin and in the inner depressions, although less so in the Galanta basin. Intense fault tectonics were current in Badenian and Sarmatian times.

Evidence of the Styrian Tectonomagmatic Epoch in the Backdeep

In the Backdeep in Czechoslovak territory (the Ipel'ská and Lučenecká kotlina depressions, the southeastern part of the Podunajská nížina lowland) there are various manifestations of tectonics. During the Ottnangian in Ipel' and Lučenec depressions continental sedimentation occurred, which at the beginning of Karpathian times was replaced by marine sedimentation, at first in semi-detached bays but later in the open sea environment. In the Karpathian the distant silicic volcanism was active. The end of Karpathian times saw the immersion, denudation and faulting activity. At the beginning of the Badenian subsidence occurred again and resulted in a new transgression according to a different plan.

During the Badenian regional uplift produced sea regression and sedimentation in local depressions under continental conditions.

In the Badenian and/or in the Sarmatian fault activity took place. In the area of the later Gabčíkovo basin of the Backdeep the pattern was the same as in the intramontane Galanta basin.

Evidence of Moldavian and Rhodanian Tectonomagmatic Epochs

During the Moldavian and Rhodanian tectonomagmatic epochs in the western Carpathians the late molasse developed in the backdeep. The molasse basin of the backdeep annexed the intramontane basins and depressions. The late molasse also sporadically spread over into the area of the foredeep.

Evidence of tectonic activity in the basins and depressions of the late molasse is as noted below. At the end of the Sarmatian subsidence was slower and sedimentation was interrupted. The beginning of the Pannonian saw subsidence, sometimes extensive, extension of the area of sedimentation in the Gabčíkovo basin, then reduction of the area of sedimentation and a decreasing intensity of subsidence in intramontane basins, in comparison with the Sarmatian.

Slight tectonism affecting the evolution of the late molasse was observed in the inner Carpathians. Uplift and regression and/or slower sedimentation might have proceeded during or at the end of the Pontian. In the Vienna Basin the sedimentation was also slower in the Dacian when renewed subsidence and lacustrine-fluvial sedimentation proceeded in other intramontane basins. The Dacian subsidence also caused lacustrine-fluvial accumulation at the northern margin of the former Buda Palaeogene basin of the back molasse ('the Poltár Formation'* in the Lučenská and Rimavská kotlina depressions).

In the Dacian the faults of the Hornomoravský úval graben in the Foredeep were revived and lacustrine sedimentation proceeded in a small graben. During the Dacian, in the Gabčíkovo basin the subsidence became intensified and the rate of sedimentation increased. At the end of the Dacian, the subsidence was replaced by uplift and regression. In the Hornomoravský úval graben and in the area of the Lučenská and the Rimavská kotlina depressions the lacustrine sedimentation ended. The sedimentation was also interrupted in the intra-Carpathian basin (the Vienna and Galanta basins) and in the backdeep (the Gabčíkovo basin).

At the beginning of the Romanian time the new subsidence caused local revival of sedimentation. At the end of the Romanian the regional uplift of the western Carpathians took place. In the Quaternary reduced fluvial and eolian sedimentation prevailed and subsidence took place in the Gabčíkovo basin. Silicic andesite volcanism of Pannonian age occurred in eastern Slovakia and the Central Slovakian volcanic region. At the end of the Pannonian and/or in the Pontian silicic volcanism was replaced by basalic volcanism (the Štiavnica and Podrečany basalts). In Hungary it was already recorded in the Lower Pannonian (Kiskunhálas). The basaltic volcanism culminated at the end of the Pliocene (basalts of the Cerová vrchovina highland in South Slovakia) and ended in the Quarternary (Brehy near Nová Baňa). This volcanism was preceded by the opening of deep subcrustal faults. No significant vertical displacement occurred on these faults.

CONCLUSIONS

Manifestations of the Pyrenean tectonomagmatic epoch are less conspicuous in the Western Carpathians. More intense movements occurred between the Lower and Upper Oligocene (Rupelian–Egerian) in the outer Flysch Belt. The Flysch Belt was tectonically rebuilt and its units such as the Magura Group became the source of detrital material for the Krosno lithofacies.

In the period of the genesis of main molasse (Eggenburgian–Sarmatian) in the Western Carpathians the tectonics were most intense in the zone of the Outer Flysch where the nappes formed and were transported. In molasse basins sedimentation took place. Tectonic activity resulted in refolding and an interruption of sedimentation, followed by new subsidence (foredeep, intramontane basins), or in uplift and break of sedimentation

* Poltár Formation according to the new data is Pontian in age.

followed by new subsidence (backdeep). Molasse sediments are also affected by faults that were most pronounced in the Badenian and Sarmatian. Molasse sedimentation in intramontane basins and in the backdeep was associated with silicic volcanism and later on andesite volcanism. Volcanic activity culminated in the Badenian and in the Sarmatian simultaneously with the peak of subsidence in intramontane basins. At the time of the formation of the late molasse (Pannonian–Pliocene) tectonic activity resulted in intense depression of the backdeep. Subsidence in basins was interrupted by uplift and regression. Molasse sedimentation was associated with basaltic volcanism and silicic and andesitic volcanism was waning.

The evolution of the main and late molasse was associated with regional uplift of the Carpathian region, resulting in progressive degradation of the sedimentary environment from marine through brackish to freshwater.

REFERENCES

Adam, Z. and Dlabač, M. (1969). Vysvětlivky k mapám mocnosti a litofaciálního vývoje Podunajské Nížiny. Zborník geol. vied, *Západné Karpaty*, **11**, 156–71.

Andrusov, D. (1965). Geológia československých Karát III, *Slov. akadémia vied*, 1–392.

Andrusov, D. (1968). Grundriss der Tektonik der nördlichen Karpaten, *Vydav. Slov. akadémie vied*, 1–188.

Andrusov, D. (1974). In M. Mahel' (ed.), *Tectonics of the Carpathian Balkan Regions*, Bratislava; Geol. ústav D. Štúra.

Bartow, J. A. (1978). Oligocene continental sedimentation in the Caliente Range area, California. *J. Sed. Petrology*, **48**, 75–98.

Buday, T. *et al.* (1961). Nafta a plyn v čsl. Karpatech, *Knihovna Ústř. ústav geologický*, **38**, 1–158.

Buday, T. *et al.* (1967). *Regionálni geologie ČSSR II*, vol. 2, Ústř. ústav geologický, Academia Praha, pp. 1–651.

Bystrická, H. (1979). Rupelian in the Banská Bystrica Basin, *Geol. zborník, Geologica Carpathica* **30**/1, Bratislava, 109–122.

Cicha, I., Hagn, H. and Martini, E. (1971). Das Oligozän und Miozän der Alpen und Karpaten. Ein Vergleich mit Hilfe planktonischer Organismen. Mitt. Bayer. St.-Samml, *Paläont. hist. Geol.*, **11**, 279–93.

Gašparík, J. (1978). Nové nálezy eggenburgu v Západných Karpatoch (Turčianska kotlina), Geol. práce, Správy 71, Geol. ústav D. Štúra, Bratislava, 89–92.

Chmelík, F. *et al.* (1967). Vysvětlující text k základní geologické mapě 1:50 000 list M-33-107-B (Kroměříž), MS Archív Ústř. úst. geol. Praha.

Leško, B., Ďurkovič, T., Gašparíková, V., Kullmanová, A. and Samuel, O. (1978). Nové poznatky o geológii Myjavskej pahorkatiny na základe vrtu Lubina-1, Geol. práce, Správy 70, Bratislava; Geol. úst. D. Štúra, 35–56.

Mahel', M. (ed.) (1974). *Tectonics of the Carpathian Balkan regions*, Bratislava: Geol. ústav D. Štúra, pp. 1–453.

Matějka, A. and Roth, Z. (1956). Geologie magurského flyše v severním povodí Váhu mezi Bytčou a Trenčínem, *Rozpravy Ústř. úst. geol.*, **22**, 1–332.

Nemčok, J. (1961). Vznik a výplň depresií v magurskom flyši na východnom Slovensku, *Geol. sborník Slov. akad. vied*, **12**, 2, 175–190.

Pícha, F. (1979). Nesvačilské souvrství—nový název pro vrstvy autochtonního paleogénu na jižní Moravě, *Čas. Mineral. Geol.*, **24**, 3, 305–11.

Pícha, F., Hanzlíková, E. and Cahelová, V. (1978). Fossil submarine canyons of the Tethyan margins of the Bohemian Massif in southern Moravia, *Věst. Ústř. úst. geol.*, **53**, 5, 257–78.

Roth, Z. *et al.* (1962a). Vysvětlivky k přehedné geologické mapě ČSSR 1:200 000 M-33-XXIV-Olomouc, *Ústř. úst. geol.*, 1–226.

Roth, Z. *et al.* (1962b). Vysvětlivky k přehledné geologické mapě ČSSR 1:200 000 M-34-XIX-Ostrava, *Ústř. úst. geol.*, 1–292.

Roth, Z. in Buday, T. *et al.* (1967) *Regionálni geologie ČSSR II*, vol. 2, *Ústř. ústav geologický*, Academia Praha.

Samuel, O. (1973). Paleogeographical outline and manifestations of orogegenetic phases in the

Paleogene of the West Karpathians in Slovakia and in the adjacent part of the Hungarian Midmountains. Geol. prace, Správy 60, Geol. ústav. D. Štúra, Bratislava, pp. 55–83.

Seneš, J. *et al.* (1978). First working version of time correlation tables of the Thethys-Paratethys Neogene. IGCP, project No. 25, MS, Slovak Academy of Sciences, Bratislava.

Stráník, Z. in Buday, T. *et al.* (1967) ***Regionálni geologie ČSSR II***, vol. 2, ***Ústř. ústav geologický***, Academia Praha.

Stráník, Z. *et al.* (1964). Vysvětlující text k základní geologické mapě 1:25 000 list M-33-106-D-c (Šitbořice), MS Archív Ústř. úst. geol., Praha.

Stráník, Z. (1983). Outer Carpathians in South Moravia. In Samuel, O. and Gašpariková, V. (1983). 18th European Colloquy on Micropaleontology (excursion guide). Geol. úst. D. Štúra, pp. 167–175, Bratislava 1983.

Stráník, Z. (1985). Le faciès du Miocène inférieur de la zone de Flysch et son rapport au Miocène de l'Avant-fosse Carpathique et du bassin de Vienne en Moravie du Sud. Proceeding reports of the XIII-th Congress of KBGA, Part I, pp. 66–69, Geol. Inst. Cracow.

Stráník, Z. and Molčíková, V. (1980). Untermiozän nordöstlich von Pouzdřany, ***Věstník Ústř. ústav geologický***, **55**/2, 93–104.

Špička, V. *et al.* (1976). Kompletní geologické zhodnocení opěrného hlubinného vrtu Jarošov-1, MS Archív Ústř. ústav. geol., Praha.

Vass, D. (1976). Molasové panvy a globálno-tektonický model Karpat (Molasse basins and global tectonic model of Carpathians), Zborník referátov vedeckej konferencie: Československá geológia a globálna tektonika (Smolenice 1976), Bratislava: Edičné stredisko Geol. úst. D. Štúra, pp. 111–17.

Vass, D. (1979). Genesis of inner Molasse basins in the western Carpathians in light of leading function or mantle in the earth's crust development, in *Czechoslovak Geology and Global Tectonics*, Bratislava: Veda, vydavatel'stvo SAV, pp. 183–97.

Vass, D. *et al.* (1980). Štruktúrny vrt ŠV-8 (Dolné Semerovce, Ipel'ská pahorkatina), Reg. geol. Zip. Kuprt 14, pp. 1–106.

Global Correlation of Tectonic Movements
Edited by Yu. G. Leonov and V. E. Khain

10

The Role of Middle and Late Alpine Tectonic Phases in the Geological History of Hungary

G. CSÁSZÁR, J. HAAS, J. HALMAI, G. HÁMOR AND L. KORPÁS*

This paper has resulted from activities carried out in connection with IGCP Project No. 107. It is based on two types of maps of the late Alpine tectonic phases. The maps of the first type present a palaeogeological portrayal of the time of the maximum erosion associated with the phases; the maps of the second type depict the deposits of the transgression which followed each phase. The most important geological events of the middle and late Alpine tectonic phases are shown in Fig. 1.

AUSTRIAN PHASE

The map of the Austrian phase in Hungary (Fig. 2) shows three major tectonic units: the central Alpine unit; the Hungarian Highland unit; the southern Hungarian unit. The central Alpine unit is composed of the Unter- and Ober-Ostalpin nappe systems, with the Penninicum. The southeastern boundary of the unit is determined by the Raba-Roznava line which is interpreted as a dislocation.

The Hungarian Highland unit is separated by the Balaton-Darno line into two zones. Consisting of epicontinental Mesozoic strata the Transdanubian Central Mts are characterized by southern Alpine features in the southwest part (Biancone facies) and by the northern Alpine features in the northeast part (the Rossfeld Beds). It looks as if the two strongly different facies were deposited close to each other in two separate sedimentary basins of the same tectonic unit. During this phase, the unit, prior to being strongly denuded, was block-faulted, slightly folded and imbricated.

The map of the cover rocks (Fig. 3) shows an Early–Middle-Albian transgression of southwesterly direction. In the pelagic Igal-Bükk zone of the folded structure, possibly affected by nappe tectonics, a pelagic shaly siltstone formation of uncertain age occurs between the Liassic and Upper Cretaceous. The zone and the unit itself are bounded in the south by the central Hungarian dislocation.

Characterized by the intra-Carpathian and Apuseni connections, the southern Hungarian unit includes three zones: Mecsek; Villány; and Békés-Codru. The folded-imbricated structure of the unit can in the main be attributed to the effects of the Austrian phase. Areas of continuous sedimentation, however, are likely to have existed in both the Mecsek and the

* Magyar Állami Földtani Intézet, H 1442 Budapest, P.f. 106, Hungary.

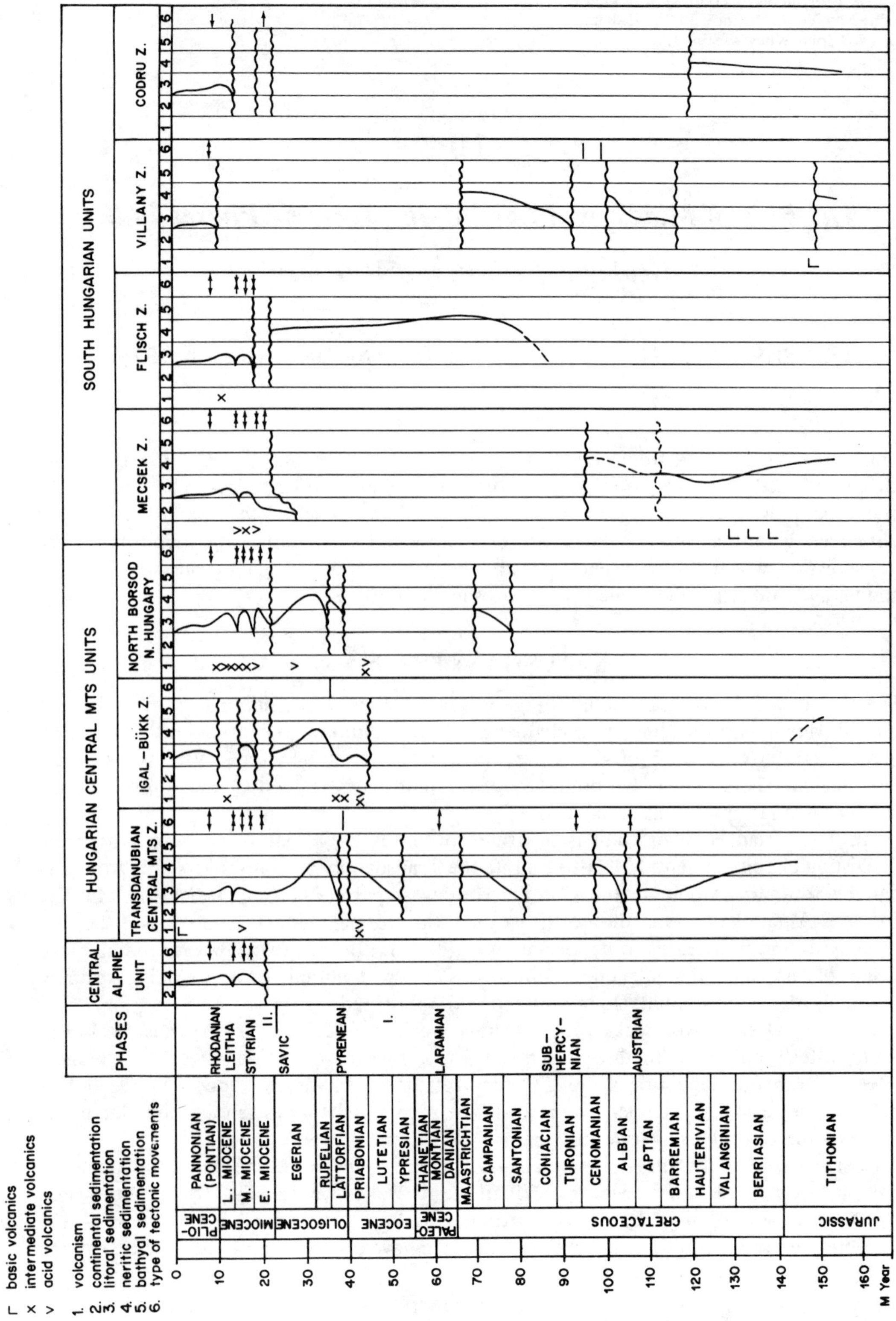

FIG. 1 Middle and late Alpine tectonic phases in Hungary.

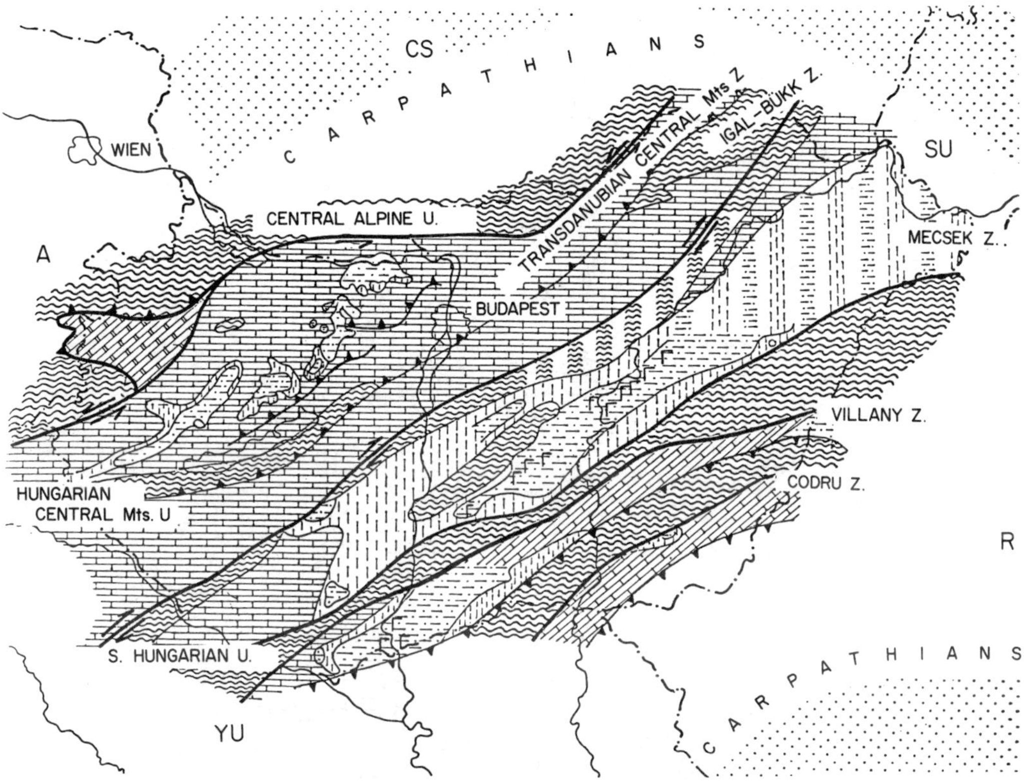

FIG. 2 Austrian phase. I. Early Albian surface.

1. Formations
   - Middle Miocene
   - Lower Miocene
   - Oligocene
   - Eocene
   - Upper Cretaceous
   - Lower Cretaceous + Aptian
   - Jurassic–Lower + Middle Cretaceous
   - Jurassic–Cretaceous
   - Jurassic
   - Triassic
   - Mesozoic
   - Pre-Mesozoic
     - supposed extension
2. Tectonic elements
   - Horizontal movement
   - Reverse fault
   - Non-differentiated
   - Inactive
3. Metamorphism
   - Epimetamorphism
4. Magmatism
   - Intrusive centre
   - Effusive centre
   - Basic volcanics
   - Intermediate volcanics

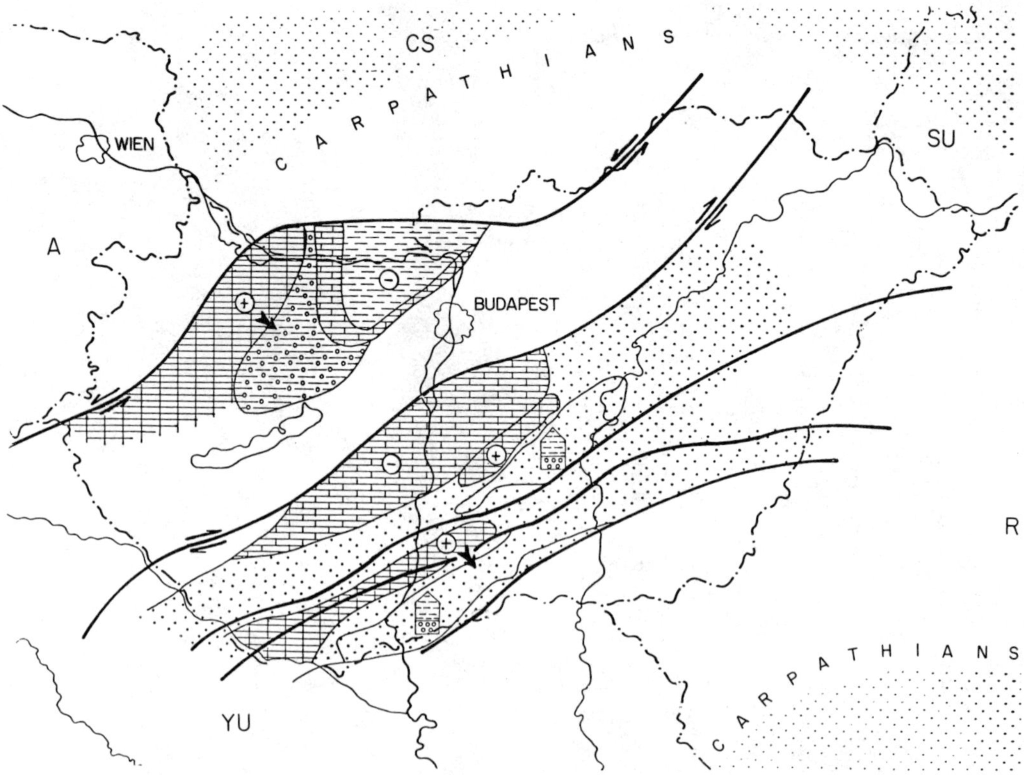

FIG. 3 Austrian phase. II. Middle Albanian transgression.

1. Geofacies
   - Source area
   - Areas of accumulations
     - Terrestrial (continental)
     - Epicontinental detrital
     - Epicontinental calcareous
     - Pelagic
     - Flysch and flysch-like sedimentation
       - supposed extension
   - Transport directions
2. Tectonic elements
   - (a) Horizontal movement
     - Normal fault
     - Non-differentiated
   - (b) Continuous sedimentation without any change in lithology
     - Continuous sedimentations with changes in lithology
3. Magmatism
   - Intrusive centre
   - Effusive centre
   - Intermediate volcanics
   - Silicic volcanics
4. Intensive uplift
   - Intensive subsidence

Villány zones, where predominantly calcareous and marly sediments are succeeded by sandstones and marls. An Early Cretaceous alkaline basalt volcanism is conspicuous in both zones. Fig. 3 suggests an overall transgression in the Mecsek and Villány zones. Because of the later denudation it is impossible to assess the Békés-Codru zone over its Hungarian area.

The compressional nature of the Austrian phase is also evident in Hungary, where the two plates seem to have collided.

SUB-HERCYNIAN PHASE

The sub-Hercynian phase starts in both the Hungarian Highland range and the southern Hungarian units with block-faulting and regional emergence at the end of the Cenomanian. The result of the denudation is visible on the pre-Senonian map (Fig. 4). The new regional subsidence began in the Early Senonian and reached its maximum in Late Senonian time (Fig. 5).

The transgression, which produced the epicontinental (terrestrial-marine) sediments with thick platform carbonate formation in the Transdanubian Central Mts zone, arrived from the southwest. In the northeastern Igal-Bükk zone, after a large break in sedimentation, the deposition of coarse detrital rhythmic sediments of more or less Gosau type set in. Within the southern Hungarian unit the Mecsek and Villány zones in the Danube-Tisza interfluve

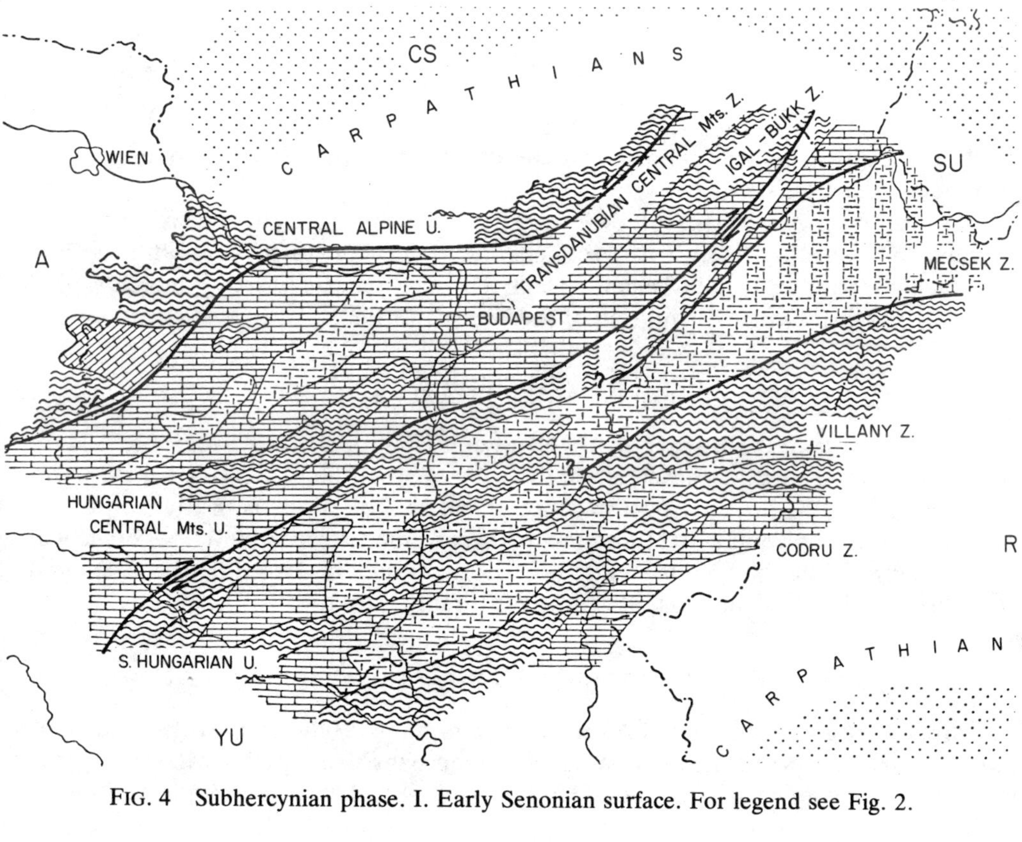

FIG. 4 Subhercynian phase. I. Early Senonian surface. For legend see Fig. 2.

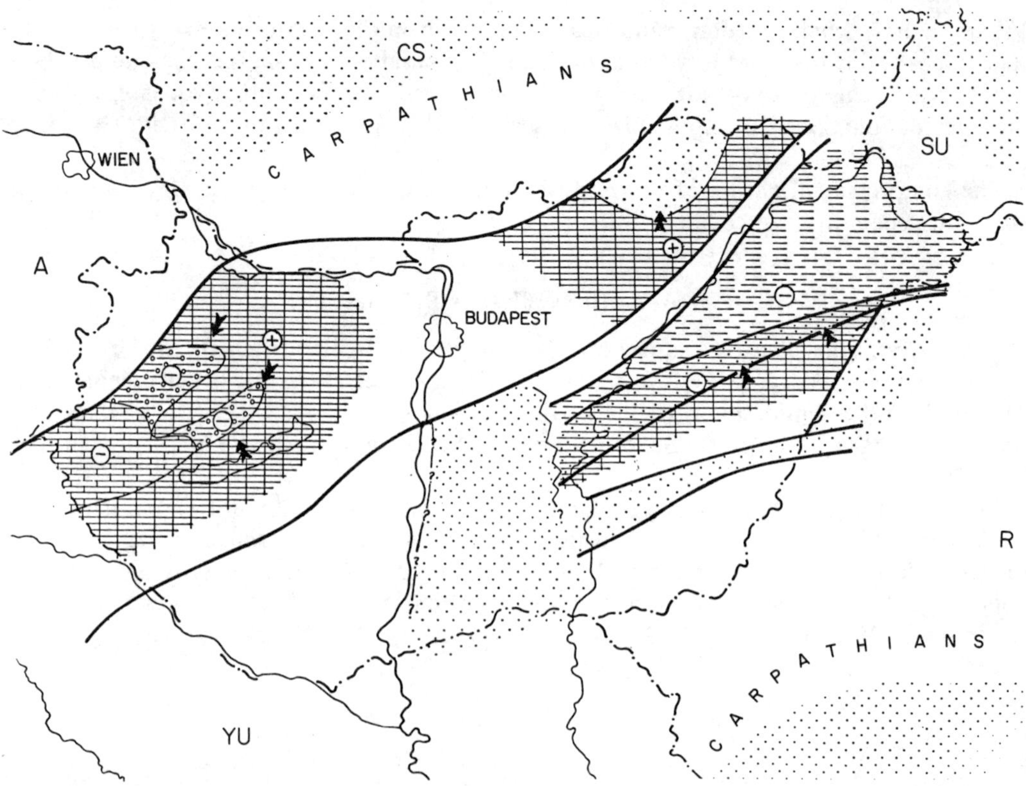

FIG. 5 Subhercynian phase. II. Early Campanian transgression. For legend see Fig. 3.

lost their separate character—possibly an early effect of the Vardar system. Here the epicontinental detrital sediments overlie crystalline rocks.

In the northeast part of the Mecsek zone, on the one hand, basinal red beds were deposited; on the other hand, flysch sedimentation showing connections with the intra-Carpathian flysch of Romania began. The flysch may indicate extreme mobility of the contacting margins during the suturing of the two major units.

LARAMID PHASE

The Laramid phase in the Hungarian Highland unit is introduced by a regression. During emergence, between the Maestrichtian and the Sparnacian, block-faulting was associated with slight tilting (Fig. 6). A new subsidence began at the end of the Early Eocene. The culmination of the epicontinental transgression akin to the Cretaceous cycle was reached at the beginning of the Priabonian (Fig. 7). In the southern Hungarian unit only the continuation of flysch sedimentation can be recorded.

PYRENEAN PHASE

The Pyrenean phase in the central Alpine part of northern Hungary provoked the start of epicontinental molasse sedimentation. In the Transdanubian Central Mts unit regional emergence was combined with block-faulting (Fig. 8). The new subsidence was also

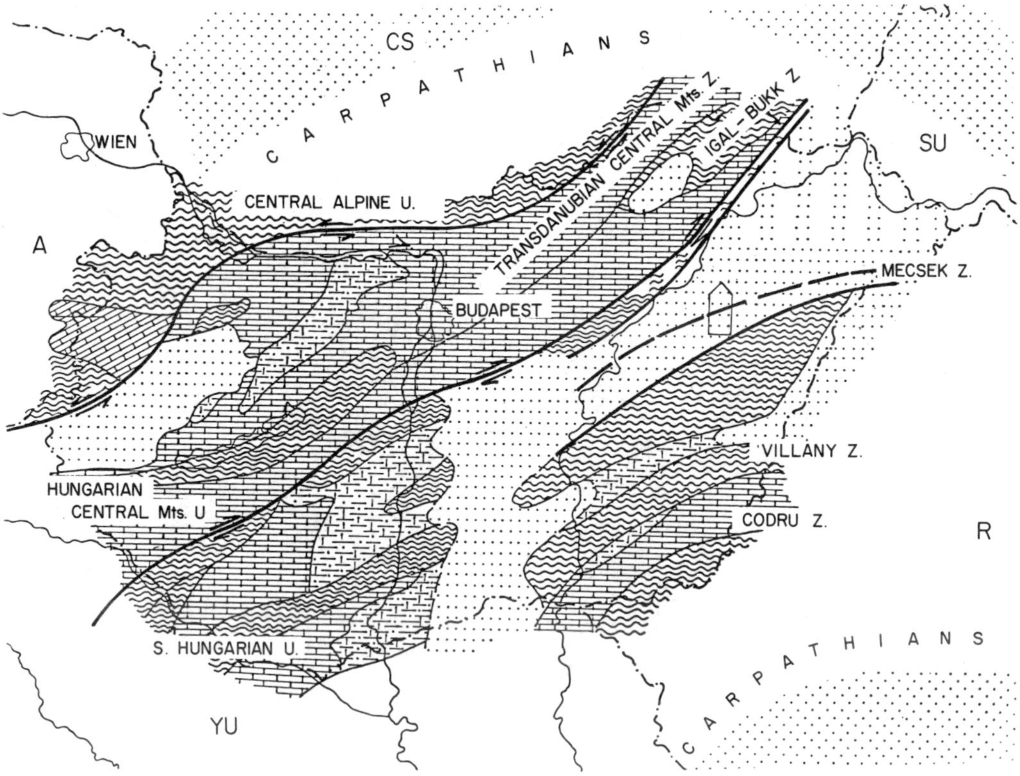

FIG. 6 Laramid phase. I. Pre-Eocene surface. For legend see Fig. 2.

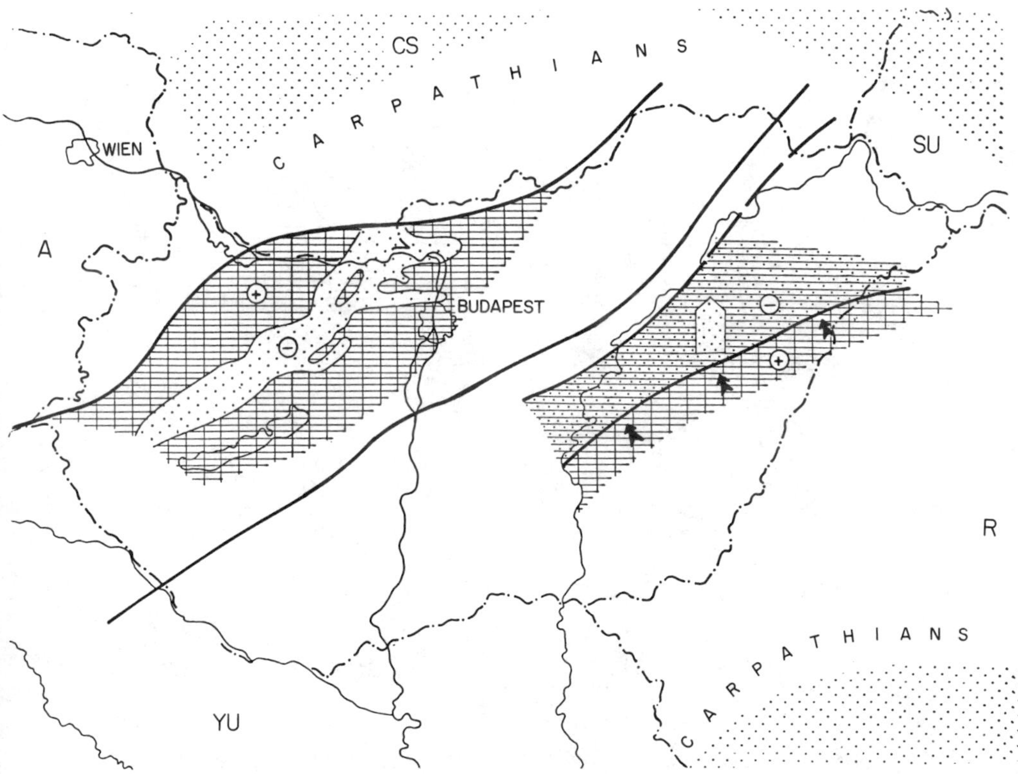

FIG. 7 Laramid phase. II. Middle-Eocene transgression. For legend see Fig. 3.

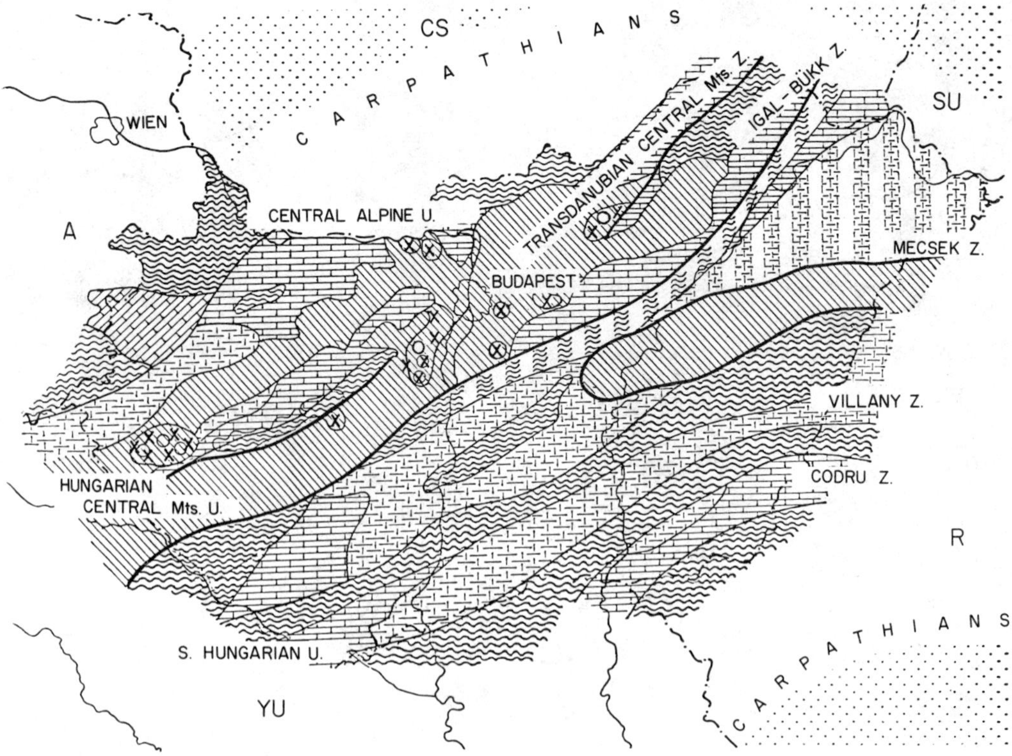

FIG. 8 Pyrenean phase. I. Pre-Oligocene surface. For legend see Fig. 2.

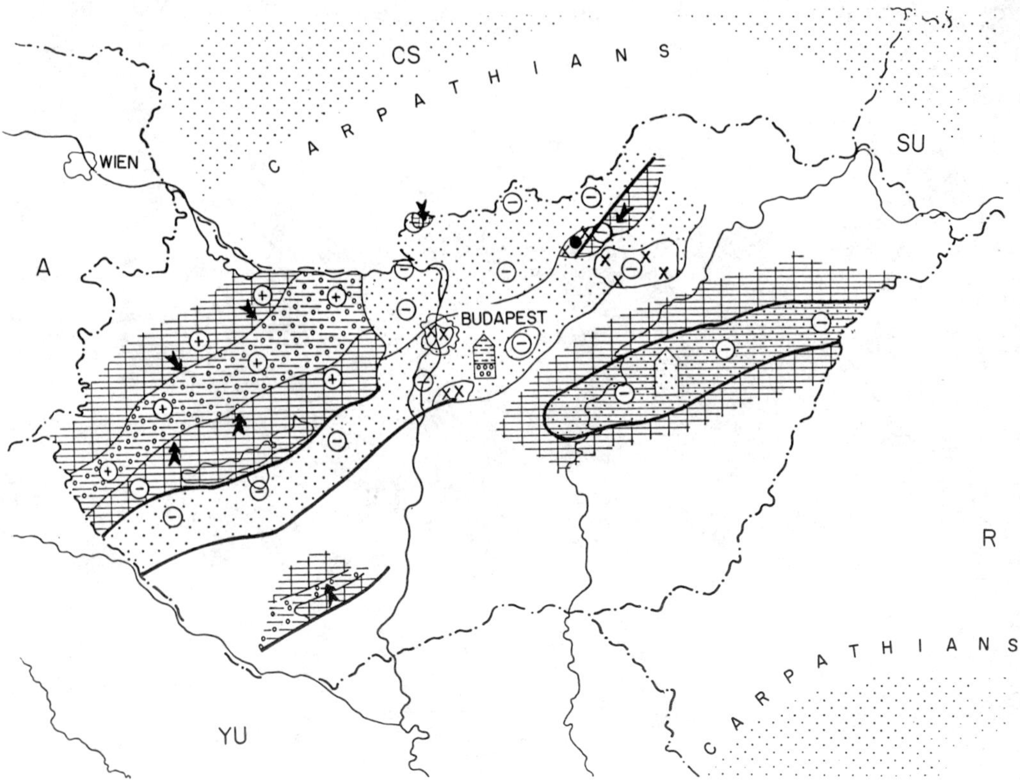

FIG. 9 Pyrenean phase. II. Rupelian transgression. For legend see Fig. 3.

manifested here by the molasse sedimentation in Early–Middle Oligocene times (Fig. 9). In the Igal-Bükk zone terrigenous carbonate sediments pass laterally with a sharp change into molasse deposits. A volcanic zone of andesitic type was formed at the contact of the two zones. In the southern Hungarian unit flysch sedimentation continued.

SAVIC PHASE

The Savic phase was the time when the Alpine-Carpathian-Dinaric system began to emerge. In the compressive subphase the folding processes progressed and were succeeded by thrust sheet movements (Fig. 10). The remnants of the Tethys were converted into the Paratethys. The Carpathian basin lost its direct communication with the Mediterranean Sea. In the southwest part of it continental sediments were found, while marine sediments were deposited in the northeast foredeep. This marine incursion issued from the western basin of the Mediterranean Sea and, skirting the Alps, it converged on the sedimentary basins of the Hungarian territory (Fig. 11).

During the tensile subphase of the orogeny the deep fractures widened into open fissures which served as channels for the ascent of silicic volcanics, at the Eggenburgian–Ottnangian boundary. The Savic phase played a decisive role in the replacement of the structural belts, heretofore mainly longitudinal, stretching from northeast to southwest, by a partly transverse northwest to southeast trend. At this time the Vardar zone became an emergent ridge.

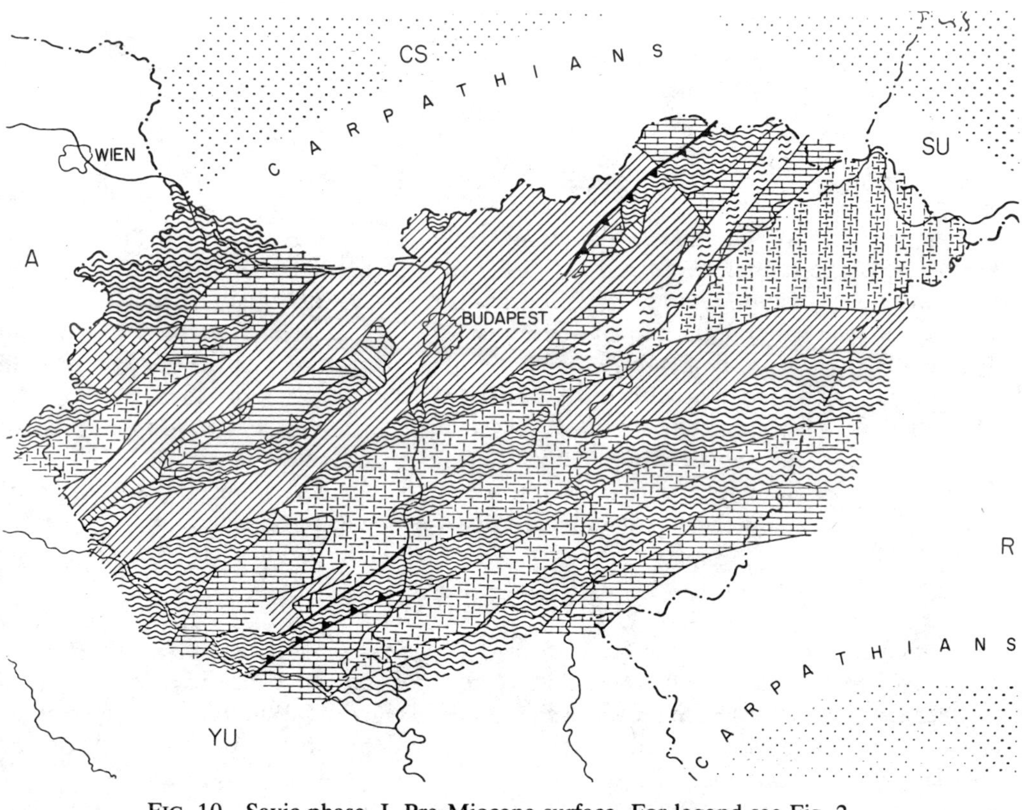

FIG. 10 Savic phase. I. Pre-Miocene surface. For legend see Fig. 2.

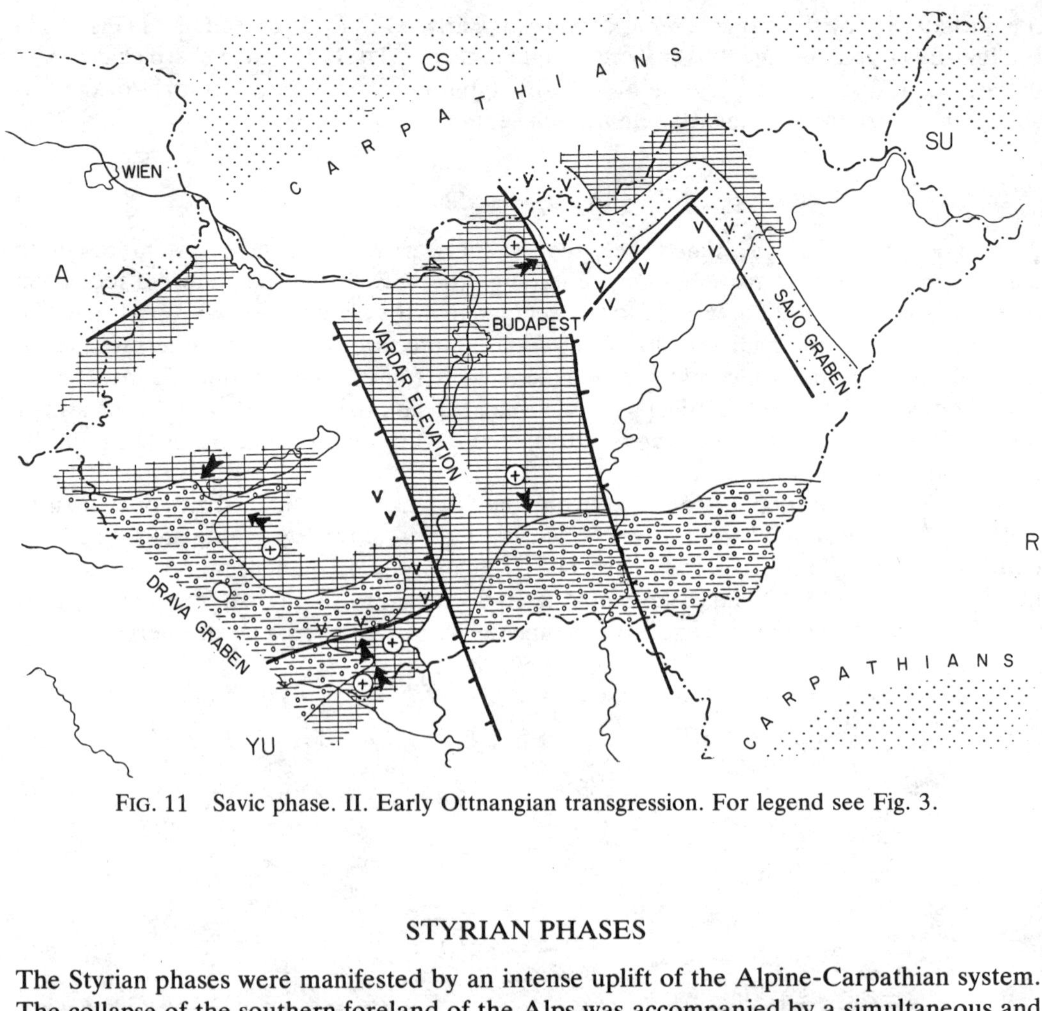

FIG. 11 Savic phase. II. Early Ottnangian transgression. For legend see Fig. 3.

STYRIAN PHASES

The Styrian phases were manifested by an intense uplift of the Alpine-Carpathian system. The collapse of the southern foreland of the Alps was accompanied by a simultaneous and drastic subsidence of the Dinarides and of areas within the Carpathian region (Fig. 12).

Two large sedimentary grabens of southwest to northeast trend were formed within Hungarian territory (Fig. 13). Being an offshoot of the Adriatic basin the northern graben had a marine communication with the Vienna basin across southern Slovakia, while the southern graben was connected with the Transylvanian basin. Northeastward transgressions took place in Karpatian and Early Badenian times.

In the intra-Carpathians, the uplift of terrestrial areas and vigorous andesitic-rhyolitic volcanism brought about the so-called intra-Carpathian volcanic chain.

LEITHAIAN PHASE

The Leithaian phase represented the decisive stage in the formation of the Alpine-Carpathian-Dinarid system; a phase during which mountain-size nappe masses were thrust one upon another and dislocated towards the foredeep (Fig. 14). With the emergence of the Carpathian arc the only communication with open seas was restricted to the Negotinian Gate via the Lower Danube.

Once again, transverse fractures of northwest to southeast direction define the structural

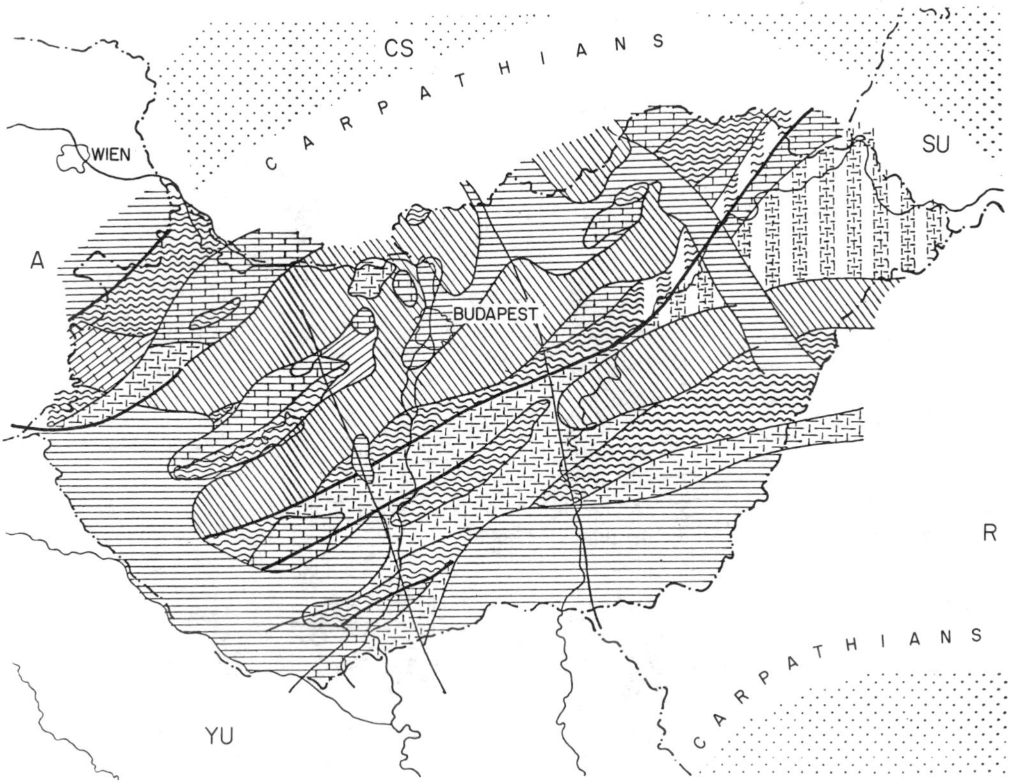

FIG. 12 Styrian phase. I. Pre-Karpatian surface. For legend see Fig. 2.

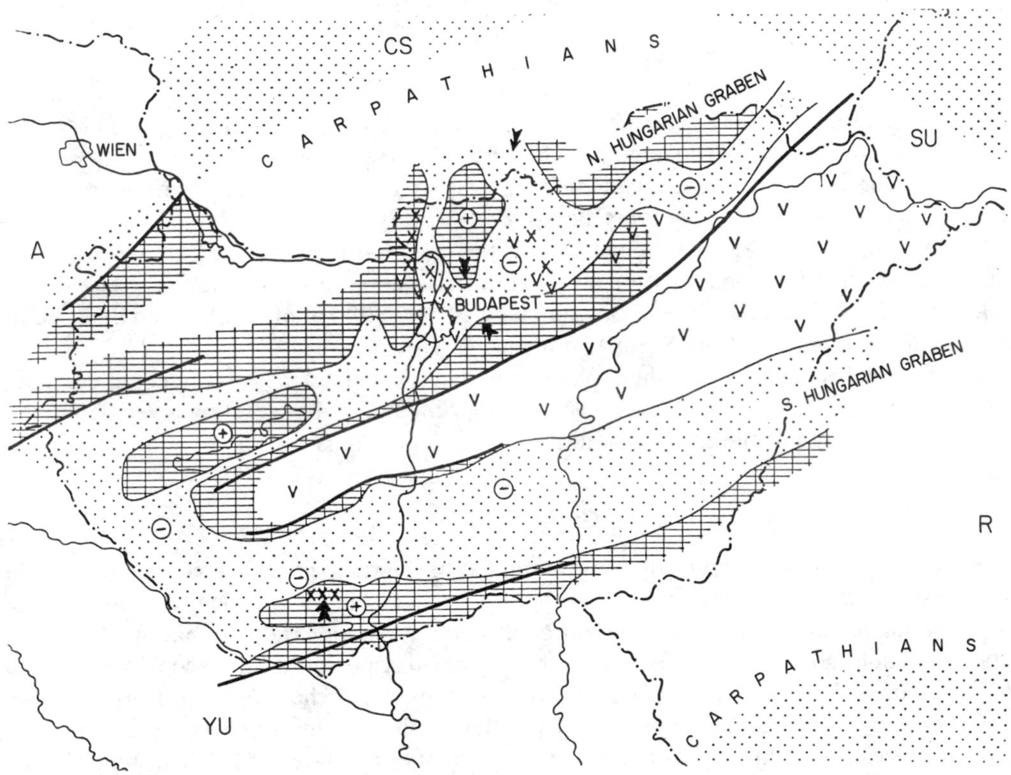

FIG. 13 Styrian phase. II. Karpatian transgression. For legend see Fig. 3.

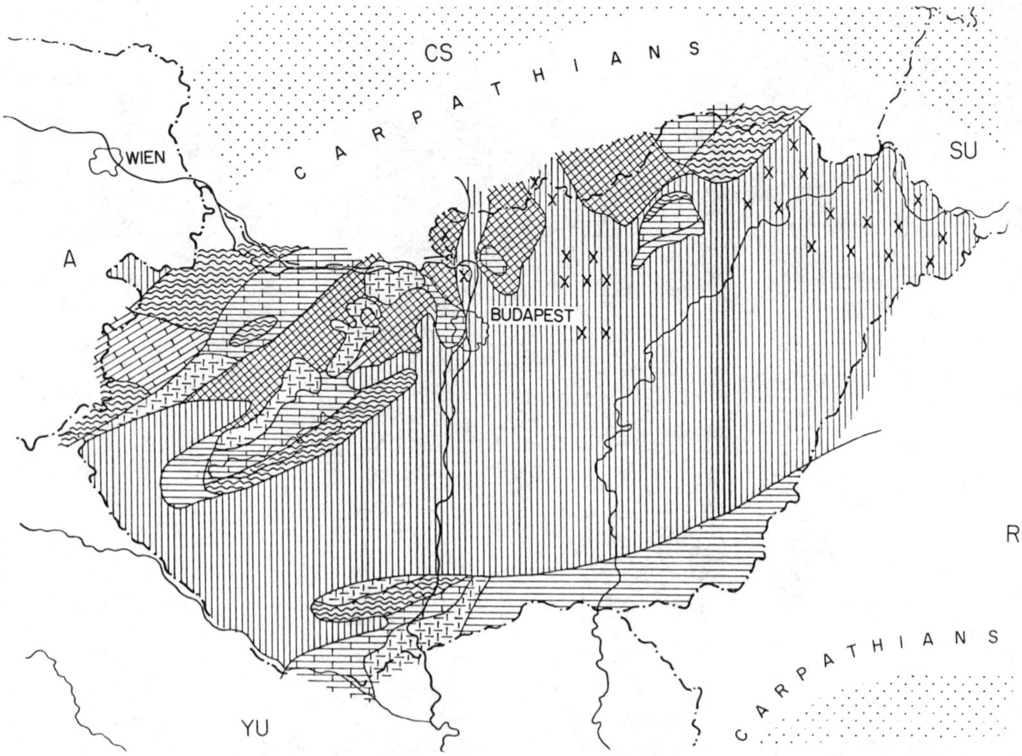

FIG. 14 Leithaian phase. I. Early Badenian surface. For legend see Fig. 2.

pattern of the region: the Drava graben, the Vardar graben and the Szatmár graben became more and more distinctly delineated and the Little Hungarian Plain (depression) came into existence. The grabens became recipients of pelagic sediments, while on the ridges between them epicontinental carbonates were deposited (Fig. 15).

In the extensional subphase the collapse of the deep basins was accelerated, so that the Sarmatian–Lower Pannonian sediments became up to 2000–3000 m thick. The last strong volcanic activity also took place in this subphase (2000–3000 m of rhyolite and andesite in the Tokaj Mountains and the Szatmár graben).

With the Rhodanian phase the isolation of the Carpathian basin was completed. Its marine communication was totally lost and detrital materials were removed from the emergent margins and the inselbergs to fill up the depressions.

CONCLUSION

The tectogenic history of Hungary representing a considerable part of the Carpathian basin includes two stages (Fig. 1). The first is characterized by the independent development of three major tectonic units; the different geological features of their formation suggesting that they could not have been formed in the present-day position. The second stage began in the period between the Pyrenean and Savic phases, when the three units under consideration occupied their present-day position. This was the time when the transverse fractures became predominant or at least as important as the longitudinal. This resulted in a

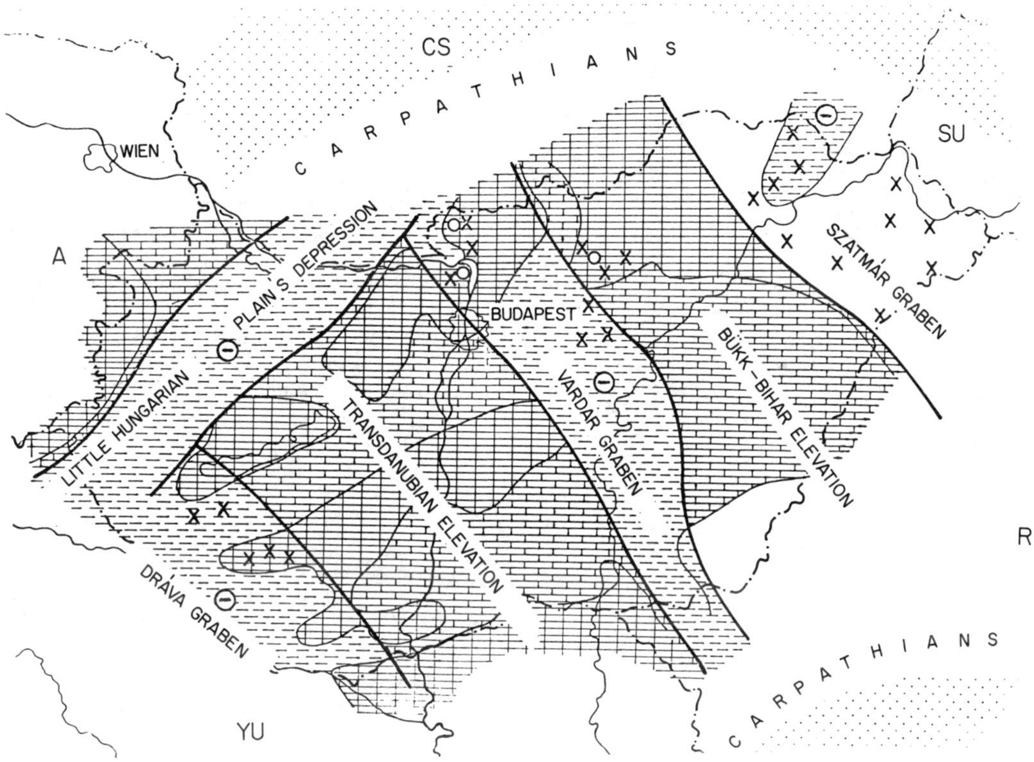

FIG. 15 Leithaian phase. II. Late Badenian transgression. For legend see Fig. 3.

chequer board fault block pattern; that is, a structure in which very deep intramontane basins had formed among the inselbergs. Accordingly, the Hungarian basin as such consists of a series of graben-like structures superimposed on various plate fragments.

REFERENCES

Balázs, E., Báldi, T., Dudich, E., Gidai, L., Korpás, L., Radócz, Gy., Szentgyörgyi, K. and Zelenka, T. (1980). Structural and facies study on the Eocene–Oligocene boundary formations in Hungary. *Discussiones Palaeontologicae*, **25**, 13–46. (In Hungarian with English summary.)

Bendefy, L. (1968). Contribution to knowledge on the internal structure of the Pannonian Massif. *Földrajzi Közlöny*, **16** (92) 4, 289–313. (In Hungarian with English summary.)

Bossellini, A. and Hsü, K. J. (1973). Mediterranean plate tectonics and Triassic palaeogeography, *Nature*, **244**, 5412, 144–6.

Császár, G. and Haas, J. (1979). Review of the facies and palaeogeography of the Cretaceous in Hungary, *Aspekte der Kreide Europas. IUGS*, ser. A, **6**, 413–24.

Géczy, B. (1973). Plate tectonics and palaeogeography in the East Mediterranean Mesozoics, *Acta Geol. Acad. Sci. Hung*., **17**, 421–8.

Hámor, G. (1978). Die Orogenphasen des Badenian, in *Chronostratigraphie und Neostratotypen. Miozän M4 Badenien*, Bratislava: Veda, p. 109.

Horváth, F. and Channell, I. E. T. (1977). Further evidence relevant to the African/Adriatic promontory as a palaeogeographic premise for Alpine orogeny, in *Structural History of the Mediterranean Basins*, Symposium in Split 1976, pp. 133–42.

Horváth, F. and Stegena, L. (1977). The Pannonian basin: a Mediterranean interarc basin, in *Structural History of the Mediterranean Basins*, Symposium in Split 1976, pp. 333–40.

Körössy, L. (1970). Entwicklungsgeschichte der Neogenen Becken in Ungarn, *Acta Geol. Acad. Sci. Hung*., **14**, 2, 421–9.
Laubscher, P. H. (1971). Das Alpen-Dinariden Problem und Palinspastik der Südlichen Tethys, *Geol. Rundschau*, **60**, 3, 813–33.
Laubscher, H. and Bernoulli, D. (1977). Mediterranean and Tethys, in *Structural History of the Mediterranean Basins*, Symposium in Split 1976, pp. 129–32.
Mahel, M. (ed.) (1974). *Tectonics of the Carpathians—Balkan Regions*, Bratislava: Veda.
Szádeczky-Kardoss, E. (1971). Theses on the study introducing a discussion entitled 'The Carpathian-Dinarid area from the point of view of the new global tectonics', Geonómia és Bányászat 4, 82.
Varga, I. and Grecula, P. (1980). Major tectonic dividing zones on the inner side of the western Carpathians, *Földtani Kutatás*, **23**, 3, 17–22. (In Hungarian.)
Vörös, A. (1977). Provinciality of the Mediterranean Lower Jurassic brachiopod fauna: causes and plate-tectonic implications, *Palaeogeography, Palaeocl., Palaeoec*., **21**, 1–16.
Wein, Gy. (1969). Tectonic review of the Neogene-covered areas of Hungary, *Acta Geol. Acad. Sci. Hung*., **13**, 1, 399–436.

Global Correlation of Tectonic Movements
Edited by Yu. G. Leonov and V. E. Khain

11

Main Stages in the Evolution of the Polish Carpathians During Late Palaeogene and Neogene Times

NESTOR OSZCZYPKO* AND KAZIMIERZ ŻYTKO**

INTRODUCTION

The work presented here has been prepared as a contribution to IGCP 107 Project 'Global correlation of epochs of tectogenesis'. The present paper deals with the northernmost part of the Carpathians segment of the Alpine orogenic belt. This part is divided into two structural units: internal Carpathians (Tatra Mountains and Podhale flysch) and external or flysch Carpathians including also the Pieniny klippen belt. The internal Carpathians were folded between the Cenomanian and Coniacian, and the flysch Carpathians in Miocene times (Książkiewicz, 1977). During the Late Cretaceous the flysch basin of the external Carpathians began to split into a number of smaller basins, which later gave rise to the development of main tectonic units. From south to north these are (Fig. 3A): the Magura unit (its internal part is represented by the Pieniny klippen belt), the pre-Magura zone (Dukla-Grybów unit and the pre-Magura scale), the Silesian unit, the sub-Silesian unit, the Skole unit and the Boryslav-Pokucie unit. The last, unrecorded in Poland, is fairly well recognized in the Ukrainian Carpathians. During the Neogene time a foredeep filled with molasse deposits was formed mainly on the platformal basement, north of the Carpathians. The molasse of the internal part of the foredeep now constitutes the Stebnik-Sambor unit.

Besides the intra-Cretaceous orogenic movements the important tectonic events occurred at the boundary between the Cretaceous and the Palaeogene in the Pieniny klippen belt (Birkenmajer, 1970; Książkiewicz, 1977) and in the Krynica zone forming the internal part of the Magura unit (Żytko, 1977). In this report the authors have concentrated on the effects of the later movements. These effects are shown within the present structural framework of the Carpathians with no palinspastic restoration. The maps presented in Figs 1–3 reveal regional tendencies in the orogenic evolution of different parts of the geosyncline and its foreland during the earliest Oligocene (post-Pyrenean time—Fig. 1), the Early Miocene (post-Savian time—Fig. 2), the Badenian–Early Sarmatian (post-Styrian time—Fig. 3). The subdivision of the Neogene follows regional stages of central Paratethys (Senès, 1979), but the general subdivisions of the Early Miocene are also used.

* Institute of Geological Sciences, Jiagiellonian University, Oleandry 2a, 30-063 Kraków, Poland.
** Geological Institute, Carpathian Branch, Skrzatowl, 31-560 Kraków, Poland.

THE END OF THE EOCENE AND BEGINNING OF THE OLIGOCENE (PYRENEAN MOVEMENTS)

After the mid-Cretaceous orogenic movements the internal Carpathians were subjected to intensive subaerial erosion, whereas in the external Carpathians the uninterrupted sedimentation mainly of the flysch facies took place. Already in the Early Senonian, transgression covered parts of the internal Carpathians (western Slovakia), and the evidence from the Tatra Mountains and Podhale indicates that transgression commenced there in the Early Lutetian (Książkiewicz, 1977). During Lutetian–Priabonian times shallow-water sedimentation took place and conglomerates and Nummulitic limestones, succeeded by mudstones, were deposited. The Upper Eocene sediments of the external Carpathians became more diverse. In the Magura basin, thick sequences of sandy flysch were deposited, while the sediments of more outer basins are dominated by thin green claystones.

In the regions under consideration, the Pyrenean movements were marked mainly by great changes in the type of argillaceous sediments which, to a greater or lesser degree were affected by tectonic deformations. In the internal Carpathians, these movements initiated a rapid subsidence and sedimentation of the Podhale flysch which is now considered as a correlative of the Krosno Formation. In the external Carpathians the orogenic activity brought about a change in sediment type and the unification of pelagic deposition, resulting in the formation of the *Globigerina* Marl and the overlying dark bituminous Menilite Shale.

In the Magura basin, the Pyrenean movements continued to cause the sedimentation of sandy flysch. In the southeastern part of this basin, where lower Oligocene sediments are now preserved, tectonic deformation resulted in an angular unconformity (Książkiewicz and Leško, 1959; Oszczypko, 1973). All this orogenic activity caused unified conditions of sedimentation in the Magura basin (Krynica, Bystrica and Rača zones). In eastern Slovakia the boundaries between these zones are unconformably overlain by deposits of the latest Eocene (Variegated Shales, *Globigerina* Marl) and Early Oligocene age (Malcov Formation—an equivalent of the Krosno Formation —Leško and Samuel, 1968). In the western part of the Polish Carpathians, in the vicinity of Wielka Racza, in the Magura unit (Rača zone) there is a known relict of the variegated shales of the latest Eocene age (Żytko, 1962). It is therefore possible that at the Eocene–Oligocene boundary the Magura unit was already formed and partly advanced northwards (Fig. 1).

In the pre-Magura zone, the Pyrenean movements were reflected in the sedimentation of thick, sandy flysch complexes (Mszanka Sandstone, Sub-Cergowa Marl and Cergowa-Krosno Sandstone). These complexes also originated in the internal eastern part of the Silesian basin (Iwonicz Fold, Bukowica Fold), indicating the same facies in the latter area as in the pre-Magura zone. The great contrast in the facies development of the Upper Cretaceous–Eocene deposits between the Dukla (pre-Magura) and Silesian areas, the disappearance of variegated shales and marl facies of the Late Cretaceous age known from the pre-Magura scale (the Soła River catchment-area in the west) and from olistholiths in the Silesian unit (Ślączka, 1963) all suggest that the Pyrenean activity there resulted in a tectonic shortening between these two sedimentary basins. Additional evidence comes from the fact that in eastern Slovakia, the Dukla unit is pushed far forward on to thick, dark sediments of Late Eocene—Early Oligocene age as revealed by the borehole Zboj no. 1 (Cieszkowski *et al.* 1981). It is noteworthy that during the latest Eocene and Early Oligocene times the Dukla basin was affected by strong synsedimentary movements, being a prerequisite for the formation of initial folds (Ślączka, 1969; Żytko, 1977).

Similarly, in the Silesian, sub-Silesian and Skole basins, beginning with the Oligocene there originated bituminous, muddy sediments containing a horizon of cherts. Later on in the pre-Magura zone flysch deposits appeared (Krosno Formation). The Krosno Formation began to form at first in the southern part and gradually developed towards the outer part of the flysch trough. The large areas of the outer part of the Skole basin were covered from the latest Oligocene onwards by the Krosno-type flysch. Through much of the Oligocene time these areas were dominated by slow sedimentation of bituminous clays. It is worth mentioning that during the latest Eocene in the outer part of the Skole basin olistostromes appeared (Popiele Formation—Dżułyński and Kotlarczyk, 1965) and during the earliest Oligocene in the Skole basin a rapid reversal in palaeocurrents from west to eastwards took place (Żytko, 1961; Sikora, 1971).

From the above pattern the authors have distinguished the following features of the Early Oligocene palaeogeography (Fig. 1): areas of rapid subsidence and high sediment accretion rate (B2) characterized by the zones of synsedimentary folds (B5), areas in which the Pyrenean movements caused only facies changes (B3) and the Magura basin which is characterized by a regional decline in the sediment accretion rate (B1). There is a possibility that the Pyrenean orogenic activity was responsible for the beginning of thrust movements in the Magura and Dukla units (B9).

EARLY MIOCENE OROGENIC MOVEMENTS* AND THEIR CONSEQUENCES

The paroxysmal phase of the Pyrenean movements was in Oligocene times succeeded by flysch sedimentation. It disappeared gradually, starting in the south (Podhale flysch, Malcov Formation in the Magura region) in response to the northward migration of the depocentre marked by sandy flysch and overlying marly, post-flysch sediments, which terminated flysch sedimentation in all Carpathian units (Ślączka, 1969). In the outer parts of the Silesian unit and in the Skole unit flysch sedimentation was still in existence until the Early Miocene. This is indicated by the U/Zr radiometric age, $20.5 \pm 0.9 - 15.6 \pm 1.1$ m.y., obtained for tuffite intercalations in the upper part of the Krosno Formation (Van Couvering *et al.*, 1981). Furthermore the Early Miocene age of upper levels of the Krosno Formation is provided by the micropalaeontological data (Nowak, 1979). Flysch succession in the Polish Carpathians is terminated by this formation.

Thus the emergence in the northern part of the basin occurred in Early Miocene times. The sedimentation persisted in the outermost zone—that is, in the Boryslav-Pokucie area. The Polanica Formation, corresponding in age and lithofacies to upper parts of the Krosno Formation (Pishvanova and Gruzman, 1980; Gruzman, 1981) is here followed upwards by the Vorotyshche Formation containing evaporites and a thick Sloboda Conglomerate complex. The clasts of the conglomerate consist mainly of rocks derived from the platform, although the clasts from the Palaeogene flysch are also significant. The consequence of the younger than previously assumed age of the Krosno and Polanica Formations is that the Vorotyshche Formation should be in general, related, to the Burdigalian instead of the Aquitanian as was traditionally maintained. It seems, therefore, that during Early Miocene times, at the end of sedimentation of the Krosno and Polanica Formations, a longitudinal fault divided the basin giving rise to the later structural separation between the Skole and Boryslav-Pokucie units.

The Vorotyshche Formation also occurs in the Stebnik-Sambor unit, forming its oldest known segment. It is possible that at the end of the Vorotyshche time a successive,

*Tectonic activity in the flysch Carpathians, here described was previously related at the boundary between the Oligocene and Miocene and often called Savian phase. The latter term has become ambiguous because recent discoveries showed that the effects of this orogenic activity are younger than previously assumed. Therefore, the term Savian phase is not used here.

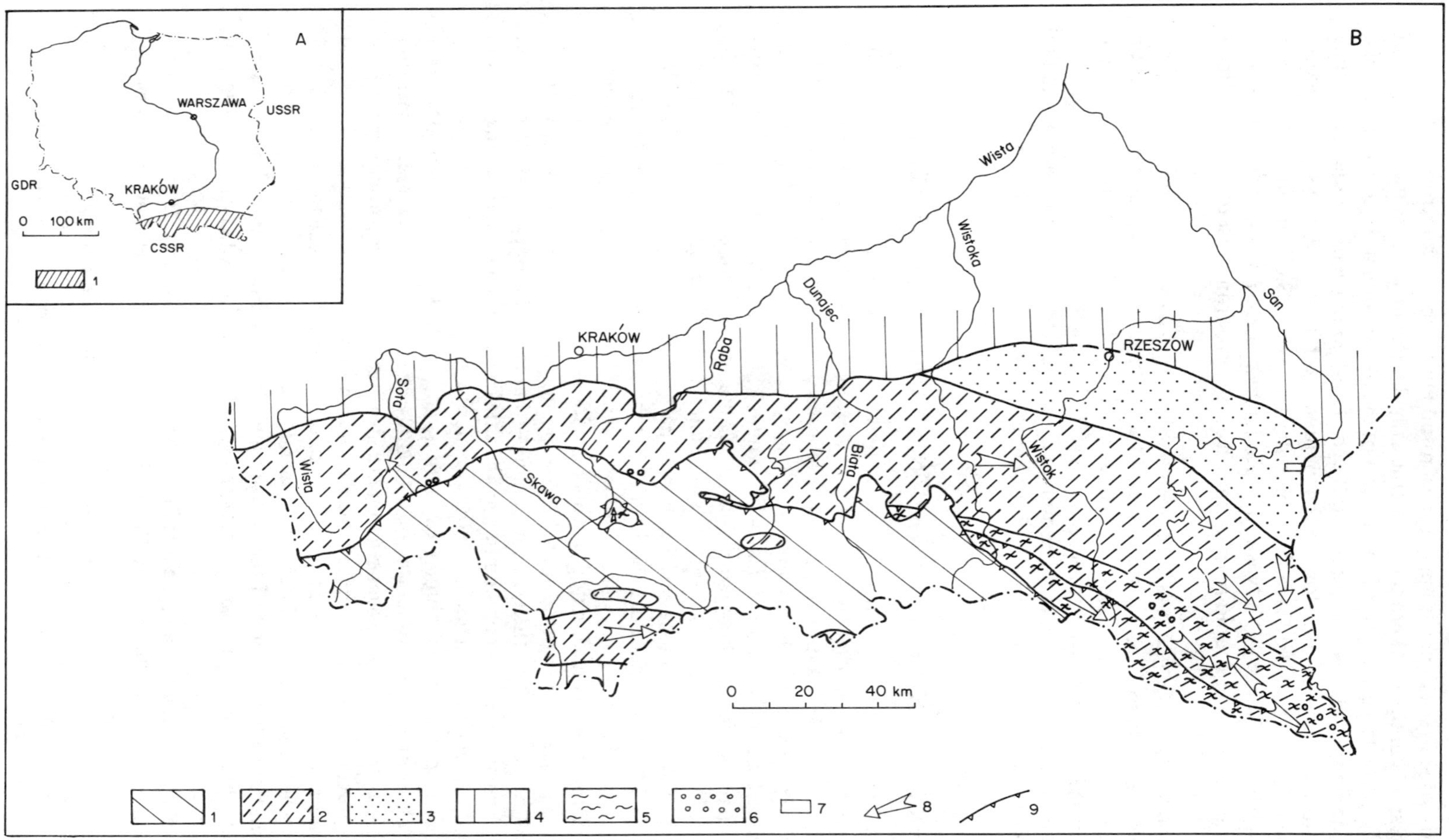

FIG. 1
A Localization map: 1 – Polish Carpathians.
B Map of tectonic activity at the beginning of the Oligocene: 1 – the end of sedimentation (?) or shallowing of the sedimentary basin; 2 – lateral spreading of sedimentation (Podhale) and rapid subsidence; 3 – continuous sedimentation with changes in lithology; 4 – areas without information; 5 – synsedimentary folding; 6 – blocks; 7 – olistoliths in latest Eocene; 8 – direction of detritic material transport; 9 – front of initial nappes.

paroxysmal stage of the Early Miocene movements resulted in the formation of the successive, more outer, longitudinal fracture which broke the molasse basin into two. This terminated the deposition in the Boryslav-Pokucie area, whereas in the Stebnik-Sambor region the molasse sedimentation remained uninterrupted. In the latter region the Stebnik and Baliche Formations were formed, which contain intercalations of evaporites in their lower parts. It is possible that molasse deposits of the Stebnik-Sambor unit were laid down both on the geosynclinal substratum (above the flysch) and on the platform of the foreland. The autochthonous molasse deposits of the Early Miocene age have not been recognized in the foredeep of the northern part of the East Carpathians.

Contrary to the above, in the East Alps and the West Carpathians the Lower Miocene (pre-Badenian) molasse is known to rest on the platform basement in the foredeep (Roth, 1978). The data presented by Brix *et al.* (1977) and by Cicha *et al.*, (1967) indicate that, east of Vienna, younger and younger Lower Miocene deposits are found to the north. This fact can be considered as a continuity of the orogenic migration of the flysch geosyncline in the foredeep.

The pre-Badenian, autochthonous molasse deposits possibly corresponding to the Stebnik and Baliche Formations and resting on the platform basement, have also been encountered in the Polish outer part of the West Carpathians in the deep Sucha IG 1 borehole (Fig. 3), in Sucha and Stryszawa Formations (Ślącz a, 1977) and in the adjacent Czechoslovakian part of the Carpathians (Jurkova, 1976; Roth, 1978). The lower portions of these molasse deposits (Sucha Formation) contain a great quantity of the flysch material (blocks and olistoliths) derived from the sub-Silesian and Silesian units (Palaeocene red and dark shales). This constitutes evidence for a much younger, Early Miocene age of thrusting of the flysch units from their area of deposition. It also indicates that during this time the Silesian unit was already thrust over the sub-Silesian, and both were pushed forward, far on to the platform.

The above data indicate the existence of at least two paroxysmal stages of Early Miocene orogenic movements in the marginal part of the East Carpathians. The first stage derived from the Baliche Formation, at the base of the Carpathian overthrust, between Raba and the Dunajec river (Połtowicz, 1972).

The above data indicate the existence of at least two paroxysmal stages of Early Miocene orogenic movements in the marginal part of the eastern Carpathians. The first stage (pre-Vorotyshche) brought about the emergence and folding of the Skole-Skiba unit, whereas the second stage (post-Vorotyshche) climaxed in overthrusting of these units on to the Boryslav-Pokucie area (folds), closing the molasse sedimentation in that area. In the West Carpathians these thrust movements caused the northwards displacement of previously formed flysch units (nappes), prior to the sedimentation of the Sucha Formation.

It is necessary to add that in the Krosno and Polanica Formations clastic material already appears in places (olistoliths) derived from the Carpathian flysch and its littoral facies. This could indicate an earlier emergence of southern areas within the Carpathian flysch geosyncline, and it corroborates the fact of the northward migration of orogenic compression (synsedimentary folds) towards the platform. This process, which was initiated in the Palaeocene, occurred continuously until Late Miocene times. It began in the Krynica zone of the Magura unit and was continued up to the outermost part of the flysch basin (Żytko, 1977).

During Early Miocene times weak andesitic volcanism took place in the Pieniny klippen belt and in adjacent parts of the Magura unit (Birkenmajer and Nairn, 1969).

The map shown in Fig. 2 is thought to reflect the main results of the orogenic processes above described which took place in the Early (pre-Badenian) Miocene. The map shows: an

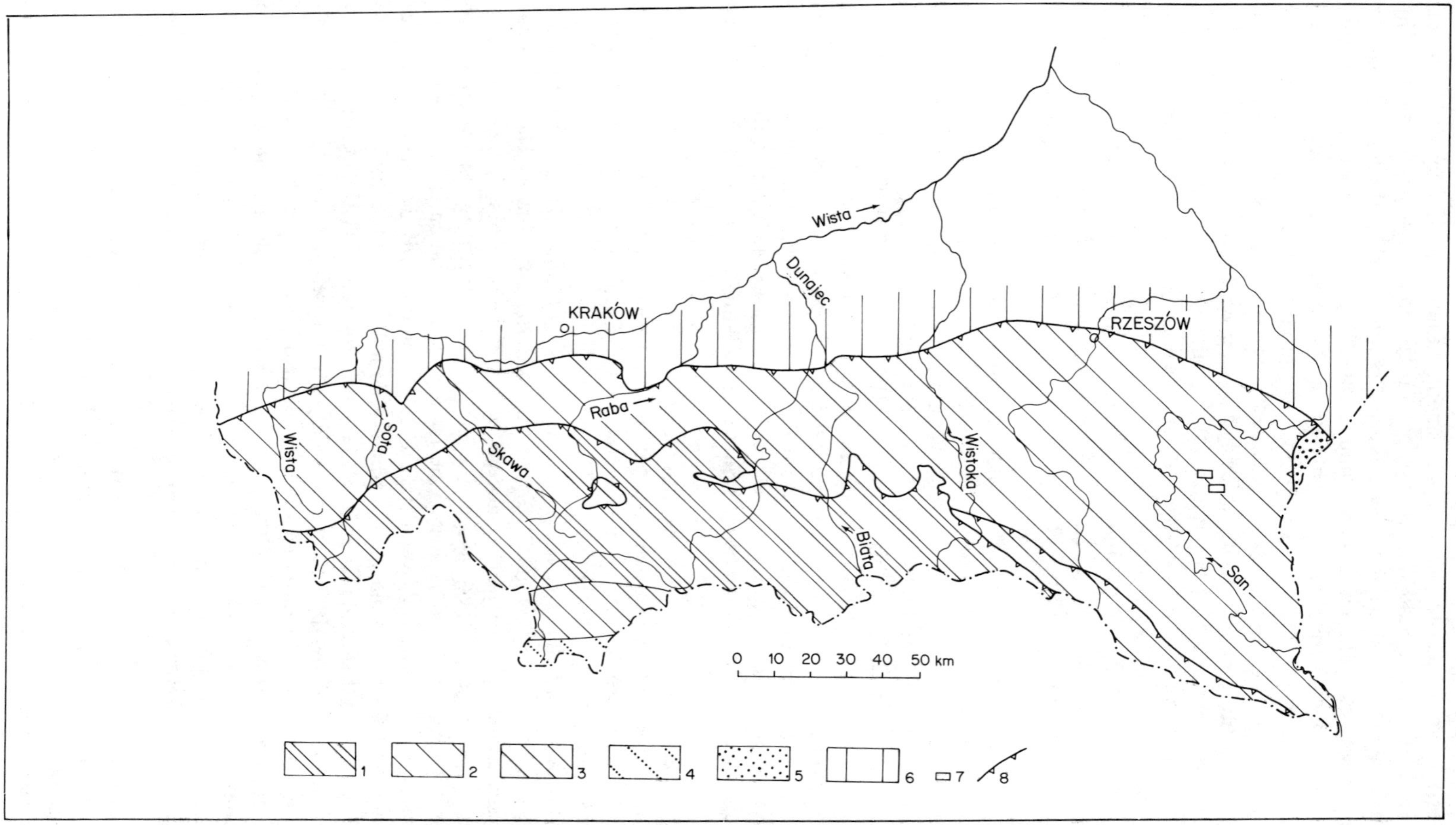

FIG. 2 Map of tectonic activity in the Early Miocene.
1 – continuation of alpinotype folding and thrusting movements; 2 – the end of sedimentation, Alpinotype folding and initial thrusting; 3 – Alpinotype folding without overthrusts; 4 – uplifted areas; 5 – continuation of sedimentation with change of flysch into the molasse sediments; 6 – areas without information; 7 – olistoliths (in final stage of sedimentation); 8 – front of main overthrusts (partly regenerated).

area of tectonic uplift (Tatra Mountains); a region of weak folding (Podhale); and an area subjected to continued folding and thrusting (Magura and Dukla units). In the latter region sedimentation came to an end in Early Oligocene times. Further to the north, Early Miocene movements resulted in the breaking off of sedimentation and folding of the flysch (Silesian, sub-Silesian and Skole units). The area shows possible facies changes from flysch to molasse (Stebnik unit). The Boryslav-Pokucie unit, if present in Poland, lies under overthrust flysch masses. The Early Miocene tectonic activity climaxed in the formation of the Silesian, sub-Silesian and Skole overthrusts.

BADENIAN AND EARLY SARMATIAN OROGENIC MOVEMENTS AND THEIR CONSEQUENCES

As briefly mentioned above, in the Early Miocene the Carpathians were folded and raised above sea level. Marine sedimentation took place only in a narrow zone of the foreland. During Karpatian times this marine basin expanded northwards, involving larger platform areas and forming the shallow foredeep of the West Carpathians (Jurkova, 1976; Ślączka, 1977; Roth, 1978). The Stryszawa Formation of Karpatian age (Ślączka, 1977) yields evidence of changing environmental conditions from brackish to fully marine. This formation can probably be correlated with the Baliche Formation known from the Stebnik-Sambor unit in the east (Ney, 1968; Gluszko and Kruglov, 1971; Połtowicz, 1972). In the Karpatian time, alluvial and paludal deposits with lignites began to accumulate in the lower parts of the Carpathians (Oszczypko, 1973).

During Early Badenian time marine sedimentation existed in the foredeep. In the Stebnik-Sambor basin, occupying part of the foredeep, the Baliche Formation was followed by molasse sediments of the Przemyśl Formation (Ney, 1968). At that time the marine area enlarged considerably, gradually involving the greater parts of the platform (Alexandrowicz, 1965; Ney *et al.*, 1974). In the substratum of the outer West Carpathians, the Lower Badenian molasse, that begins in places with the Dębowiec Conglomerate, rests on the Stryszawa Formation. There is an erosional gap at the base of Badenian deposits, as has been recorded in the Sucha IG 1 borehole (Ślączka, 1977). The deposits are overthrust by flysch. Farther west, in northeastern Moravia, the Early Badenian basin assumed a more northern position than the Karpatian one. They were separated by the Sławków-Cieszyn Ridge (Jurkova, 1976). The clayey-sandy sediments of Early Badenian age (Skawina Formation), however, most often rest directly either on the platform basement or on the eroded flysch (Alexandrowicz, 1962; Oszczypko, 1973). Their maximum thickness of 500–700 m is attained west of the Raba river, under the overthrust of the Carpathians and immediately north of their edge.

The middle segment of the Badenian succession is represented by anhydrites, gypsum and locally rocksalt (Rybnik and Zory in Upper Silesia and the Wieliczka-Pilzno area). The space distribution of the rocksalt indicates a shift of the axial zone of the foredeep towards the northeast. This shift can be related to the northward movement of the Carpathian nappes and to the partial withdrawal of the Early Badenian sea from the folded flysch. The width of the basin thus distinctly diminished. In the Late Badenian, the axial zone of the molasse basin shifted further towards the northeast. The maximum thickness of the Upper Badenian deposits (up to 1000 m) occurs at the present margin of the Carpathians between the cities of Tarnów and Rzeszów. In the east, within the Stebnik-Sambor unit Upper Badenian sediments are represented by a coarse clastic molasse—the Radycz Conglomerate, composed mainly of flysch-derived gravels (Ney, 1968). In Early Sarmatian times the shift of the depositional zone continued. The zone of maximum subsidence, characterized

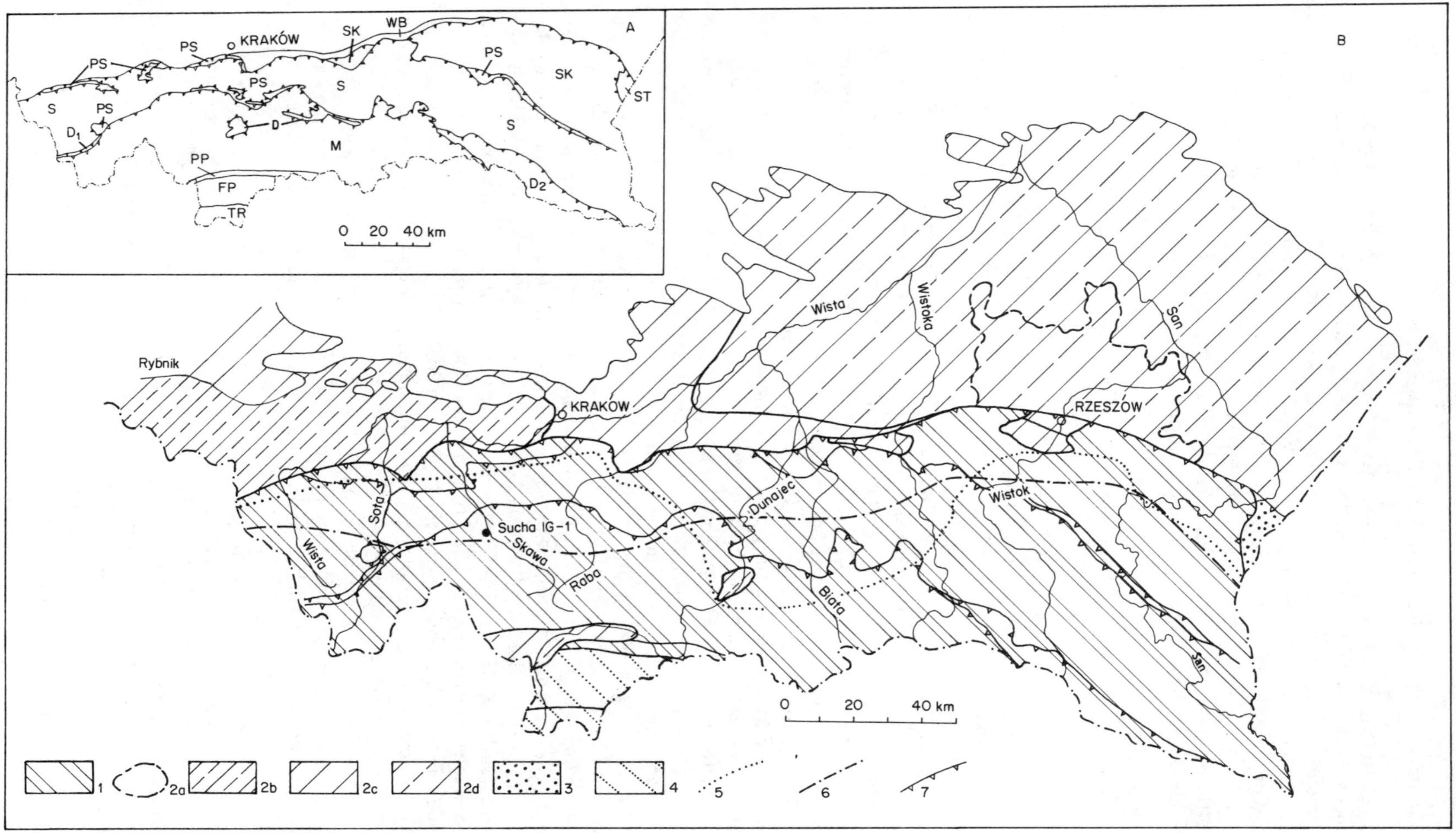
A
B
PS
KRAKÓW
SK
WB
SK
ST
S
M
D
D1
D2
PP
FP
TR
0 20 40 km
Rybnik
KRAKÓW
RZESZÓW
Wisła
Wisłoka
San
Wisłok
Dunajec
Biała
Raba
Skawa
Soła
Sucha IG-1
1
2a
2b
2c
2d
3
4
5
6
7

by a sediment infill of up to 2500 m, occurred outside the present edge of the overthrust flysch masses. During Early Sarmatian time areas of littoral accumulation expanded northeastwards (Ney *et al*., 1974). In the south, in the marginal part of the Skole unit, in places Badenian and Lower Sarmatian deposits occur, lying within Miocene erosional remnants above the flysch strata.

Upper Badenian and Lower Sarmatian sediments of the foredeep are mainly represented by shales and mudstones with subordinate sandstone intercalations and, in places, sandstone complexes.

During the Sarmatian continental coarse clastics also accumulated in some places. They are now known to occur in the Orawa-Nowy Targ Valley, in Podhale (Birkenmajer, 1978). The sedimentation here persisted until the Pleistocene (Fig. 3).

The Badenian sedimentation was accompanied by a very weak volcanic activity manifested by rare and thin tuffite layers. The U/Zr radiometric age of the tuffites from the Upper Badenian deposits ranges between 11.9–12.0 ± 0.8 and 14.1 ± 0.9 m.y. (Van Couvering *et al.*, 1981).

It therefore follows that the main features in the orogenic development of the Carpathians from Karpatian until Early Sarmatian times consisted of the encroachment of the molasse basin on the platform and a simultaneous diminishment of sedimentary areas in the south, due to the overthrusting of flysch units (Ney, 1968; Oszczypko and Ślączka, 1979).

During Early Badenian and Karpatian times a westsouthwest–eastnortheast trending foredeep existed on the platformal basement of the Polish western Carpathians. It reached Moravia in the westsouthwest, while its eastern prolongation was the Stebnik basin displaying northwest-southeast orientation.

In Middle Badenian times, the axial zone of the foredeep in the western Carpathians was marked by a linear belt of the rocksalt deposits between the vicinity of Rybnik (Upper Silesia) and Tarnów. This more or less, westnorthwest-eastsoutheast axial zone was dominant until the Latest Badenian. In this region between Racibórz-Opawa (Upper Silesia) and Tarnów this zone is represented by thick Late Badenian deposits. During the Early Sarmatian, the axial zone acquired a northwest-southeast orientation, the trend characteristic for the East Carpathians. The western extremity of the Sarmatian foredeep followed the present course of Wisła river between the mouths of the Dunajec and San rivers (Fig. 3). This right lateral rotation of the axial zone of the foredeep is related to an eastward shift of folding and thrusting processes along the margin of the orogen.

The western part of the Polish Carpathians is thrust over the Lower Badenian deposits, the middle part is pushed forwards on to the Middle and Upper Badenian, while the eastern part occurs directly either on the Lower Sarmatian or on folded molasse of the Stebnik-Sambor unit (Wdowiarz, 1976). As was revealed by boreholes, the magnitude of the

---

FIG. 3

A Tectonic sketch—map of the Polish Carpathians: TR – Tatra units; FP – Podhale flysch; PP – Pieniny klippen belt; M – Magura nappe; D – pre-Magura zone (pre-Magura scale – D_1; Dukla unit – D_2); S – Silesian nappe; PS – sub-Silesian nappe; SK – Skole nappe; ST – Stebnik-Sambor nappe; WB – Badenian folds.

B Map of tectonic activity in Badenian–Early Sarmatian time: 1 – weak folding and overthrusts, main tectonic transport; 2 – spreading of molasse sedimentation in Early Badenian (Rzeszów area—in Late Badenian – 2a) and beginning of regional subsidence in Early Badenian (2b), in the Middle and Late Badenian (2c), in Early Sarmatian (2d); 3 – continuous molasse sedimentation with changes in lithology and thickness; 4 – uplifted areas; 5 – supposed southern margin of Early Badenian transgression on the Carpathian flysch; 6 – supposed southern margin of Badenian deposits lying under the Carpathian flysch; 7 – overthrust.

overthrust is at least 30–40 km (Fig. 3). Manifestations of folding involved gradually younger and younger sediments, from the vicinity of Kraków up to the city of Tarnów (Książkiewicz, 1977). In this area the cores of anticlines consist of the flysch mantled by Lower and Middle Badenian deposits. Both eastward and westward of this area, folds die out, passing gradually into a narrow zone of Miocene schuppen which is wedged up between the autochthonous Miocene and the Carpathian overthrust. The deposits of folded Miocene do not occur everywhere under the Carpathian overthrust but are preserved as detached blocks between the autochthonous Miocene and the flysch.

The thrust movement of the flysch rocks was probably directed towards axial zones of the foredeep. During Karpatian and Early Badenian times thrusting took place towards the northwest, during the Middle and Late Badenian it was towards the north, and in the Early Sarmatian towards the northeast. It seems that these thrust processes occurred in a continuous manner since there is no evidence to suggest periods of increased tectonic activity. During Badenian times, synsedimentary block movements affected the substratum of the foredeep. They terminated in the Early Sarmatian (Połtowicz and Starczewska-Popow, 1974). Some of the manifestations of the Badenian–Sarmatian orogenic activity described above, are schematically shown in Figure 3.

CONCLUSIONS

During the late evolutionary stage of the Carpathian geosyncline, in the Palaeogene and Neogene, the northern Carpathians were affected by continuous orogenic processes, as already considered by Tokarski (1980). The migration of depocentres of at first flysch sediments and later molasse deposits was directed towards the margin of the geosyncline. It began in the Palaeocene and persisted until the Early Sarmation time. The forward shift of compression, accompanied by a parallel shift of synsedimentary folds, took place continuously from the Palaeocene until the Early Miocene, involving increasingly more outer parts of the flysch zone. The sedimentation gradually came to an end, beginning in the south. The initial thrust process was not contemporaneous; it commenced first in the south and during the Oligocene and the Early Miocene it progressively involved more northerly areas. Thrusting can probably be related to the continuous push exerted by the African on the Eurasian plate (Dewey *et al*., 1973).

In the course of these continuous processes, two distinct stages of increased orogenic activity occurred. First at the Eocene–Oligocene boundary the Pyrenean movements resulted in the regional replacement of green shales by brown bituminous pelites, as well as in local angular unconformities within the uninterrupted succession of the Magura area. They were also manifested by vertical movements of the sea bottom, indicated by the change in the pattern of palaeocurrents. A geanticlinal zone, formerly present between the Dukla and Silesian areas, was engulfed by the orogen. The Magura and Dukla overthrusts began to form in the southern area of the flysch geosyncline.

Second, during the Early Miocene rapid folding and emergence took place in the remaining, outer, part of the flysch Carpathians (Silesian, sub-Silesian and Skole areas). The sedimentation continued only in a narrow, marginal zone of the geosyncline (Boryslav-Pokucie and Stebnik-Sambor areas) though there was a change in facies from flysch to molasse. The Silesian, sub-Silesian and Skole overthrusts started to develop. The latter overthrust terminated the deposition in the Boryslav-Pokucie area. A relict basin still existed in the Stebnik-Sambor area, probably being linked with the molasse foredeep of the West Carpathians.

After the Early Miocene paroxysmal movements, the continuous orogenic processes occurred again. They resulted in the migration and clockwise rotation of the axial zone of the foredeep of the Polish Carpathians. During the Karpatian time, a shallow, west-southwest trending foredeep existed in the foreland of the West Carpathians. At the end of the Early Badenian a wide area of the platform was flooded by a shallow sea. During the Middle and Late Badenian the axial zone of the foredeep changed its orientation to a west-northwest trend. During the Early Sarmatian it assumed a northwest-southeast trend and the western extremity of the foredeep approached the present course of the Wisła river.

The marine regression from the foredeep of the northern Carpathians could possibly have been due to the change of the direction of compression between the African and Eurasian plates, some 9 m.y. ago (Dewey *et al*., 1973).

REFERENCES

Alexandrowicz, S. W. (1962). Microfauna and stratigraphic position of Miocene clays from Iwkowa and Żegocina (Western Carpathians), *Bull. Acad. Pol. Sci.*, Sèr. Sci. gèol. geogr., **10**, 161–6.

Alexandrowicz, S. W. (1965). L'évolution de l'avant-fosse des Carpathes et la stratigraphie du Miocène de la Pologne Méridionale, *Carp. Balk. Geol. Ass. VII Congr.*, Part II, 1, pp. 189–93.

Birkenmajer, K. (1970). Pre-Eocene fold structures in the Pieniny klippen belt, Carpathians of Poland (English summ.), *Stud. Geol. Pol.*, **31**, 1–77.

Birkenmajer, K. (1978). Neogene to early Pleistocene subsidence close to Pieniny klippen belt, Polish Carpathians, *Stud. Geomorph. Carpatho-Balcanica*, **12**, 17–28.

Birkenmajer, K. and Nairn, A. E. M. (1969). Palaeomagnetic studies of Polish rocks. III. Neogene igneous rocks of the Pieniny Mountains, *Rocz. Pol. Tow. Geol.*, **38**,4, 475–89.

Brix, F., Kröll, A. and Wessely, G. (1977). *Die Molassenzone und deren Untergrund in Niederösterreich*, Hamburg-Wien: Erdöl-Erdgas Zeit.

Cicha, J., Schmidt-Thomè, P., Janoschek, R. and Prey, S. (1967). Problems of flysch and molasse complexes, Internat. Geol. Congr. XXIIIth Session, Guide to Excursion 26 AC, Prague, pp. 3–56.

Cieszkowski, M., Durkovic, T., Jawor, E., Koràb, T. and Sikora, W. (1981). A new tectonic unit in the Polish and Slovak Carpathian flysch, Abstract of the 12th Congr. of CBGA, Bucharest, pp. 118–20.

Couvering van, I. A., Aubry, M. P., Berggren, W. A., Bujak, J. P., Naeser, C. W., Wieser, T. (1981). The Terminal Eocene Event and the Polish Connection, *Paleogeogr. Paleoclim. Paleoecol.*, **36**, 3/4, 321–362.

Dewey, J. F., Pitman, W. C. III., Ryan, W. B. F. and Bonnin, J. (1973). Plate tectonics and the evolution of the Alpine system, *Bull. Geol. Soc. America*, **84**, 3137–80.

Dżułyński, S. and Kotlarczyk, J. (1965). Sedimentation of the Popiele beds in the Polish Carpathians, *Rocz. Pol. Tow. Geol.*, **35**,1, 103–5.

Glushko, V. V. and Kruglov, S. S. (1971). Geologicheskoje strojenije i gorjuchyje iskopajemyje Ukrainskih Karpat, *Trudy UkrNIGRI*, Vyp. **XXV**, Moskva: Nedra, pp. 342 (Russian).

Gruzman, A. D. (1981). Planktonnyje foraminifery oligocena-miocena Skibovoj zony Ukrainskih Karpat, Abstracts of the 12th Congr. of CBGA, Bucharest, pp. 70–1.

Jurkova, E. (1976). The structure of the Carpathian foredeep and flysch nappes in north-eastern Moravia (English summ.), *Časopism pro mineralogii a geologie*, **21**,4, 349–62.

Książkiewicz, M. (ed.) (1977). The tectonics of the Carpathians in Pożaryski W, in *Geology of Poland*, vol. IV, *Tectonics*, Warszawa: Geolog. Inst., pp. 476–620.

Książkiewicz, M. and Leško, B. (1959). On the relation between the Krosno and Magura flysch, *Bull. Acad. Pol. Sc.*, Sèr. Sci. gèol. geogr., **7**,10, 773–80.

Leško, B. and Samuel, O. (1968). *The Geology of the Eastern Slovakian Flysch* (English summ.), Bratislava: Vydav. Slov. Akad. Vied, pp. 245.

Ney, R. (1968). The role of the 'Cracov Bolt' in the geological history of the Carpathian foredeep and in distribution of oil and gas deposits (English summ.), Pr. geol. Kom. Nauk Geol. PAN Oddz.w Krakowie, 45,82p., Kraków.

Ney, R., Burzewski, W., Bachleda, T., Górecki, W., Jakóbczak, K., and Słupczyński, K. (1974). Outline of palaeogeography and evolution of lithology and facies of Miocene layers on the

Carpathian foredeep (English summ.), Pr. geol. Kom. Nauk Geol. PAN Oddz.w. Krakowie, 82,65p., Warszawa.

Nowak, W. (1979). Niektóre jednostki lito—i biostratygraficzne i ich znaczenie dla korelacji najmłodszych utworów fliszu karpackiego (Polish) *Kwart. geol.*, **23**,4, 928–30.

Oszczypko, N. (1973). The geology of the Nowy Sącz Basin, The Middle Carpathians (Enlgish summ.), *Biul. Inst. Geol.*, **217**, 101–97.

Oszczypko N. and Ślączka, A. (1979). Relations between flysch and molasse deposits in the northern Carpathians, *Veröff. Zentralinst. Physik Erde*, **58**, 209–19.

Pishvanova, L. S. and Gruzman, A. D. (1980). Planktonic Foraminifera of Polyanitskaya Suita of the Precarpathian foredeep (English summ.), *Paleont. Sbornik*, **17**, 27–33.

Połtowicz, S. (1972). Layers of Balice to the west from Dunajec (English summ.), *Geofizyka i Geologia Naftowa,* **11–12**, 363–75.

Połtowicz, S. and Starczewska-Popow, A. (1974). The development of the Carpathian foredeep between Tarnów and Przemyśl, Polish eastern Carpathians, *Rocz. Pol. Tow. Geol.*, **48**,4, 495–516.

Roth, Z. (1978). Geology of the Moravian margin of the platform and its relation to the structure of the Carpathian Mountains *Časopis pro mineralogii a geologii*, **23**,4, 349–56.

Senès, J. (1979). Géochronologie des stratotypes des étages du Miocène inferieure et moyen de la Paratethys centrale utilisables pour la correlation globale, *Geol. zbornik*, **30**,1, 99–108.

Sikora, W. (1971). Tectonic phases in the geosyncline of the Polish outer Carpathians, Sav. Geol. Društ. Tektonska Komisja KBGA, Beograd, pp. 65–83.

Ślączka, A. (1963). Variegated Shales from Miłkowa and geology of the adjacent area, *Kwart. geol.*, **7**,2, 295–305.

Ślączka, A. (1969). Final stages of geosynclinal development in the SE part of the Polish Carpathians, *Acta Geol. Acad. Sci. Hung.*, **13**, 331–5.

Ślączka, A. (1977). Rozwój osadów miocenu z otworu wiertniczego Sucha IG-1 (Polish), *Kwart. geol.*, **21**,2, 404–5.

Tokarski, A. (1980). Dynamics of outer Carpathian Tertiary orogenesis, *Publ. Inst. Geophys. Pol. Acad. Sci.*, A-8, **130**, 129–42.

Wdowiarz, S. (1976). On the relation of the Carpathians to the Carpathian foredeep in Poland (English summ.), *Przegl. geol.*, **6**, 350–7.

Żytko, K. (1961). Occurrence of Variegated Marls in the Upper Cretaceous and Eocene of the Skole unit (Flysch Carpathians) (English summ.), *Kwart. geol.*, **5**,3, 594–601.

Żytko, K. (1962). Stratigraphy of the Magura unit in the southwestern part of the Beskid Żywiecki (Flysch Carpathians), *Bull. Acad. Sci. Pol.*, Sèr. sci. gèol. geogr., **9**,3, 167–77.

Żytko, K. (1977). Uwagi o paleogeńskich ruchach tektonicznych w Karpatach zewnętrznych (Polish), *Kwart. geol.*, **21**,4, 938–40.

Global Correlation of Tectonic Movements
Edited by Yu. G. Leonov and V. E. Khain

12

Himalayan Orogeny

B. N. RAINA U. C. PATI AND N. SRIMAL*

INTRODUCTION

This paper presents an attempt at a synthesis of the data available on the geology and structure of the Himalayas in order to build a pattern of pulses of the Himalayan orogeny which uplifted this range into the highest mountains of the world. The range is nearly 2400 km long and approximately 320 km wide, and has a complicated history of structure and tectonics. This, together with the fact that parts of it lie in Pakiston, Nepal, Bhutan and Tibet, for which details are not available, while the data available from India are insufficient, renders any attempt at a synthesis of the tectonic history very difficult. However, despite this, efforts have been made during the IGCP Project 107 to lay a foundation for such a study so that in due course, with the availability of more data, a more exact picture may emerge. The authors are only too conscious of the tentative nature of their correlations and inferences and welcome suggestions and criticism from the national and international fraternity of geological scientists engaged in the study of the Himalayas.

GEOLOGICAL SETTING

Longitudinally the Himalayas are divisible into a number of belts, with characteristic lithologies, structures and tectonics arranged from north to south (Fig. 1). These are briefly discussed here. Figs 2–8 show sections across different parts of the Himalayas.

The Tertiary Belt of Foothills

The outermost hills of this range consist of Tertiary and quaternary sediments, with simple structures and little induration. The foothills comprise the molassic sediments of the Siwalik Group, which has yielded a prolific vertebrate fauna, ranging from Middle Miocene to Pleistocene. The belt extends from the northwest eastwards up to Bhutan with remarkable continuity and uniformity in lithology.

The Lesser Himalayas

The Lesser Himalayas represent the most complicated setting of stratigraphy and structure and it is here that most controversies arise. To the south this zone is marked by the Main

* Geological Survey of India.

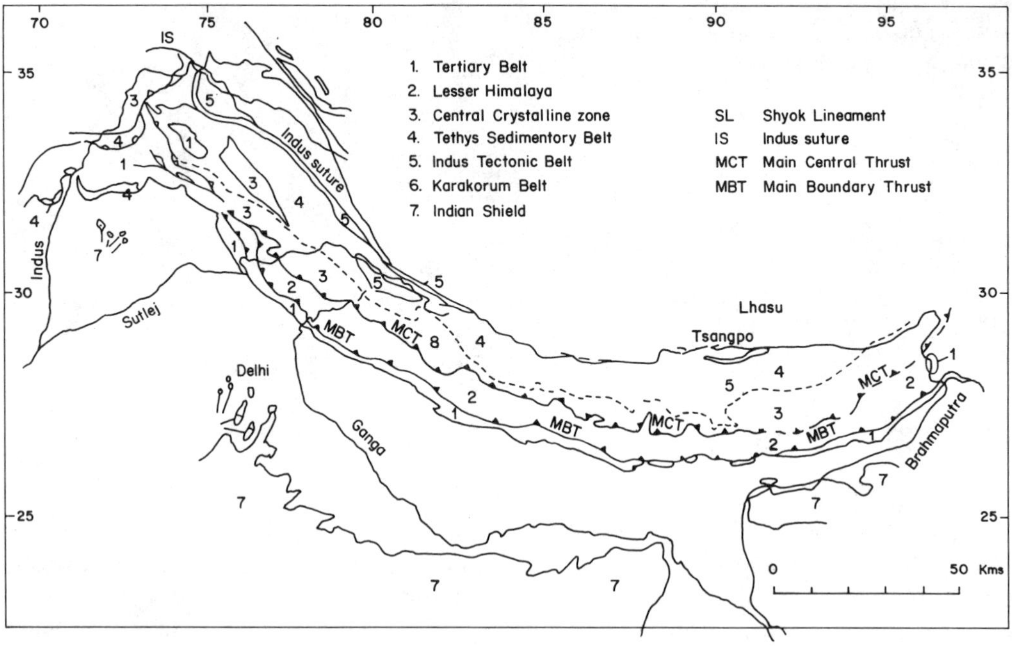

FIG. 1 Main tectonic units of Himalaya system.

Boundary Thrust and in the north by the Central Himalayan Crystalline Group, which is thrust southwards on the sedimentary rocks along the so-called Main Central Thrust but which has now been found to be a zone of two tectonic planes, which are sometimes mutually intersecting, Pati and Rao (1983). Broadly speaking, the Lesser Himalayas can be divided into two tectonic belts—one which is regarded as comprising autochthonous rocks and the other which is generally considered as allochthonous.

The autochthonous belts

These comprise the autochthonous folded belt of Kashmir and the Thanamandi-Pira belt (Raina and Krishnaswamy, 1976) which comprise volcanics, agglomerates (Late Palaeozoic) and limestones with outliers of Permian, Triassic and Eocene rocks. In Himachal the Shali and the Deoban are considered to be the autochthonous belts, while in U.P. Himalayas the Deoban-Garhwal belt is taken to be parautochthonous. The stratigraphy of this belt has been worked out in detail, but no precise clues as to its age have been discovered. Some bryozoans have been found in the limestones, a middle formation of this group, which suggest a Palaeozoic age (Raina and Dungrakoti, 1975). But these bryozoans have been found only in thin sections and their precise identification has so far not been possible.

In the eastern sector of Sikkim and Bhutan, the Buxa Group of rocks and Lower Gondwana have been identified as windows representing an autochthonous belt. In the Arunachal area of the eastern sector, the Tenga Group, the Bichom Formation and some Gondwana sediments perhaps represent the autochthonous to parautochthonous units. The almost impenetrable forest of this region makes geological work difficult. Lithologically and stratigraphically the Tenga Group of the Arunachal area seems to be identical to the Deoban-Garhwal Group of the U.P. Himalayas. However, in the Dihang valley of the Arunachal Pradesh, the Tenga Group is unconformably (?) overlain by a sequence of

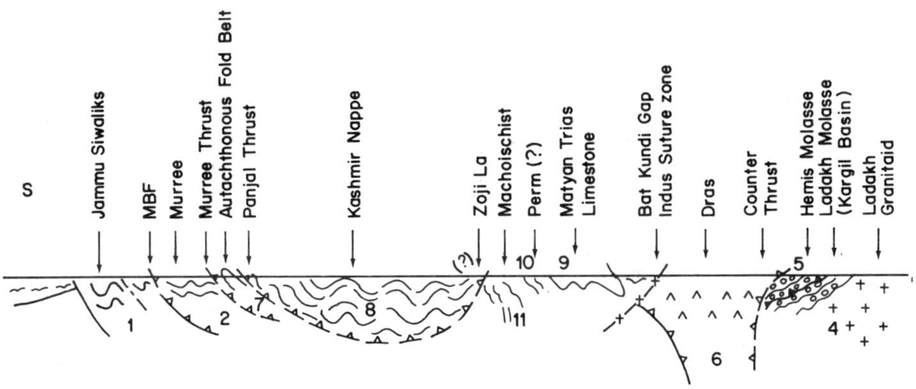

1. SIWALIK GROUP
2. MURREE GROUP
3. LADAKH MOLASSE
4. LADAKH GRANITOID
5. HEMIS MOLASSE
6. DRAS OPHIOLITE BELT
7. AUTOCHTHONOUS FOLD BELT PERMIAN - EOCENE
8. KASHMIR NAPPE PRE CAMBRIAN - JURASSIC
9. MATYAN LIMESTONE - TRIAS
10. PERMIAN BEDS
11. MACHOI SCHIST (PALAEOZOIC ?)

FIG. 2

volcanics, minor quartzites and slates and siltstones, known as the Inkiyong Group (Jain *et al.*, 1975). These slates and siltstones are lithologically identical to the Simla slates–Morar-Chakrata beds of the H.P. and U.P. Himalayas. However, the slates of the Inkiyong Formation have yielded well-preserved dicotyledonous Angiosperm plant fossils of Tertiary age.

The allochthonous belts

In Kashmir, the Kashmir nappe, representing a sequence from the Precambrian Salkhala to the Jurassic, is found thrust southwards over the Thana-Mandi-Pira belt as a synformal nappe along the Panjal-Salkhala thrust. The Palaeozoic and Mesozoic rocks in the nappe belong to the Tethyan sedimentary belt, commonly found to the north of the Great Himalayas. In Kashmir, and in the Garhwal area of U.P. they have been brought forward to more southerly areas than commonly found in the Himalayas.

In Himachal Pradesh, the Jutogh nappe, the Chail nappe and rocks of the Krol belt represent allochthonous masses which overlie the autochthonous to parautochthonous sedimentary rocks of the Lesser Himalayas. Some of these nappe units are in dispute and since all of them are completely devoid of fossils the precise stratigraphic correlation and references to their tectonic positions are necessarily controversial. In the U.P. Himalayas, the allochthonous belts of the H.P. Himalayas, the Jutogh, the Chail and the Krol continue southeastward. In addition some crystalline nappes are found such as the Almora nappe, the Bageshwar-Baijnath and Askot crystalline klippes. These crystalline allochthons are considered to be the erosional remnants of a large crystalline sheet which covered the U.P. Himalayas almost up to the Main Boundary thrust. However, the most important allochthonous unit of the U.P. Himalayas is the Bijni tectonic unit of the Garhwal area, which has brought Palaeozoic rocks of the Tethyan sequence over the Krol allochthonous belt and is overlain by the tectonic klippe of Amri crystallines. As mentioned earlier, this is the only other known occurrence of Tethyian rocks south of the great Himalayas apart from the

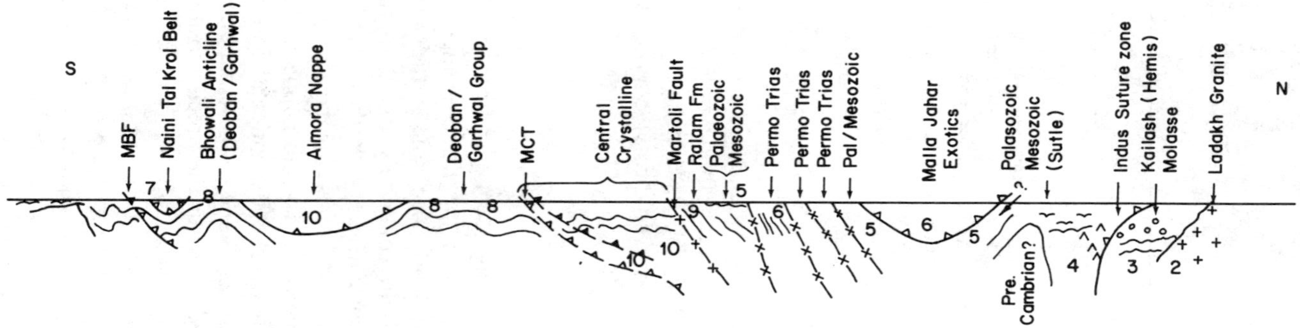

1-SIWALIK GROUP; 2-LADAKH GRANITOIDS; 3-KAILASH MOLASSE; 4-INDUS OPHIOLITE BELT; 5-PALAEOZOICS 6-MALLA JOHAR EXOTICS – EXOTICS OF PALAEOZOIC MESOZOIC IN OPHIOLITES; 7-NAINITAL KROL BELT (PALAEOZOIC-MESOZOIC?); 8-DEOBAN GARHWAL GROUP (PALAEOZOIC?); 9-RALAM FORMATION; 10-PRE-CAMBRIAN

FIG. 3

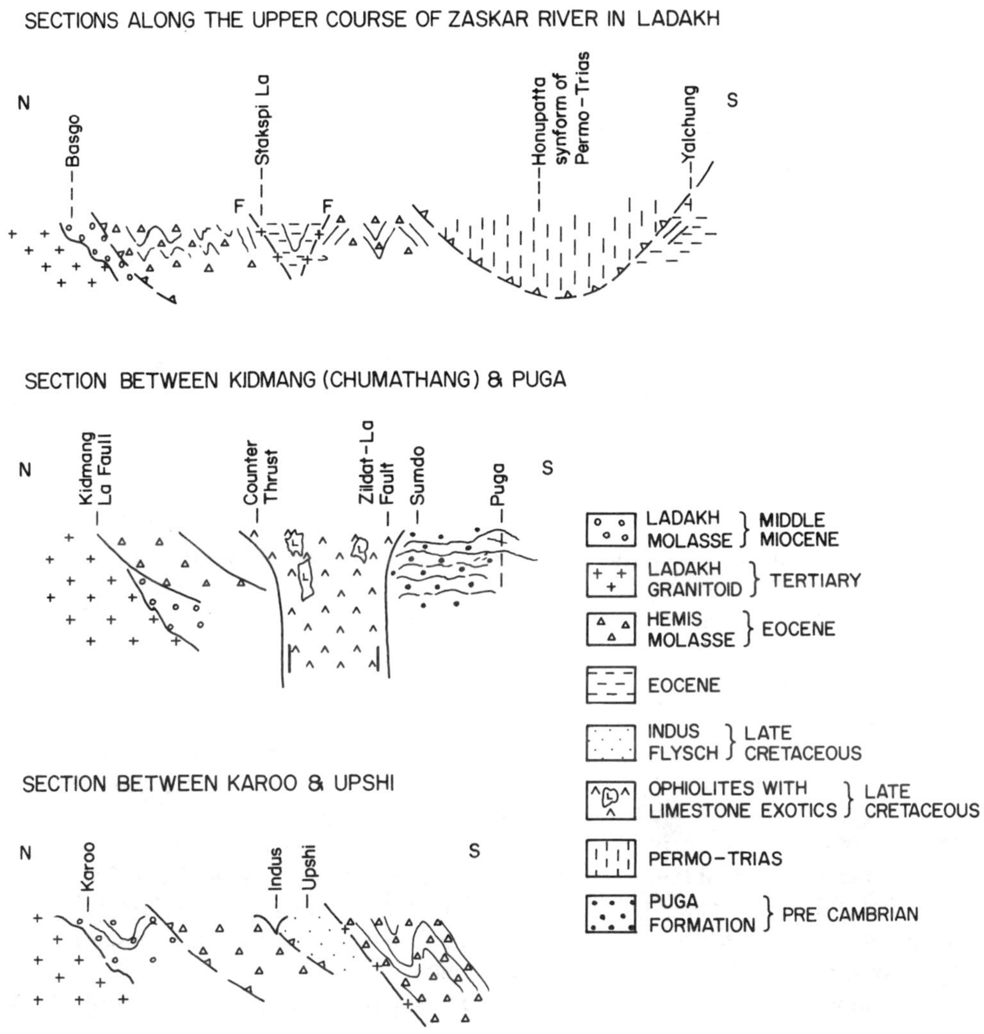

FIG. 4

Kashmir nappe. Exact mechanics of this klippe are not yet clearly understood. No clear-cut nappe unit has so far been reported from Arunachal Pradesh, except perhaps in the Along area.

The Central Crystalline Zone

Stretching across the axial zone of the Himalayas is a large expanse of crystalline rocks, commonly grouped under the blanket heading of the Central Crystalline or the Central Himalayan Crystalline Group. As mentioned earlier, these are partially thrust southward over the sedimentary belts of the Lesser Himalayas and towards the north they form the basement of metasediments over which the Phanerozoic rocks of the Tethys belt were deposited. No detailed study of the lithology, stratigraphy and structure of these has yet been carried out.

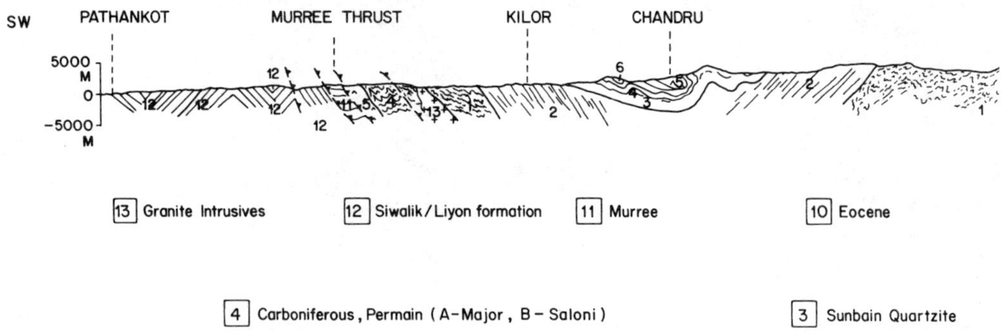

FIG. 5

Controversies exist regarding the age and origin of these metamorphic rocks and associated granites. This crystalline complex was earlier regarded as Archaen. During recent work in U.P. it has been found that the southern part of this complex comprises rocks equivalent to Chail and Jutogh in the Simla area. It has also been found that much of the gneiss occurring in this complex is of intrusive origin and has invaded rocks of various ages of different geological periods, as for instance, the Tertiary granites of the Kedarnath and Badrinath areas and the Palaeozoic granites of Chirpatiakhal. Most of the geochronological dates from this area have given very young ages for these rocks. This, together with the study of structural details, has led some authors (Powell and Conaghan, 1973a) to doubt whether any Precambrian 'gneiss' is present. However, other geochronological dates from this belt have given Precambrian ages (Bhanot *et al.*, 1978, 1977; Crawford, 1974). Hence it seems that this great complex comprises elements of Archaen granites, granulites, orthogneisses and schists, metamorphosed Precambrian sediments and metamorphosed sediments and intrusive granites and gneisses of later periods.

The crystalline axial zone continues from the Dihang in the east to Nunkun in the west, forming the Great Himalayas with its peaks mostly composed of granites, gneisses or metasediments. In the west crystalline rocks on an equally massive scale are again exposed northwest of the Kashmir Valley in the Nanga Parbat region. The intervening area between the Nunkun peak and Nanga Parbat is occupied by the sediments of the Tethys zone. This together with the fact that the Nanga Parbat crystallines have a trend transverse to the Great Himalayas has led many geographers and geologists to doubt the continuity of the Great Himalayas beyond Nunkun.

The Tethyan Sedimentary Belt

The Central Himalayan Crystalline Group forms the basement over which the sediments comprising a complete sequence from the Precambrian to the Cretaceous was deposited and which is commonly referred to as the Tethys Himalayas. Since these rocks are richly fossiliferous, it has been possible to build their stratigraphy with the help of fossils. However, the contact between the Central Himalayan Crystalline Group and the Tethyan sediments in places such as, for example the Spiti-Zasker area seems gradual, while elsewhere as in the Kumaun Himalayas it is marked by a fault.

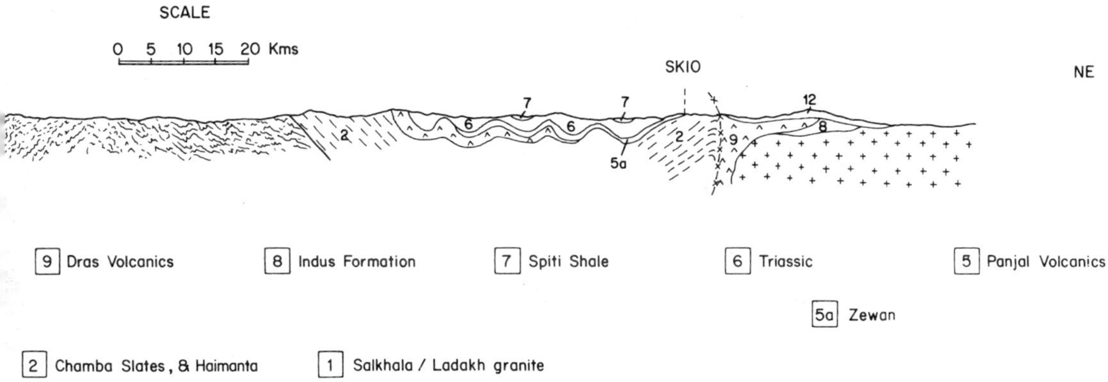

The late Carboniferous marine regression is detected from the deposition of the agglomeratic slate sequence in the Kashmir basin, Talai pebbly sandstone in the Chamba-Bhadarwah basin, volcanic flows in the Zaskar basin and a dimictite known as Ganmachidam Formation in the Spiti area. As land conditions prevailed in the Kashmire area, Gondwana flora developed, which explains the occurrence of Gondwana strata in the Kashmir basin. There was another marine transgression in the later part of Early Permian, which became widespread during Triassic times. Sedimentation stopped in the Chamba-Bhadarwah and Kashmir basins during the late Triassic to the Early Jurassic periods. In the Spiti-Zaskar basin the sea began regressing during the Middle Jurassic. In the Kumaun area marine sedimentation ceased during the Middle and Late Cretaceous. It was at this period that the first pulse of the Himalayan orogeny was initiated and was followed by Eocene sedimentation in the narrow Indus tectonic belt. The Tethys sediments or Kshir-Sagar Tethys basins (Raina and Krishnaswamy, in press) show only broad simple folds in the Spiti, Zaskar and Kashmir area, but occur as compressed and imbricated slices in the Kumaun area.

The Indus Tectonic Belt

In the Ladakh district of Jammu and Kashmir, from Hanle in the southeast to Nanga Parbat area in the northwest, a suite of mafic and ultramafic rocks and sediments ranging in age from the Late Cretaceous to the Mio-Pliocene are exposed in a narrow linear belt following the trend of the Indus river. Further east this zone is traced to Kuamun and Tibet along the Tsangpo river. Towards the west this belt is considered to join the mafic-ultramafic sequences of Chitral and Gilgit. It has been suggested that this zone marks the suture between the Indian and Asian continental plates along which the Indian continent was subducted below Tibet in a continent-continent collision in Late Cretaceous–Tertiary times (Gansser 1964, 1966; Dewey and Bird, 1970).

The mafic and ultramafic sequences with interbedded radiolarian chert horizons have generally been termed the Dras Volcanics. The associated sediments were earlier called the Indus Flysch (De Terra, 1935; Gansser, 1964). Recent workers (Tewari, 1964; Gansser, 1977; Frank *et al.*, 1977; Shankar *et al.*, 1976; Srikantia and Bhargava, 1978) have further subdivided the sediments of this zone. The sedimentary rocks are mainly of two types, a

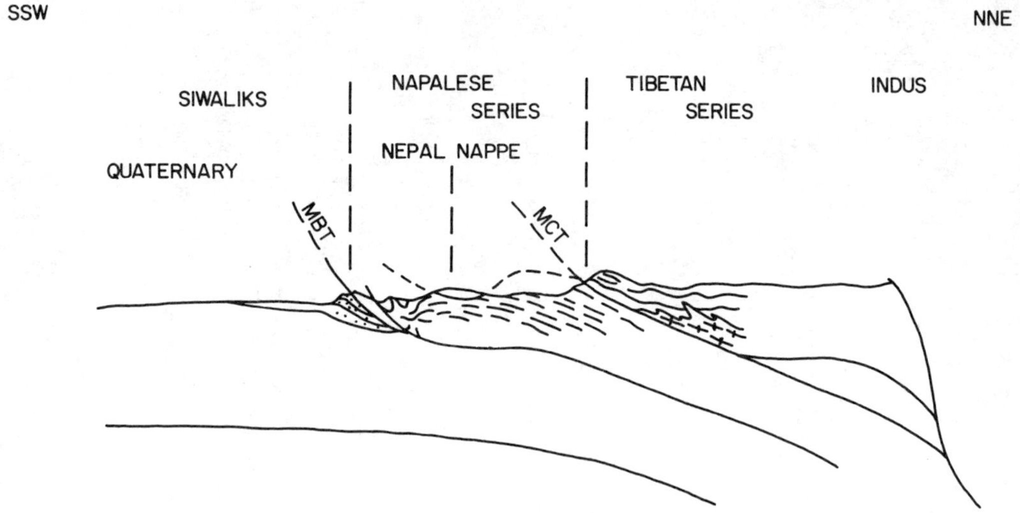

FIG. 6

flysch and a molasse. Frank *et al*., (1977) have found flyschoid sediments of two ages, one Triassic and the other Cretaceous–Palaeogene. The Tertiary molassic sediments have likewise been subdivided into two—a northern autochthonous and a southern parautochthonous belt (Frank *et al*., 1977). The Dras Volcanics are thrust over by the Late Palaeozoic–Mesozoic sediments of the Zaskar basin in the south. To the north these are thrust over the flyschoid sediments along the Counter thrust. This in turn is thrust over the molassic sediments which lie unconformably over the Ladakh granitoid complex. Gansser (1977) does not consider the Dras Volcancis to be associated with the consumption of the Indian plate. He has reported ophiolitic melanges with limestone exotics. The flyschoid sediments were deposited in the foredeeps formed in front of the newly uplifted landmass.

The Karakorum Belt

The rocks of the Karakorum have been divided into two tectonic belts by the batholithic Karakorum granitoid complex which forms the axis of the Great Karakorum range. To the south, exposed between the Ladakh granitoid complex and the Karakorum granitoid complex is a series of volcanic-sedimentary rocks which were earlier termed the 'Shyok Ophilites'. Recent work in this area has helped in the recognition of four distinct formations which are: (1) Khardung Volcanics of the Early Tertiary age consisting of silicit-intermediate volcanics and volcaniclastics; (2) the Hundiri Formation of Middle to Late Cretaceous (Eocene?) age and of flyschoid character, consisting of chert, limestone and rhythmically banded sandstone-slate sequence; (3) the Biagdang Formation of Cretaceous age consisting of a thick pile of basalt, gabbro and ultramafics with minor interbedded chert bands; (4) the Aranu Formation of Permo-Carboniferous age consisting of slate, sandstone-limstone horizons that have locally been migmatized. The Ladakh granitoid complex is intrusive in the Khardung Volcanics which have been thrust over by the Hundiri Formation along the Khalsar thrust. The Hundiri Formation is underlain by the Biagdang Formation with a normal contact. The Biagdang Formation is exposed in the core of a northwest to southeast trending antiform with its axis aligned along the Shyok river. This antiformal fold has caused a repetition of the Hundiri Formation and the Khardung

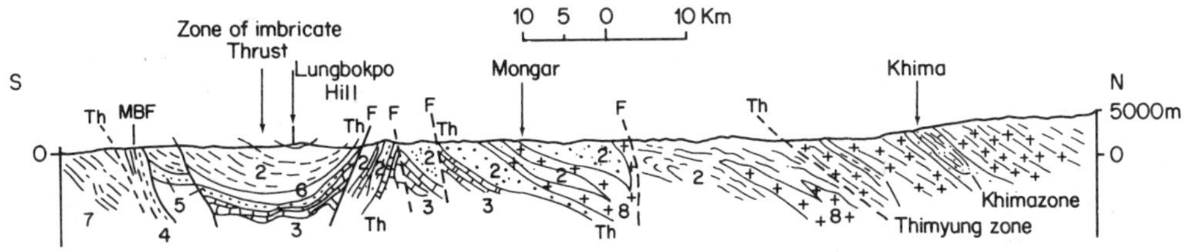

1. Thimpu Farmafion – Tethyan Belt
2. Shumar Farmafion
3. Buxa Farmafion
4. Gondwana
5. Thungsiling Quartzite
6. Diurz Boulder Slate
7. Granite Gneiss
8. Siwalik – Tertiary Belt

(Items 2–7: LESSER HIMALAYAN BELT)

(After Jangpangi, 1978)

FIG. 7

Volcanics in the Saltoro Hills in the northern bank of the Shyok river. The Khardung Volcanics, repeated in the Saltoro and the Great Karakorum Range along the Nubra Valley, are thrust over by a narrow band of the Aranu Formation, itself overthrust by the Karakorum batholith. This thrust zone marks a major regional lineament in the area and has been designated as the Shyok lineament.

To the north of the Karakorum batholith lies the Karakorum Tethyan basin consisting of Upper Palaeozoic to Tertiary rocks. In the eastern Karakorum the Karakorum granitoid intrudes into the Permo-Carboniferous limestones containing bryozoans and bivalves. These limestones are the oldest formations recorded in the area and are overlain by a thick sequence of slate and sandstone which in its upper part contains bands of Late Permian fusilinid limestone. This is overlain by a continuous sequence of buff and grey limestones ranging from the Triassic to the Cretaceous (Aghil Limestone). This Late Palaeozoic–Mesozoic platform sequence is unconformably overlain by a Tertiary red molassic sequence. The youngest sediments are represented by Quaternary gravels and claystones which form the high level plains of Dipsang.

In the western and central Karakorum, Desio (1974) has worked out a similar stratigraphy consisting of the Darkot Group of Carboniferous–Permian age overlying the intrusive Karakorum batholith. The Darkot Group is overlain by the Gujhal Dolomite which is homotaxial with the Aghil Limestone.

MAJOR STRUCTURAL AND TECTONIC FEATURES OF THE HIMALAYAS

The significant structural and tectonic features of the Himalayas are as follows:

(1) The fold/reverse fault system is very important in the Lesser Himalayan region and has been observed in Tertiary as well as pre-Tertiary formations.

(2) The Lesser Himalayas and partly the Tertiary belt show cross-structures transverse to the regional strike. The structures are north northeast to south southwest to northeast to southwest in the western Himalayas, north to south in the Central Himalayas, and northwest to southeast in the eastern Himalayas. These probably reflect basement reactivation during orogenic times (Anon. G.S.I., 1976).

(3) Post-thrusting folding and faulting and later erosion has brought to the surface a number of window structures like the Larji-Rampur window in Kulu Himalaya, the Pharat window in Garhwal Himalaya and the Rangit window in Darjeeling-Sikkim Himalayas.

(4) Although the Lesser Himalayan thrusts might have been embryonic in Pre-Miocene times the actual transportation probably did not take place before the Miocene as is shown by the detailed structural studies in some areas. The Lesser Himalayas consists of several thrust sheets piles up one above the other. But the sequence of their order in the field need not necessarily reflect the sequence of their emplacement.

Evidence is available to indicate that the upper thrust sheet is directly in contact with the Eocene below, while the lower thrust sheet has overriden Pleistocene sediments (Anon. G.S.I., 1976). It has also been recently found that the Jutogh thrust might be pre-Chail in age (Pati, 1978). Although there is broad agreement on the importance of thrust structures in the Himalayas, the controversy regarding the root zones of the thrust sheets is not yet satisfactorily resolved. While some workers consider the root zones of the Lesser Himalayas to lie concealed below the southward moving Central Crystallines sheet others consider some of the thrust sheets to be parautochthons which have moved over their own basement.

(5) Although a number of orogenic episodes have been postulated for the Himalayan region (Valdiya, 1964; Srikantia, 1977; Dhoundial *et al*., 1977; Saxena, 1971) imprints of the earlier phases are yet to be established for the whole of the region. This is further complicated by the fact that although there are a number of conglomeratic horizons in the Himalayas, no angular unconformity has so far been recognized. Even the unconformity at the base of the Eocene is not in the nature of an angular unconformity, though the underlying formations belong to greatly differing ages. The obliteration of earlier structural and metamorphic imprints by Tertiary orogenic episodes has added to this confusion.

(6) The Main Central thrust represents a schuppen zone along which the Central Himalayan Crystallines of the axial belt have moved southward at different times.

(7) The Main Boundary thrust which runs the length of the Himalayas was earlier thought to be in the nature of a boundary fault delineating the basin of deposition of the Siwalik sediments. Work in the last few decades by the Geological Survey of India and the Oil and Natural Gas Commission has shown that it is neither a boundary fault nor a single tectonic plane. The Main Boundary fault thrust represents a tectonic zone along which rocks older than the Siwalik Group have overriden Siwalik sediments.

(8) The Indus suture zone represents the furthest and the northernmost tectonic feature of the geological province of Himalaya and is considered to mark the northern limit of the Indian plate. The Indus suture displays a variety of vergence, northward in Spiti and Kailas, southward in Amlang La and Kiogar and ambivergent in Rupshu. The nature of the Tsangpo suture, which is regarded as the eastward extension of the Indus suture, is as yet very little known. The Palaeozoic–Mesozoic rocks of the Zaskar basin are thrust over the Dras Volcanics in the south. In the north the Dras Volcanics are thrust over the flyschoid sediments along the back-thrust. The tectonics of this belt are very complex with different units having been thrust over each other.

(9) A major regional lineament can be traced at the base of the Great Karakorum range along the Nubra and Shyok river in the eastern Karakorum. To the southeast

this lineament crosses the Ladakh range and continues further east in Tibet. To the northwest it passes into Gilgit and beyond. In the Nubra Valley the batholithic Karakorum granitoid complex is thrust along this lineament over the Late Palaeozoic Aranu Formation and the Khardung Volcanics. This lineament is marked by a zone of shearing and mylonitization as well as geothermal activities manifested by a string of hot springs such as those of Changlung and Panamik.

(10) On both the extremities of the Himalayas, the gentle pattern of the regional structures of Kashmir and the Arunachal Himalayas abruptly veers towards the south. These are generally known as the western and eastern syntaxes. The western syntaxis was first described by Wadia (1931) and later studied in detail by Calkins *et al.* (1969). The northeastern syntaxis is poorly known due to difficult terrain conditions. Nandy (1976), however, on the basis of his studies in the Mishmi hills and surrounding areas does not agree to the existence of a syntaxial bend in the eastern Himalayas.

Although there is a number of recent publications with detailed structural analyses of local areas (Ray and Naha, 1971; Ghosh *et al.*, 1974; Vashi and Merh, 1974; Powell and Conaghan, 1973; Sinha Ray, 1973; Sarkar and Shrish, 1976, and Srimal and Pati, 1981 among others) the work has been sporadic and patchy. A synthesized picture for the Himalayas is yet to emerge. Structural studies in different sectors of the Himalayas have revealed three to four-fold trends. Whereas workers such as Srikantia *et al.* (1976), Valdiya (1964), Saxena (1971) and Srikantia (1977) have postulated pre-Tertiary orogenic episodes in the Himalayan region, Conaghan and Powell (1973) argue against any pre-Jurassic deformation of Himalayan rocks. It is difficult to conceive that this region had suffered no deformation earlier than the Tertiary. Evidently the evidence of earlier deformational and metamorphic phases is more or less completely masked by the Tertiary orogeny which was of a very high intensity and global in nature.

Three main phases of folding are widely recognized in the western Himalayas. In order of antiquity they are: (1) northeasterly plunging reclined folds; (2) northwest southeast trending Himalayan folds; and (3) northeast-southwest to north-south cross-folds. Because of obliteration of stratification in varying degrees by later deformation and metamorphism, isoclinal nature and high amplitude–wavelength ratio, the early folds are recognizable only with great difficulty.

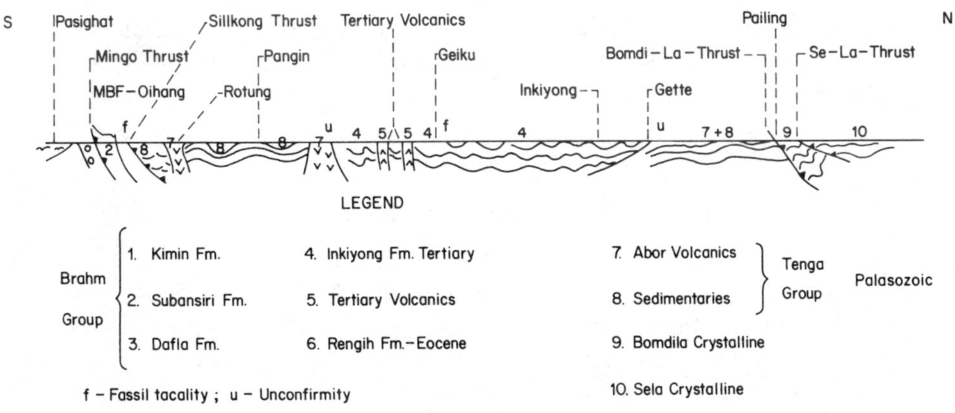

FIG. 8

MAGMATIC ACTIVITY ASSOCIATED WITH THE HIMALAYAN OROGENY

Dating of the Himalayan igneous rocks so far has been patchy and scanty and as a result it is difficult to get a comprehensive picture of the magmatic activity associated with different stages of the evolution of the Himalayas. An attempt has been made here to synthesize the data available so far and to fit them into a unified picture.

Silicic Igneous Activity

Two main types of granitic rocks are found in the Himalayas: a highly foliated rock composed mainly of quartz, feldspar and biotite and a younger non-foliated rock composed of muscovite, black tourmaline or hornblende in addition to quartz and feldspar. However, the presence of tourmaline, hornblende or garnet is not restricted to granites of Tertiary age and they are also found in the older granitoids (Pascoe, 1973).

Field evidence indicates that the Nela Valley granite of the Tehri district, U.P., the Burzi La, Ladakh and Karakorum granites of Kashmir, the granites of the Lahaul area, the Gangotri-Kedarnath-Badrinath granitoids of the Garhwal Himalayas, the tourmaline or hornblende rich granitic rocks of the Everest region in southern Tibet, north central Nepal and northeastern Bhutan and the leucogranites of the high Himalayas of Napal are all associated with Himalayan orogenic episodes. The Early Tertiary calc-alkaline suite of island arc volcanic rocks of Khardung in the Karakorum area is also associated with the Himalayan orogeny.

Basic Igneous Activity

Although acid igneous activity associated with the Himalayan orogeny is seen in different sectors of the Himalayas, basic igneous activity associated with the Himalayan orogeny is limited to certain localized belts. These are the Cretaceous–Eocene mafic rocks of the Dras and Shyok area with zones of ultramafic bodies, the dolerite dykes intruding into the Late Cretaceous–Miocene Ladakh granite and the basic sills and dykes seen within the Palaeo–Mesozoic Krol sequence and the Subathu rocks of the Himachal area.

GEOCHRONOLOGICAL DATA

Although the geochronological data for the Himalayan region are still few they can be synthesized to arrive at some general assumptions. A perusal of the geochronological data (Krummenacher *et al.*, 1978; Ashgirei *et al.*, 1977; Saini *et al.*, 1978; Jager *et al.*, 1971; Nagpal and Nagpaul, 1974; Sinha and Bagdasarian, 1977) indicates major igneous and metamorphic episodes during Late Cretaceous to Oligocene and Miocene times with a few data indicating Pliocene ages.

SEA FLOOR SPREADING

The study of the spreading history of the Indian Ocean indicates that the rate and direction of spreading has varied in the course of geological time. Twice the movement has slowed down and even stopped during the Late Eocene–Early Miocene time and twice it has resumed during the Early Oliogene and Late Miocene period (Le Fort, 1975). The movement is rotational, the centre of the rotation having changed from about 20°N, 26°E to 29°N, 27°E (Le Pichôn and Hertzler, 1968; Le Pichôn *et al.*, 1973). The rotation of centres on the western part of the Indian plate indicates a comparatively more northward move of the eastern side than of the western side of the Himalayas. Meyerhoff and Meyerhoff, (1978), Gansser (1964, 1966), Fuchs (1975 and Le Fort (1975), basing their work on geological and palaeontological data of the Himalayas, show that the crustal contraction in the Himalayas is between 300 and 600 km. From the spreading history of the Indian Ocean ridge over 35 m.y., it is estimated that 1300 km of new ocean floor has been generated. Le Pichôn and Hertzler (1968) have arrived at a mean value of 5–6 cm/year of plate convergence in the Himalayas. Le Fort therefore urges that by taking into account the uncertainties of these figures not much inconsistency arises in the figures obtained by sea-floor spreading and contraction due to crustal shortening limited to the Himalayas.

THE HIMALAYAN OROGENY AND ITS PHASES

In spite of the complicated nature of the geological structure of the Himalayas, the data so far obtained tend to support the view that three main episodes of orogenic activity have taken place in the Tertiary and that this has continued up to the Quaternary with diminishing magnitude. The tectonomagmatic activity associated with the different phases of the Himalayan orogeny is summarized in Table 1.

The model presented conceives a ridge separating the eugeosynclinal Tethys sea from the miogeosynclinal Lesser Himalayan basins. The ridge possibly, came into existence in the Precambrian, although the fossiliferous Tethyan sediments testify to an uninterrupted sedimentation. From the Precambrian to the Early–Middle Carboniferous and again from the Late Carboniferous to the Cretaceous, the model developed for the Lesser Himalayas is only conjectural, in the absence of fossils. Except for the presence of nanno fossils and polynomorphs, the absence of megafossils in the Lesser Himalayan sequences until the Late Cretaceous shell-limestone of the Krol belt is pronounced and has not been satisfactorily explained. Some workers doubt the glacial origin of the Blaini Boulder Bed (Rupke, 1974; Singh, 1979), on the basis of which it was being correlated with the Late Carboniferous Talchir Boulder Bed of Peninsular India and was being considered as an important datum for establishing the stratigraphy of the Lesser Himalayas. The few reports of fossils of Permo-Carboniferous, Jurassic–Cretaceous and Eocene ages, from the Lesser Himalayas have been explained as due to transgression in linear belts. Workers such as Crawford (1974) and Singh (1979) consider the whole of the Lesser Himalayas to consist of Precambrian rocks.

The discovery of bryozoans from the limestone band of the Deoban–Garhwal Group and Tertiary plant fossils in the Arunachal Himalayas (Dungrakoti, personal communication) from rocks similar in lithology, geological and tectonic arrangement to the Simla Slate–Morar Chakrata formations of western Himalaya goes against the view that all the Lesser Himalayan rocks are Proterozoic in age. However, a satisfactory explanation for the absence of fossils in Lesser Himalaya remains to be found.

The model also envisages that the Indus suture zone marks the northern limit of the Indian plate. Hypotheses propounding that the northern limit of the Indian plate lies in the Tien Shan area have been put forward by Crawford (1974), Kaila and Narain (1976) and Singh (1979). However, the geology of southern China and Tibet and the Karakorum area is yet to be clearly established. Unless a clear geological picture of the area is established it would be difficult to conceive the Indian plate boundary to be in the Tien Shan area, based on conjectures only, when on the other hand geological evidence points to its existence at the Indus suture zone. Similarly the Main Central thrust has also been considered to mark the suture zone (Srikantia and Bhargava, 1978; Ray, 1976; Valdiya and Gupta, 1972). As

TABLE

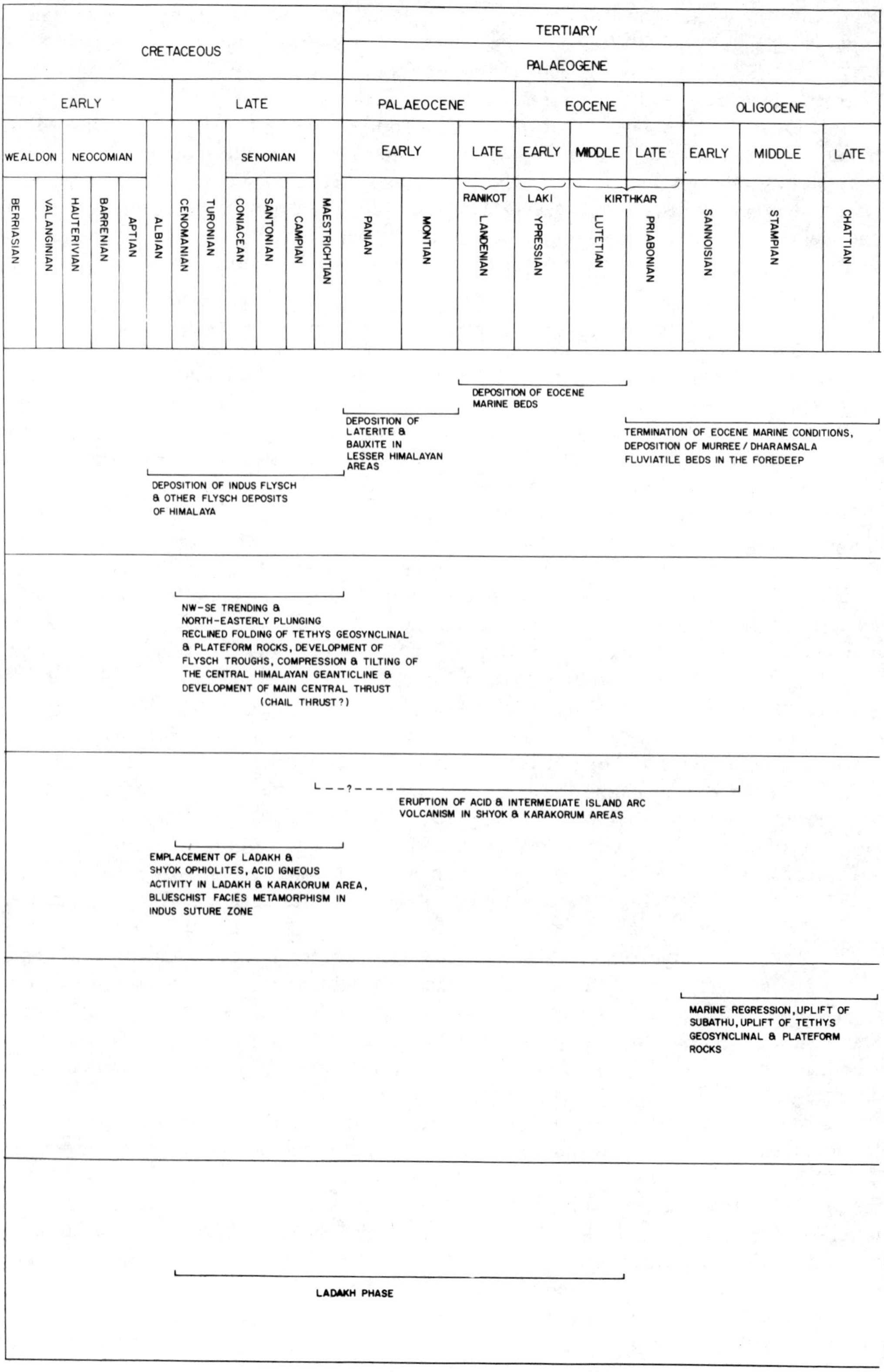
CRETACEOUS
TERTIARY
PALAEOGENE
EARLY
LATE
PALAEOCENE
EOCENE
OLIGOCENE
WEALDON
NEOCOMIAN
SENONIAN
EARLY
LATE
EARLY
MIDDLE
LATE
EARLY
MIDDLE
LATE
RANIKOT
LAKI
KIRTHKAR
BERRIASIAN
VALANGINIAN
HAUTERIVIAN
BARRENIAN
APTIAN
ALBIAN
CENOMANIAN
TURONIAN
CONIACEAN
SANTONIAN
CAMPIAN
MAESTRICHTIAN
PANIAN
MONTIAN
LANDENIAN
YPRESSIAN
LUTETIAN
PRIABONIAN
SANNOISIAN
STAMPIAN
CHATTIAN
DEPOSITION OF EOCENE MARINE BEDS
DEPOSITION OF LATERITE & BAUXITE IN LESSER HIMALAYAN AREAS
TERMINATION OF EOCENE MARINE CONDITIONS, DEPOSITION OF MURREE / DHARAMSALA FLUVIATILE BEDS IN THE FOREDEEP
DEPOSITION OF INDUS FLYSCH & OTHER FLYSCH DEPOSITS OF HIMALAYA
NW-SE TRENDING & NORTH-EASTERLY PLUNGING RECLINED FOLDING OF TETHYS GEOSYNCLINAL & PLATEFORM ROCKS, DEVELOPMENT OF FLYSCH TROUGHS, COMPRESSION & TILTING OF THE CENTRAL HIMALAYAN GEANTICLINE & DEVELOPMENT OF MAIN CENTRAL THRUST (CHAIL THRUST?)
?
ERUPTION OF ACID & INTERMEDIATE ISLAND ARC VOLCANISM IN SHYOK & KARAKORUM AREAS
EMPLACEMENT OF LADAKH & SHYOK OPHIOLITES, ACID IGNEOUS ACTIVITY IN LADAKH & KARAKORUM AREA, BLUESCHIST FACIES METAMORPHISM IN INDUS SUTURE ZONE
MARINE REGRESSION, UPLIFT OF SUBATHU, UPLIFT OF TETHYS GEOSYNCLINAL & PLATEFORM ROCKS
LADAKH PHASE

12.1

PERIOD	TERTIARY								QUATERNARY													
	NEOGENE																					
EPOCH	MIOCENE					PLIOCENE			PLEISTOCENE												HOLOCENE	
	EARLY		MIDDLE		LATE	EARLY	LATE															
	AQUITANIAN	BURDIGALIAN	HELVETIAN	TORTONIAN	SARMATIAN	PANNONIAN	PLAISANCIAN	ASTIAN	VILLIFRANCHIAN	BIBER	B.D. INTERGLACIAL	DANUBE	D.G. INTERGLACIAL	GUNZ	G.M. INTERGLACIAL	MINDEL	M.R. INTERGLACIAL	RISS	R.W. INTERGLACIAL	WURM	RECENT	SUB-RECENT

DEPOSITION

- DEPOSITION OF LOWER SIWALIK DEPOSITION OF MOLASSIC SEDIMENTS IN QIZILLANGAR AREA & LADAKH
- DEPOSITION OF MIDDLE SIWALIK
- DEPOSITION OF LOWER KAREWA
- DEPOSITION OF UPPER SIWALIK
- DEPOSITION OF UPPER KAREWA
- DEPOSITION OF QUATERNARY SEDIMENTS OF DIPSANG PLAIN, DEOSAI PLAIN, LINGZHITANG PLAIN & MORE PLAIN
- DEPOSITION OF YOUNGER TERRACES, DEPOSITION OF YOUNGEST LOAMS OF KASHMIR VALLEY

STRUCTURAL ELEMENTS

- UPLIFT OF MURREE & DHARAMSHALA BEDS, FORMATION OF SIWALIK TROUGH, EMPLACEMENT OF TECTONIC NAPPE OF LESSER HIMALAYA, SYNTAXIAL BENDING, NW-SE TRENDING HIMALAYA FOLDS HIGH LEVEL TOURMALINE GRANITES IN BADRINATH-KEDARNATH & GANGOTRI AREAS
- NE-SW TO N-S TRENDING CROSS FOLDS & FRACTURES, MAIN BOUNDARY THRUST, MURREE THRUST, REACTIVATION OF SOME OF THE TECTONIC NAPPE

METAMORPHISM AND MAGMATISM

- WIDESPREAD REGIONAL METAMORPHISM, ACID IGNEOUS ACTIVITY IN LADAKH-KARAKORUM & AXIAL BELT OF HIMALAYA, DYKES IN THE LOWER TERTIARY ROCKS

EPEIROGENIC MOVEMENTS AND ISOSTATIC READJUSTMENT

- UPLIFT & EROSION OF LOWER KAREWA
- UPLIFT OF KAREWA BEDS UPLIFT OF HUNDES MOLASSE UPLIFT OF SIWALIK ROCKS UPLIFT OF INDUS & SUTLEJ TERRACES FORMATION OF INDO-GANGA TROUGH UPLIFT OF KARAKORUM

OROGENIC PHASE

- DHARAMSALA PHASE
- SIWALIKIAN PHASE

will be discussed later, the Main Central thrust does not show the characteristics of a plate boundary.

Following Gansser (1964, 1977) the authors consider the Indus suture zone to mark the Indian plate boundary. The sequential evolution of the Himalayas during the Tertiary orogeny have been drawn after the model of Dewey and Bird (1970). Figs 9–14 show the

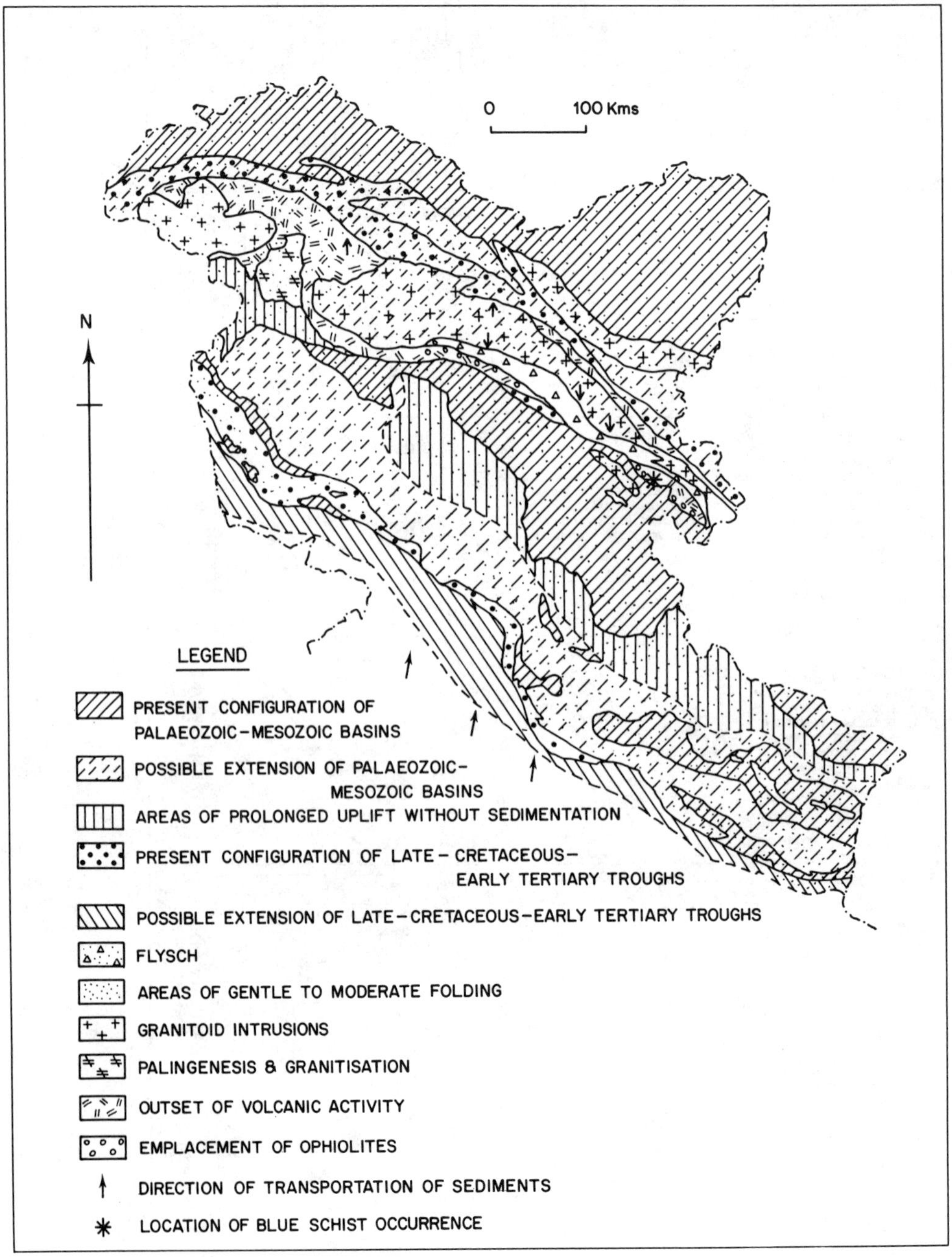

FIG. 9 Ladakh phase (Late Cretaceous–Early Eocene) in Western Himalaya.

different episodes of the Himalayan orogeny on this model, briefly discussed below. Fig. 15 shows the evolution of the Himalayas in the Tertiary.

Recent work by Bhandari *et al*. (1978) and Srimal *et al*., (1979) in the Karakorum area has established beyond any doubt that the ophiolites in that region are of Late Cretaceous age. Since these are also considered to be the continuation of the Ladakh belt in the first phase of Tertiary orogeny, the Ladakh phase is considered to be active from Late

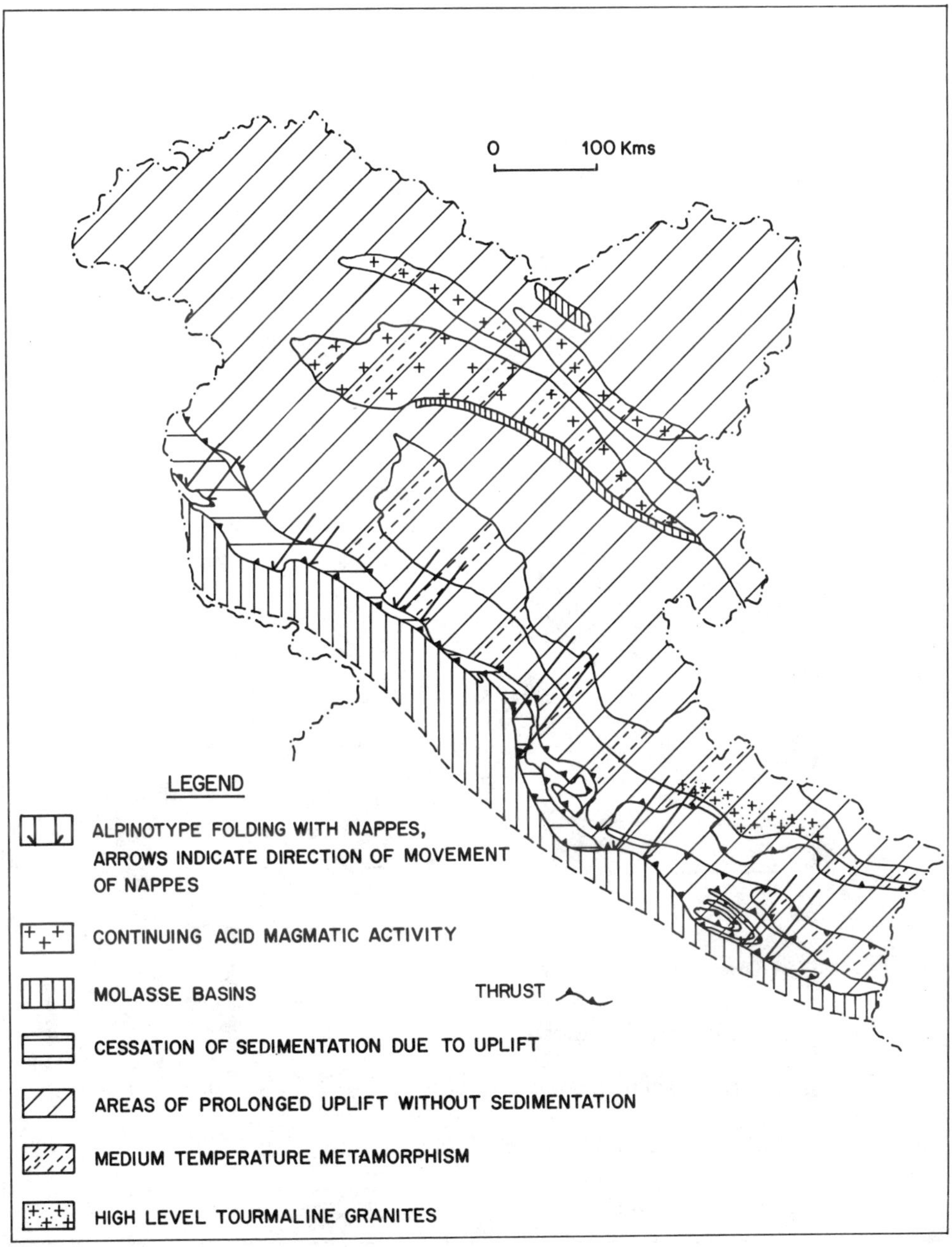

FIG. 10 Dharamshala phase (Middle Miocene) in Western Himalaya.

Cretaceous to Early Eocene times. The earth movements which culminated in the uplift of the Himalayas had begun. The tip of the Indian plate first collided in the northwestern syntaxial zone. The squeezing of the sediments caused development of reclined folds with northeast to southwest axial planes in the western Himalayas and northwest to southeast axial planes in the eastern Himalayas. With the northwestern syntaxial zone as the pivot, the Indian plate started rotating northeastwards, gradually colliding against the Asian plate. This caused shallowing of the seas, the emplacement of wedges of ophiolitic suite along the

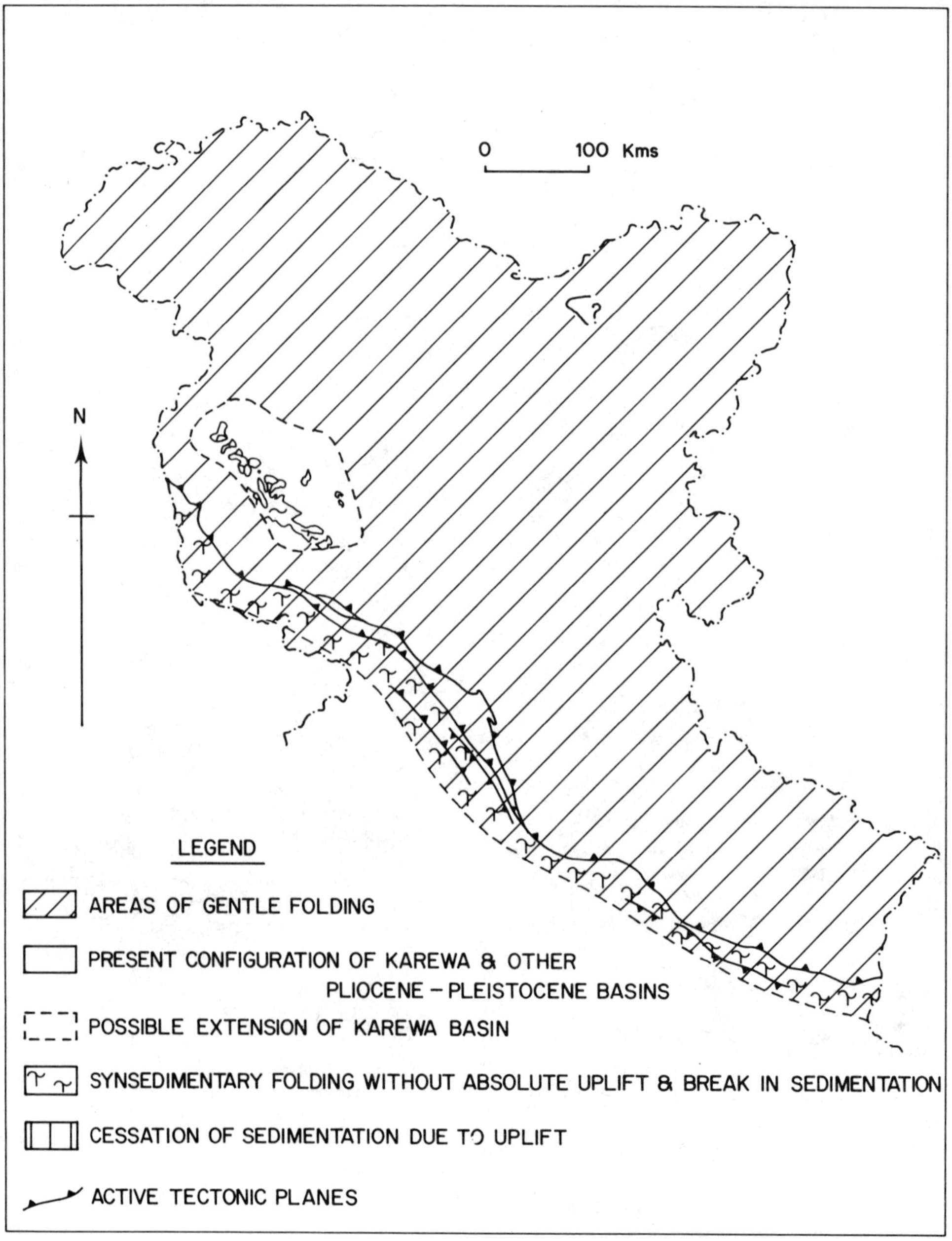

FIG. 11 Siwalikian phase (pliocene–pleistocene) in Western Himalaya.

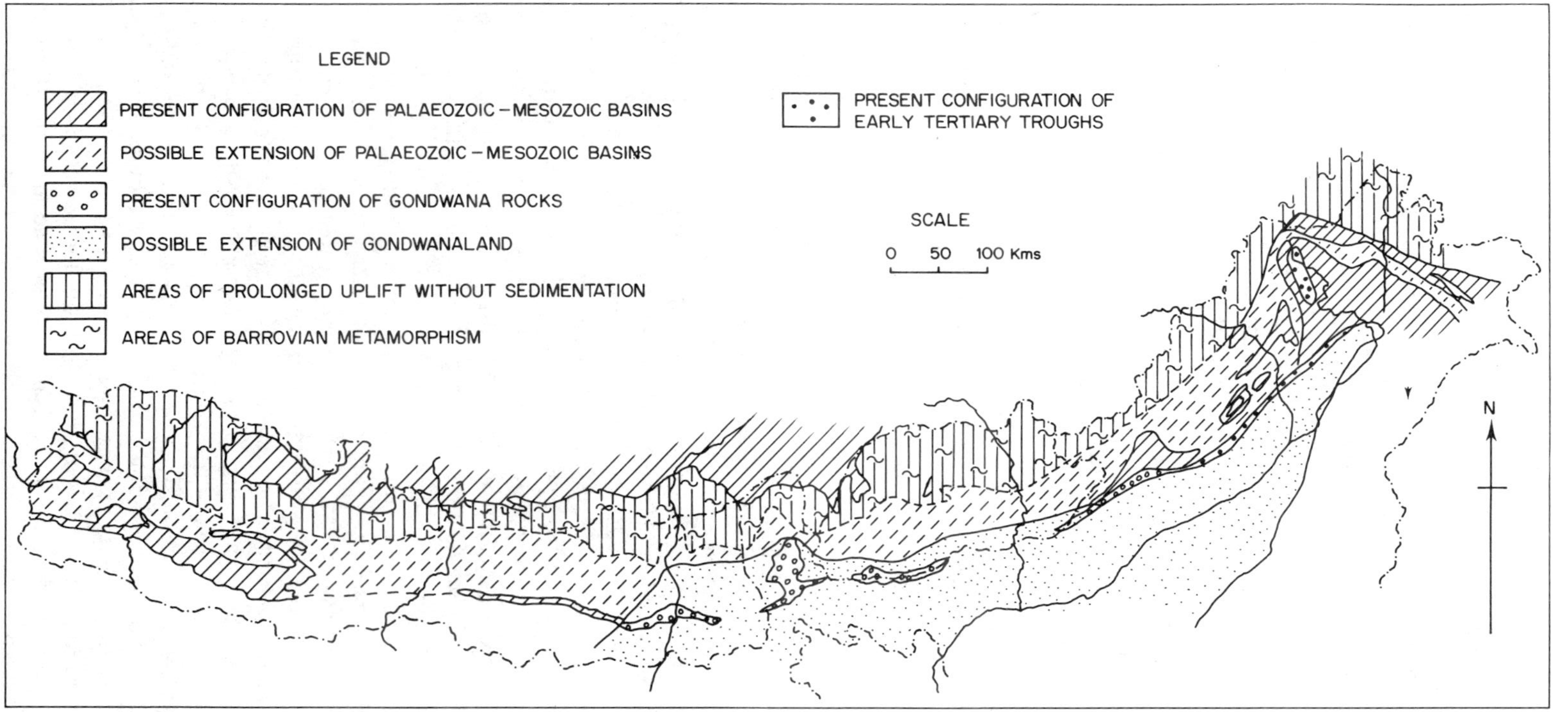

FIG. 12 Ladakh phase (Late Cretaceous–Early Eocene) in Central-Eastern Himalaya.

Indus suture zone, granitic intrusions, palingenesis and granitization in the Ladakh-Karakorum area (Figs 9 and 12). This was followed by transgression of the sea in the Eocene period. Patches of Eocene rocks are found on varied rock formations such as the Tal (Jurassic), the Krol (Permo-Trias?), Simla slates (Palaeozoic?) and Deoban rocks (Precambrian to Palaeozoic?). This indicates folding and erosion of these sedimentary basins prior to the deposition of Eocene sediments. The Tethys was squeezed into a small, narrow sea in which flyschoid sediments of the Indus suture zone were deposited.

The second phase of Himalayan orogeny, the Dharamsala, culminated in Middle Miocene times. This was the most vigorous of all the Tertiary orogenic movements. The interlocking of the Indian and Asian plates caused north-south compressive forces which generated northwest-southeast trending folds in the western Himalayas and northeast-southwest trending folds in the eastern Himalayas, Subathu, Murree and Dharamsala. Due to the underthrusting of the Indian plate below the Tibetan plate and the concurrent compression huge thrust sheets were detached from their root zones and moved southward to come to their present positions (Figs 10 and 13). Most of the thrust movements of the Himalayas are associated with this phase as also are the widespread regional metamorphism and silicic igneous activity of the area. This upheaval also affected biotic development in the Himalayas, resulting in an explosive evolution of mammalian fauna. After this movement, migratory routes to other parts of the world were severed and the development of species, which earlier had some common elements with Europe and other parts of Asia, adopted an independent course of evolution and several endogenic forms of fauna and flora evolved (Pascoe, 1973).

Consequent drainage developed and a foredeep was formed in the front of the rising Himalayas into which the Siwalik molassic sediments were deposited in the western Himalayas. In the eastern Himalayas, the collision probably took place in Middle Miocene times. This would explain the brackish water environment of deposition of the Dafla Formation of Lower Miocene age in the Arunachal Himalayas in which arenaceous Foraminifera have been found. The rocks of the Dafla Formation are lithologically different both from the Siwalik and the Murree rocks (Karunakaran and Ranga Rao, 1976). Some furrows also developed in the Tibetan side wherein the Kargil Molasse and the Kazilangar Conglomerate of the Karakorum area were deposited.

The third phase is termed the Siwalikian Phase, which was active from Late Pliocene to Middle Pleistocene times, when the molassic sediments of the western Himalayas were uplifted. The pebbly sandstone of the Subansiri Formation, which overlies the Dafla Formation in the Arunachal Himalayas suggests uplift and strong denundation of the eastern Himalayas. Whereas the major uplift in the western Himalayas was in the Eocene–Miocene period, the major uplift in the eastern Himalayas was in the Miocene–-Pliocene period as the eastern part of the Indian plate collided with the Asian plate later compared to the western part. Most of the present structures of the Himalayas developed during this phase. Tectonic creep along some of the Himalayan thrusts, translation of Upper Siwalik rocks over the subrecent gravel beds along the Satlita and foothill thrusts and the seismic unstability of the region suggest that this phase has not yet completely died down. Figs 11 and 14 show the effects of this phase.

The above synthesis is also substantiated by the seabed spreading data. In the western Himalayas underthrusting of the Indian plate below the Tibetan block due to buoyancy constraints stopped during Late Eocene time. There was isostatic readjustment and erosion of the newly risen landmass. Melting, mobilization and intrusion of granites helped the seabed spreading to resume during the Early Oliogocene and underthrusting again continued up to the Early Miocene. When intracontinental compression was not able to move

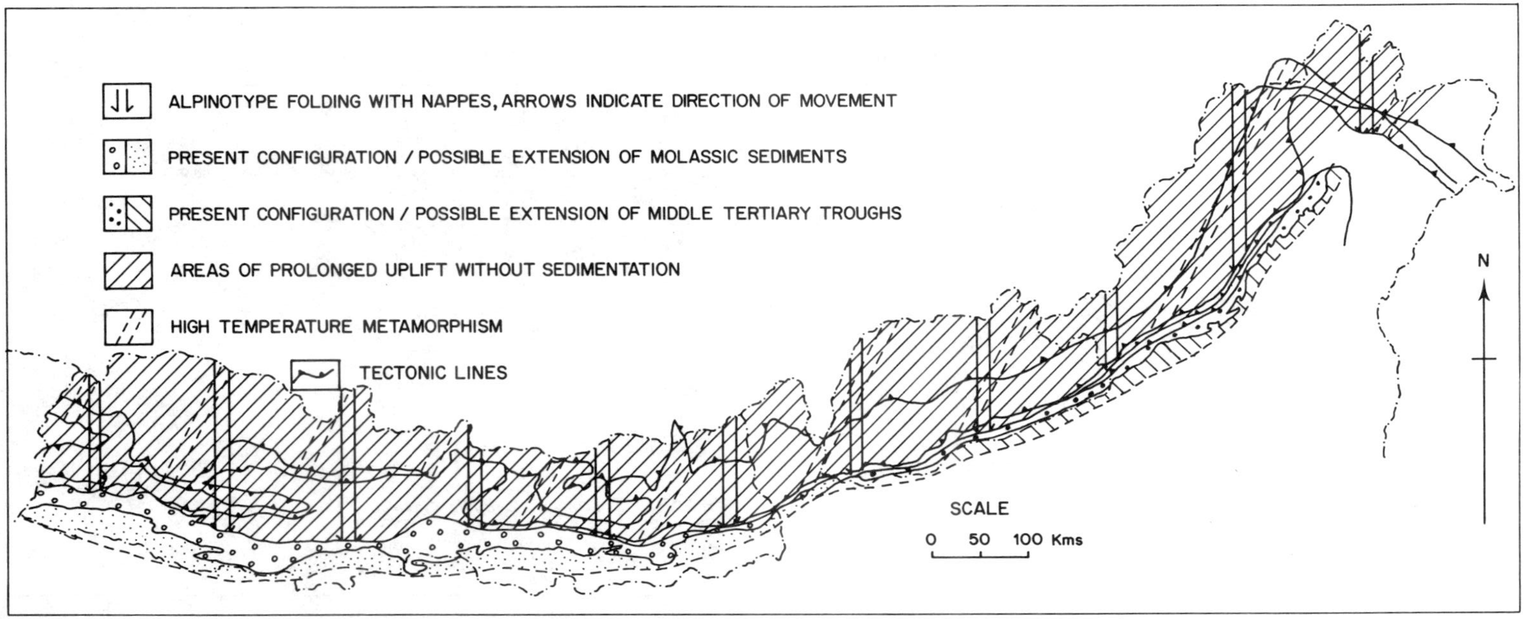

FIG. 13 Dharamasala phase (Middle Miocene) in Central-Eastern Himalaya.

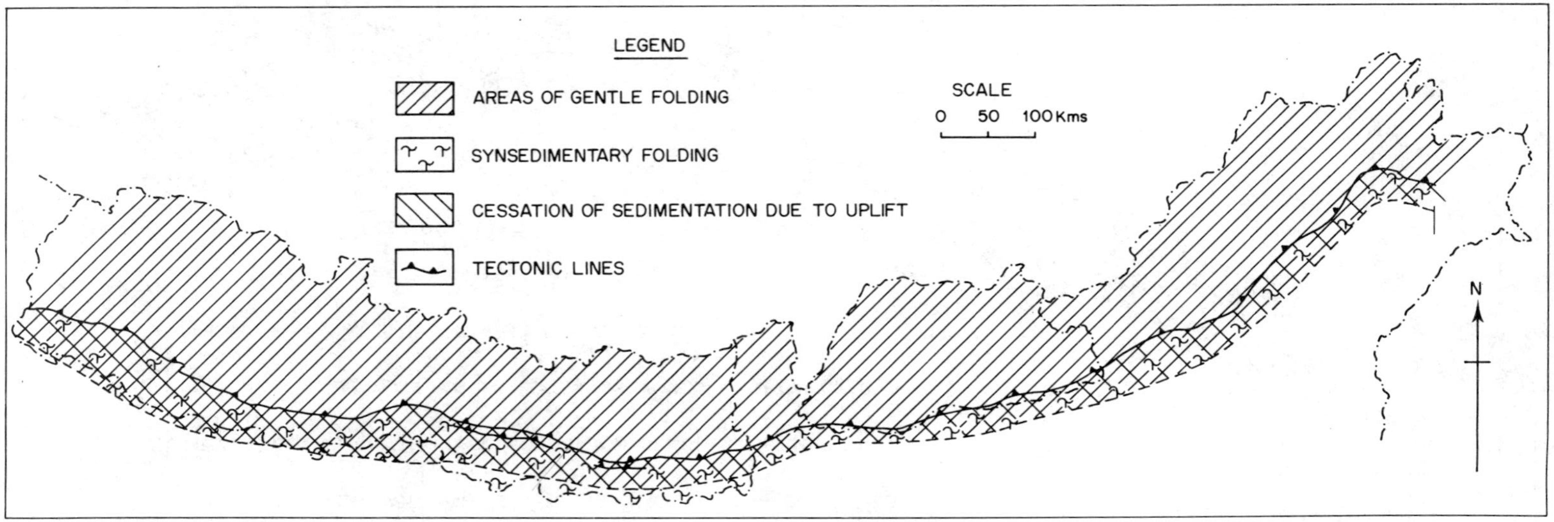

FIG. 14 Siwalikian phase (pliocene–pleistocene) in Central-Eastern Himalaya.

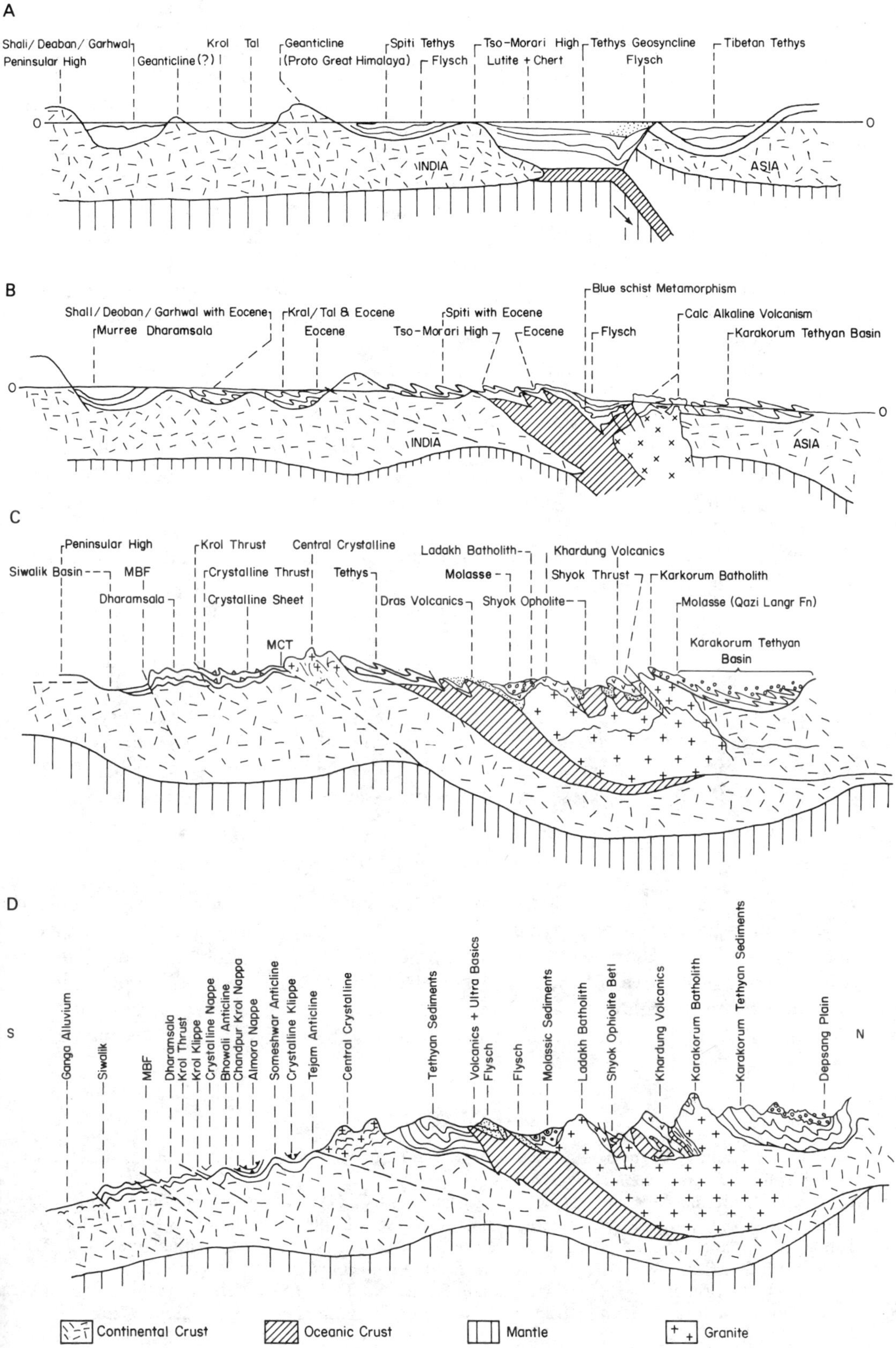

FIG. 15 Evolution of the Himalayas in the Tertiary (D-present structure).

the crust any further, movement again stopped during the Early Miocene. The compressional forces generated huge nappe sheets which later moved southwards. The authors do not agree with Le Fort (1975) and Powell and Conaghan (1973b) that the Main Central thrusts played the part taken by the Indus-Tsangpo suture earlier or with Srikantia and Bhargava (1978) who suggest that the Main Central thrust represents the site of collision. There is no geological or geophysical evidence to support this contention. The lack of ophiolitic rocks, blue schist facies metamorphism and cataclasis expected from a zone of collision does not support this view. Further, Kaila and Narain, based on their geophysical studies in the Himalayas, think that 'seismicity is not totally associated with the Main Central thrust' (1976, p. 18). It is suggested by the authors that from the Middle Miocene period onwards the northward movement of the Indian plate became virtually neglible and the most significant movement was from west to east. The small northward component of the movement was accommodated along the southward-moving nappe sheets without the involvement of subduction anywhere.

The nature of the Tsangpo suture zone in the eastern Himalayas is very little known. However, from the available evidence it can be inferred that the collision of the eastern part of the Indian plate and the Asian plate did not take place before the Miocene. Buoyancy constraints would have stopped underthrusting and subduction in the eastern part at a later date, probably in Pliocene times.

Nandy (1976) suggests that the Arakan Yoma suture zone in the Patkoi-Naga-Lusai-Arakan Yoma hill ranges represents the eastern boundary of the Indian plate. He extends the Arakan-Yoma suture through the Andaman islands to the Sunda Arc. There is no evidence of the continuity of the Tsangpo suture zone with the Arakan Yoma suture zone. Nandy (1976) considers that the Arakan Yoma suture zone represents a subduction zone in oceanic plate and island arc setting. That the oceanic crust component of the Indian plate is still being subducted below the central Burma molasse basin and the Shan-Tenasserim block is supported by the quaternary volcanoes along the basin. Underthrusting of the Indian plate under Burma at an angle of 40–60° is obtained by plotting earthquake hypocentres of the Indo-Burmese area in a vertical section (Das and Filson, 1975).

ACKNOWLEDGEMENTS

The authors express their gratitude to Shri V. S. Krishnaswamy, Director General, Geological Survey of India, for having provided this opportunity of compiling the Tertiary history of the Himalayas. The authors express their gratitude to Shri K. D. Shukla, Deputy Director General, Geological Survey of India, Northern Region, for his encouragement and suggestions. They acknowledge the help rendered by their colleagues S/Shri B. S. Jangpangi, V. K. Raina and A. K. Bhandari in the shape of stimulating discussions and constructive criticisms.

REFERENCES

Anon (1976). Geology of the Himalayas, appraisal of status and definition of problems, Geol. Surv. Ind., *Int. Him. Geol. Sem., New Delhi*.

Ashgirei, G. D, Sinha, A. K., Pande, I. C. and Mallick, B. C. (1977). A contribution to the geology, geochronology and history of regional metamorphism of Himachal Himalaya, *Him. Geol.*, **7**.

Bhandari, L. L. and Agarwal, G. C. (1967). Eocene (Subathu Series) of the Himalayan foothills of North India, *Pub. Cent. Adv. Stud. Geol., Punjab University* (**3**).

Bhandari, A. K., Srimal, N., Redcliffe, R. P. and Srivastava, D. K. (1978) Geology of parts of Shyok and Nubra valleys, Ladakh district, J. & K., India—A preliminary report, *Him. Geol.*, **9**.

Bhanot, V. B., Singh, V. P., Kansal, A. K. and Thakur, V. C. (1977). Early Proterozoic Rb/Sr whole rock age for Central Crystalline gneiss of Higher Himalaya, Kumaun, *Jour. Geol. Soc. Ind.*, **18**(2).

Bhanot, V. B., Kwatra, S. K., Kansal, A. K. and Pandey, B. K. (1978). Rb/Sr whole rock age for Chail Series of Northwestern Himalaya, *Jour. Geol. Soc. India*, **14**(5).

Calkins, J. A., Offield, T. W., and Tayab Ali, S. (1969). Geology and mineral resources of the southern Hazara district, West Pakistan and part of Western Kashmir, *U.S. Geol. Survey*.

Crawford, A. R. (1974). The Indus Suture Line, the Himalayas, Tibet and Gondwanaland, *Geol. Mag.*, **111**(5).

Das, S. and Filson, J. R. (1975). On the tectonics of Asia, *Earth & Planetary Science Letters*, **28**.

Desio, A. (1974). The Karakorum in Mesozoic-Cenozoic Orogenic belts, ed. Spencer, *Spec. Pub. Geol. Soc. Lond.* **4**.

De Terra, H. (1935). Geological studies in the northwest Himalaya between the Kashmir and Indus Valleys, *Mem. Connecticut Acad. of Arts and Science, 8, Yale North India Expedition. Article II, New Haven*.

Dewey, J. F. and Bird, J. M. (1970). Mountain belts and the new global tectonics, *Jour. Geophys. Res.*, **75**.

Dhoundial, D. P., Kakoti, N., Sangma, L., Dange, M. N., Das, A. K., Manna, S. S. and Santra, D. K. (1977). Observations on tectonic lines in Arunachat Pradesh, *Mis. Pub. G.S.I.*, **31**.

Frank, W., Gransser, A. and Trommsdorft, V. (1977). Geological observations in the Ladakh area (Himalaya): a preliminary report, *Schweiz Mineral Petrogr. Mitt.*, **57**.

Fuchs, G. R. (1975). Contributions to the geology of the northwestern Himalayas: Vienna, *Geol. Bundesan-statt Abh*, **32**.

Gansser, A. (1964). *Geology of the Himalayas*, London: Wiley.

Gansser, A. (1966). The Indian Ocean and the Himalayas—a geological interpretation, *Ecol. Geol. Helv.*, **59**.

Gansser, A. (1977). The great suture zone between Himalaya and Tibet. A preliminary account, in *Coll. CNRS 268, Ecologie et Geolgie de la Himalaya*.

Ghosh, A., Chakraborti, B. and Singh, R. K. (1974). Structural and metamorphic history of the Almora Group, Kumaun Himalaya, U.P., *Him. Geol.*, **4**.

Jager, E., Bhandari, A. K. and Bhanot, V. B. (1971). Rb/Sr age determination on biotites and whole rock samples from the Mandi and Chor granites, Himachal Pradesh, India, *Ecol. Geol. Helv.*, **64**(3).

Jangpangi, B. S. (1978). Stratigraphy and structure of Bhutan Himalaya, in *Tectonic Geology of Himalaya ed. Saklani*, New Delhi: Today & Tomorrow Publishers.

Jain, A. K., Thakur, V. C. and Tondon, S. K. (1974). Stratigraphy and structure of the Siang district, Arunachal (NEFA) Himalaya, *Him. Geol.*, **4**.

Kaila, K. B. and Narain, H. (1981). Evolution of the Himalaya based on seismotectonics and deep seismic soundings, *Geol. Surv. Ind., Misc. Pub.*, **41**.

Karunakaran, C. and Ranga Rao, A. (1979). Status of exploration for hydrocarbons in the Himalayan region—contributions to stratigraphy and structure, *Geol. Surv. Ind. Misc. Pub.*, **41**.

Krishnan, M. S. and Aiyengar, N. K. N. (1940). Did the Indobrahm or the Siwalik river exist? Rec. G.S.I., **75**(6).

Krummenacher, D., Basett, A. M., Mingery, F. A. and Layne, H. F. (1978). Petrology, metamorphism and K/Ar age determinations in Eastern Nepal, *Tectonic geology of the Himalaya, ed. Saklani*, New Delhi: Today & Tomorrow Publishers.

Le Fort, P. (1975). Himalaya: The collided range. Present knowledge of the Continental arc, *Am. Jour. Sci.*, **275**(A).

Le Pichôn, X. and Hortzler, J. R. (1968). Magnetic anomalies in the Indian Ocean and sea floor spreading, *J. Geophys. Res.*, **73**.

Le Pichôn, X., Francheteau, J. and Benin, J. (1973). *Plate Tectonics*, Amsterdam: Elsevier.

Meyerhoff, H. A. and Meyerhoff, A. A. (1978). Spreading history of the eastern Indian Ocean, and India's northward flight from Antarctica and Australia: Discussion and reply, *Geol. Soc. Am. Bull.*, **89**(4).

Nagpal, N. K. and Nagpaul, K. K. (1974). Fission track age of some Himalayan Muscovites, *Him. Geol.*, **4**.

Nandy, D. R. (1982). The Assam Syntaxis of the Himalayas—a re-evaluation, *Misc. Pub. G.S.I.*, **24**(II).

Nandy, D. R. (1976). Geological set-up of the eastern Himalayan and the Patkoi-Naga-Arakan Yoma (Indo-Burman) hill ranges in relation to the Indian plate movement, *Int. Him. Geol. Sem., New Delhi*.

Pati, U. C. (1978). Mutual relationship of Chail and Jutogh thrusts in the Tiuni area dist. Dehradun and Uttarkashi, U.P. (Abstr.) in *Ninth Sem. Him. Geol. Dehradun*.

Pati, U. C. and Rao, P. N. (1983). Geotectonics of the main central thrust of U.P. Himalaya, *Himalayan Shears*, ed. P. S. Saklani, Himalavan Books, New Delhi.

Pascoe, E. H. (1973). *A Manual of the Geology of India and Burma*, 3rd ed., Calcutta: Govt. of India Press.

Powell, C. McA. and Conaghan, P. J. (1973a). Polyphase deformation of Phanerozoic rocks of the Central Himalayan gneiss, northwest India, *J. Geol.*, **81**(2).

Powell, C. McA. and Conaghan, P. J. (1973b). Plate tectonics and the Himalayas, *Earth & Planet Sci. Lett.*, **20**(1).

Raina, B. N. and Dungrakoti, B. D. (1975). Geology of the area between Nainital and Champawat, Kumaun Himalaya, U.P., *Him. Geol.*, **5**.

Raina, B. N. and Krishnaswamy, V. S. (in press). Prefaratory note to the stratigraphic atlas of sedimentary basins of western Himalaya (in press).

Raina, B. N. and Krishnaswamy, V. S. (1982). A review of recent work in the western sector of the Himalayas, *Geol. Surv. Ind. Misc. Pub.*, **41**.

Ranga Rao, A. (1971). New mammals from Murree (Kalakot Zone) of the foothills near Kalakot, J & K State, India, *Jour. Geol. Soc. Ind.*, **12**(2).

Ranga Rao, A. (1972). Further studies on the vertebrate fauna of Kalakot, India, New Mammalian genera and species from the Kalakot Zone of Himalayan foothills near Kalakot, J & K State, India, *Directorate of Geology ONGC, India, SP Paper No. 1*.

Ranga Rao, A. (1973). Notices of the new mammals from the Upper Eocene Kalakot beds, India, contributions to the study of vertebrate fauna of Kalakot India, *Directorate of Geology, ONGC, India, SP Paper No. 2*.

Ray, S. (1976). India-in Gondwanaland and the Himalayas, *Ind. Jour. Earth. Sc.*, 236–40.

Ray, S. K. and Naha, K. (1971). Structural and metamorphic history of the 'Simla Klippe'—a summary, *Him. Geol., (1)*.

Remy, J. M. (1978). New results on lithostratigraphy and structure of West Nepal Himalaya in *Tectonic Geology of Himalaya*, ed. Saklani, New Delhi: Today & Tomorrow Publishers.

Rupke, J. (1974). Stratigraphic and structural evolution of the Kumaun Lesser Himalaya, *Sediment. Geol.*, **11**(2–4).

Sahni, A. and Khare, S. K. (1971). Three new Eocene mammals from Pajauri dist. J & K, *Jour. Pal. Soc. Ind.*, **16**.

Sajni, H. S., Nagpaul, K. K. and Sharma, K. K. (1978). Fission track ages from the Pandoh-Baggi area, Himachal Himalaya, and their tectonic interpretations, *Tectomophysics*, **46**.

Sarkar, S. N. and Shrish (1976). Tectonic analysis of a part of the folded Baijnath nappe and inner sedimentary belt in the Baijnath-Kausani-Someswar area, U.P., *Him. Geol.*, **6**.

Saxena, M. N. (1971). Geological classification and the tectonic history of the Himalayas, *Proc. Ind. Nat. Acad. Sci.*, **37A**.

Shankar, R., Padhi, R. N., Prakash, G., Thussu, J. L. and Das, R. N. (1982). The evolution of Indus Basin, Ladakh, India, *Geol. Surv. Ind. Misc. Pub.*, **41**(III).

Singh, I. B. (1979). Some thoughts on the Evolution of Himalэya and the northern limit of the Indian Shield, *Sond. aus der Geol. Bunds. Band.*, **68**.

Sinha, Roy, S. (1973). Kinematic significance of conjugate folds in the Daling metamorphites from Kalimpong hills, Sikkim Himalaya, *Him. Geol.*, **3**.

Sinha, A. K. and Bagdasarian, G. P. (1977). K/Ar dating of some magmatic and metamorphic rocks from Tethyan and Lesser Zones of Kumaun and Garhwal, Indian Himalaya and its implication in the Himalayan tectogenesis. In *Himalaya: Sciences de la terre*. éditons du CNRS.

Srikantia, S. V. (1977). Sedimentary cycles in the Himalayas and their significance on the orogenic evolution of the mountain Belt, *Himalaya, Sciences de la terre*, éditons du CNRS.

Srikantia, S. V. and Bhargava, O. N. (1978). The Indus tectonic belt of Ladakh Himalaya, its geology, significance and evolution, in *Tectonic Geology of Himalaya*, ed. Saklani, New Delhi: Today & Tomorrow Publishers.

Srikantia, S. V., Ganesan, T. M., Sinha, P. K. and Tirkey, B. (1976). Geology of a part of Zaskar basin, Ladakh, with special reference to Late Caledonian Kurgiakh Orogeny, *Abst. Sem. on Geol. of Kashmir Himalaya, Geol. Surv. Ind. Srinagar*.

Srimal, N., Bhandari, A. K. and Chakrabarty, S. (1980). Report on traverse in Karakorum region, Ladakh, J. & K, *Him. Geol.*, **10**.

Srimal, N. and Pati, U. C. (1981). Structure of Almora crystallines and underlying sedimentary sequence in Marchula–Juimar area (U.P.), *Metamorphic Tectonites of the Himalaya*, ed. P. S. Saklani, Today and Tomorrow Publishers, New Delhi.

Tewari, A. P. (1964). On the Upper Terkary deposits of Ladakh Himalayas and correlation of various geotectonic units of Ladakh with those of the Kumaon-Tibet region, *Proc. Sec. XI, XX 22nd Sess. Int. Geol. Cong., New Delhi*.

Tripathi, C., Cupta, P. D., Ghosh, R. N., Malhotra, G. and Dungrakoti, B. D. (1980). Geology of the area around Pasighat, Siang dist. Arunachal Pradesh, with special reference to its Eocene rocks. *Him. Geol.*, **10**.

Valdiya, K. S. (1964). A note on the tectonic history and the evolution of the Himalayas, *Int. Geol. Cong. 22nd Sess., Part XI, New Delhi*.

Valdiya, K. S. and Cupta, V. J. (1972). A contribution to the geology of northeastern Kumaun with special reference to the Hercynian gap in Tethys Himalaya, *Him. Geol.*, **2**.

Vashi, N. M. and Merh, S. S. (1974). Fold history of the Almora Synform, *Him. Geol.*, **4**.

Wadia, D. N. (1931). The syntaxis of the north-west Himalayas—its rocks, tectonics and orogeny, *Rec. Geol. Surv. Ind.*, **65**(2).

Global Correlation of Tectonic Movements
Edited by Yu. G. Leonov and V. E. Khain

13

On Two Tertiary Tectonic Phases in Cuba

KÁROLY BREZSNYÁNSZKY* AND MANUEL A. ITURRALDE-VINENT**

The two small-scale maps of the island of Cuba in this paper were originally presented at the meeting of IGCP project No. 107 held in Baku, 1979. They were compiled according to the agreed captions and other recommendations of the working group.

The two timespans corresponding to the Eocene–Oligocene and Oligocene–Miocene boundaries do not belong to phases of the highest tectonomagmatic activity in Cuba. In spite of this we think it will be useful to compare the tectonic events of a given time occurring in Cuba and the Eurasian continent.

THE 'PYRENEAN EPOCH' (Fig. 1)

The consolidation processes of the Cuban crust began in the Mid-Eocene time and continued during the Late Eocene–Oligocene. The latter epoch saw the process by which the earlier characteristic structural belts extending parallel to the island were losing their individuality. The longitudinal sectioning of the island changed to a transverse division caused by a series of normal and strike-slip faults. The region is devoid of any manifestation of Alpinotype deformation and vertical movements prevailed. This period of development beginning with the Middle Eocene and characterized by mainly vertical movements was called the 'platform-like period' by Iturralde-Vinent (1977).

The time, shown in the map, corresponds to the Late Eocene–Early Oligocene, to the *Globorotalia cerroazulensis* and *Globorotalia opima opima* zones, although in many cases it was very difficult or even impossible to determine exactly the age of the fauna.

The Pyrenean epoch in Cuba is characterized by a general regression which first occurred in the eastern part of the island. This process began in the Middle Eocene and reached its maximum in the Middle Oligocene. Landmasses—sites of weak faulting—were exposed, but in the western and central part of the present-day island we have to recognize more intense vertical movements, as shown by the coarse-grained conglomeratic rocks. The landmasses, forming archipelagos, were separated by marine basins. In many basins, sites of clastic-carbonate sedimentation, there was no change in facies or rate of sedimentation and only some synsedimentary folds in the western part show crustal instability.

In the Los Palacios basin (western Cuba), in connection with the horizontal movements along the Pinar fault, there was an accelerated subsidence of the basin, accompanied by the

* Central Office of Geology, Budapest, Hungary.
** Geological Institute of Cuban Academy of Sciences, La Habana, Cuba.

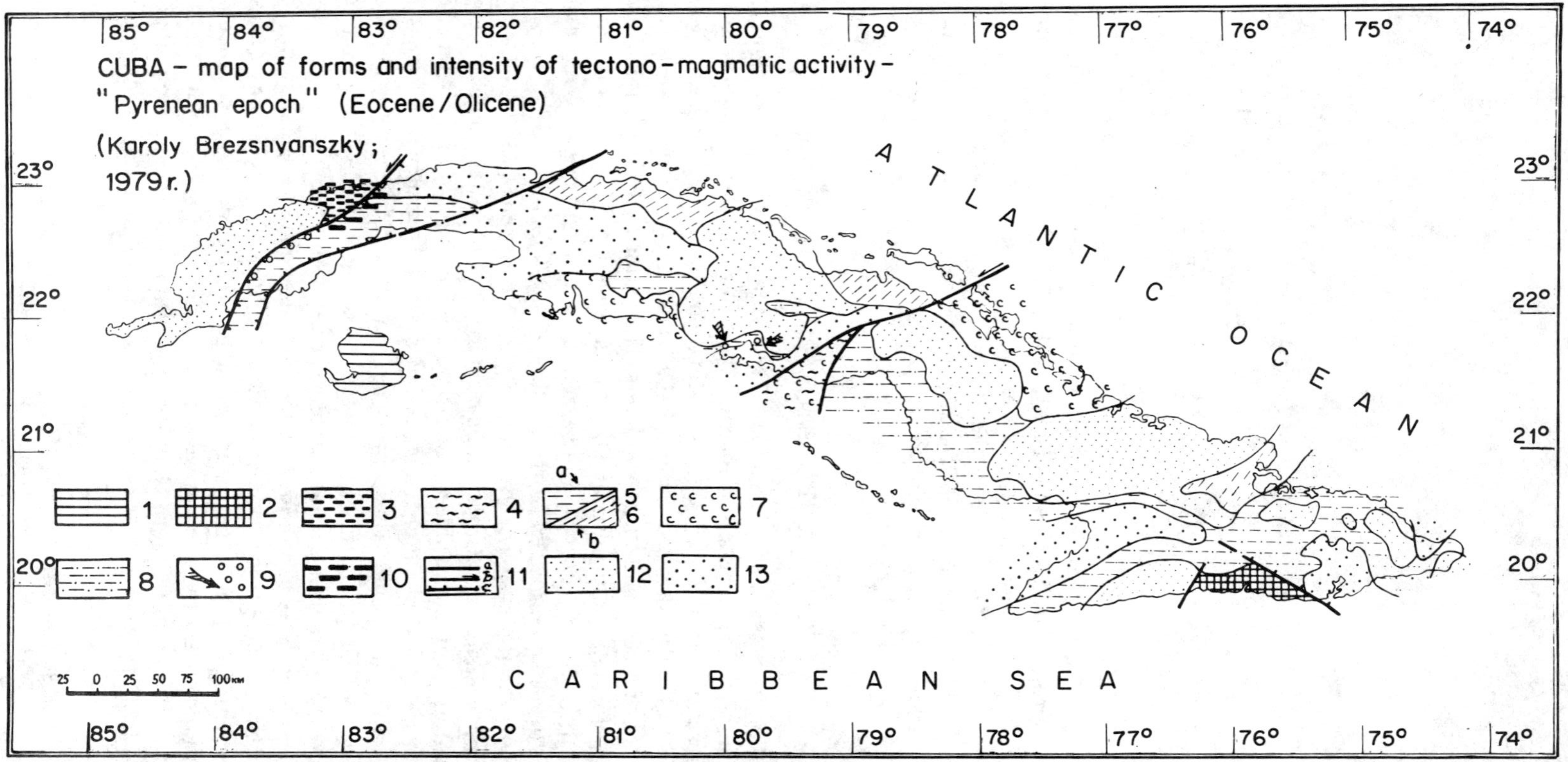

FIG. 1 Map of forms and intensity of tectonomagmatic activity: 'Pyrenean epoch' (Eocene–Oligocene boundary). For explanation of symbols see Fig. 2.

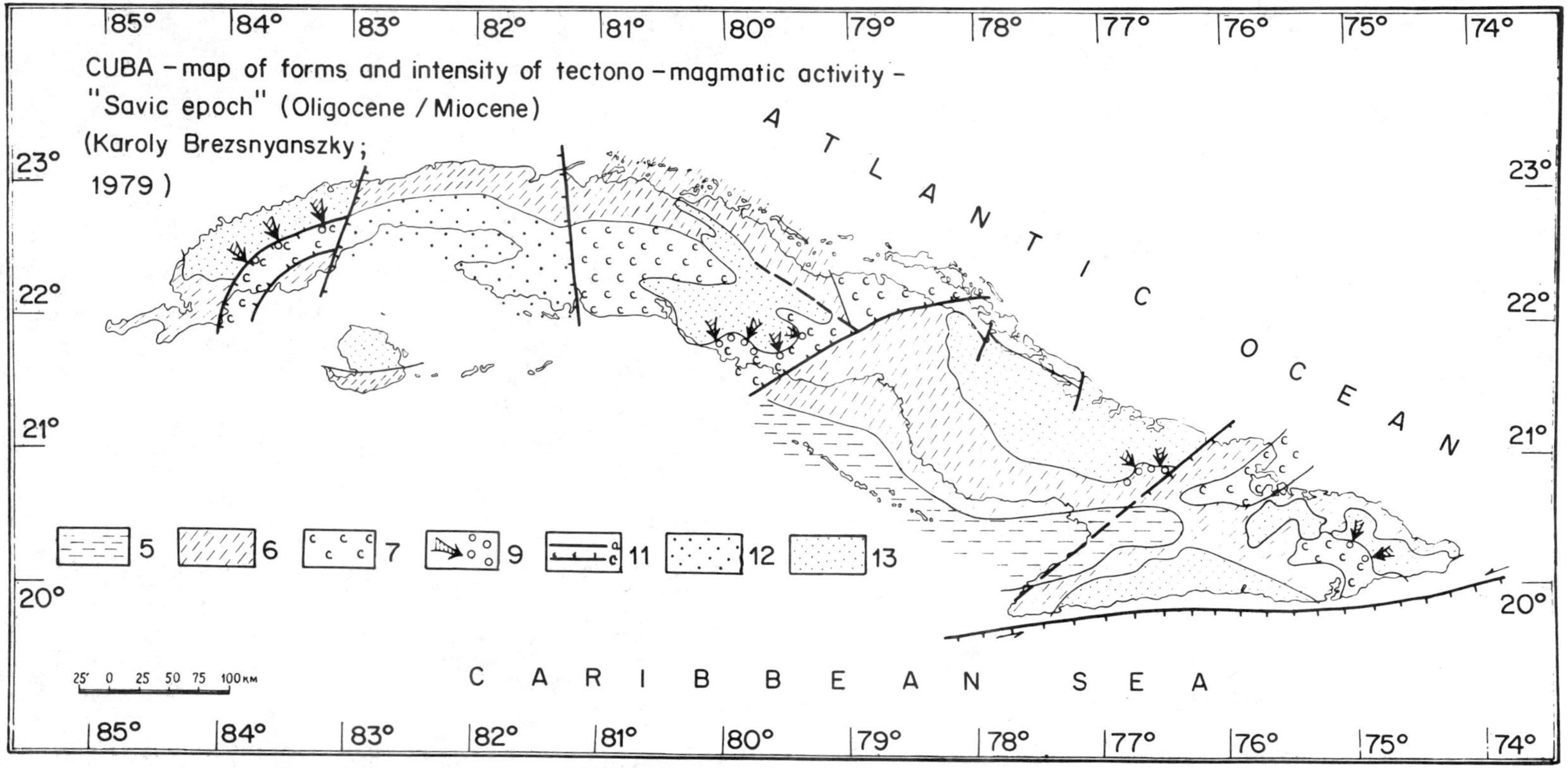

FIG. 2 Map of forms and intensity of tectonomagmatic activity: 'Savic epoch' (Oligocene–Miocene boundary).

1 – Warping and tilting; 2 – Block tectonics ('germanotype', 'Saxonian' folding): (a) gentle, (b) intense; 3 – Angular unconformity, type of deformation non-specified (magnitudes of unconformity shown in degrees); 4 – Synsedimentary folding; 5 – Pronounced acceleration of subsidence; 6 – Beginning of regional subsidence (subsidence over a vast area); 7 – Slowing down of subsidence or beginning of upward movements; 8 – Synsedimentary uplift; 9 – Coarse clastics: breccia, conglomerates, etc. (usually interpreted as intensification of uplift in source area) (source of sediments should be indicated); 10 – Zones of olistostrome complexes; 11 – (a) Fractures and fracture zones, (b) strike-slip faults, (c) normal faults; 12 – Areas of continuous sedimentation; 13 – Areas of long uplift and washout where it is impossible to distinguish movements at the timespan under review.

formation of marine olistostromes and breccias (Brezsnyánszky and Iturralde-Vinent, in press). In the central part of the present-day island the Sierra de Escambray underwent an intense vertical uplift which also changed the polarity of movements in the surrounding basins.

THE 'SAVIC EPOCH' (Fig. 2)

The period under consideration spans the Late Oligocene to Early Miocene (Aquitanian) time, *Globigerina ciperoensis ciperoensis* and *Catapsidrax dissimilis* zones.

The Savic epoch in Cuba is characterized by the culmination of a general transgression which started at the end of the Middle Oligocene. The present area of the island of Cuba is subdivided by some active, generally northeast-southwest directed faults. Each of the megablocks, separated by the above mentioned faults, has an individual character manifested by the pattern of the land and basinal areas and by the varying intensity of vertical movements. The Bartlett fault, to the South of Cuba, was at this time an active left-lateral strike-slip fault (Brezsnyánszky and Iturralde-Vinent, 1977).

The emerged landmasses generally had a low relief, but the conglomeratic series along the southern shores indicate some differential block movements. Together with the general submergence of the former landmasses the sedimentary basins, except for the Cauto basin (eastern Cuba), are affected by differential upward movements. This shows the intensification of the stabilizing process of the Cuban crust. In the Cauto basin a pronounced acceleration of submergence occurred, balanced by terrigeneous sedimentation. The clastic-carbonate sedimentation of the Pyrenean epoch contrasts with the Savic epoch in which carbonate sedimentation was predominant.

REFERENCES

Brezsnyánszky, K. and Iturralde-Vinent, M. A. (1977). Paleogeografia del Paleogeno de Cuba oriental, *Geologie en Mijnbouw*, **57**(2), 123–33.

Brezsnyánszky, K. and Iturralde-Vinent, M. A. (in press). Paleogeografia del Paleogeno de La Habana, in *Geologia de La Habana*, Cuba: Acad. Cienc.

Furrazola Bermudez, G. *et al*. (1964). *Geologia de Cuba*, Edit, Univ. 239 pp.

Iturralde-Vinent, M. A. (1972). Principal characteristics of Oligocene and lower Miocene stratigraphy of Cuba, *AAPG Bull*., **56**, 2369–79.

Iturralde-Vinent, M. A. (1977). Los movimentos tectonicos de la etapa de derarrollo plataformico de Cuba. Acad. Cienc. Cuba, Inf. Cient.-Tech. 20, 24 pp.

Global Correlation of Tectonic Movements
Edited by Yu. G. Leonov and V. E. Khain

14

Self-excited Oscillations in Geological Systems (Model Studies and Problems of Correlation)

A. V. LUKYANOV*

Geological processes develop irregularly. Their intensity changes in time, as well as in space. This is one of the principal laws of geology, on which the correlation of geological events is based. However, what causes irregular development of geological processes? Does their intensity change simultaneously all over the globe or is it merely local? These problems are fundamental to the geological science and so far have not been resolved. Yet it is clear that irregularities are both global and local and that a great many geological systems are characterized by their self-excited oscillations. Non-uniformity of evolution is the most characteristic feature of such systems. The earth is 'densely populated' with such systems, and their existence gives rise to difficulties and problems in global correlation of geological phenomena. These systems, which have their own periodicity of evolution, complicate global correlation. At the same time, however, they provide the study with original information on the specific structure of individual regions and of individual geological systems. It is at this point that modelling helps correlation.

One task in correlating geological events is to trace general patterns in the evolution of our planet and, as the ultimate goal, to find out their causes. Among these are causes of characteristic trends and cyclicity in time ('phases' and sometimes cyclic) shown by the evolution of individual regions and by the earth as a whole. Finding the causes of patterns of interest requires knowledge of the patterns themselves, as well as of particular features exhibited by these processes (both inside the earth and over its surface) which may influence the events under consideration or even predetermine their appearance. For this reason the correlation of geological phenomena has to be assisted by the simulation of geological systems and processes. Geological processes are simulated to gain better knowledge of those of their features which are impossible to study in nature. Each particular problem is studied by constructing special models according to the purpose in mind. The simulated processes are of necessity idealized and simplified. This allows one to study features of interest in their pure form, and to come to a deeper understanding of their mechanism.

In using idealization, however, one must not forget than an idealized simulated model significantly differs from a natural phenomenon which is always complicated by many incidental circumstances. For this reason the use of simulation methods for analysing specific geological objects demands caution and usually involves additional investigations in order to make sure of the applicability of the model, so as to ensure that the

* Geological Institute of USSR Academy of Sciences, Moscow.

simplifications introduced into the model do not significantly affect the object under study.

This article is included here to demonstrate some possibilities and results obtained from the simulation of geological systems with non-uniform evolution, thus causing pulses in geological processes.

Non-uniform evolution of geological phenomena can be distinctly observed in objects of very different size and significance, with, at the one end, rhythmic character of sedimentary structure and, at the other, tectonomagmatic cycles and tectonic events like separate phases of mountain-building and folding (Cycles in sedimentation, 1971; Shatskiy, 1939, 1964; Stille, 1924, 1964).

The problem of why phases, cycles or rhythms of different lengths exist is one of the major problems of tectonic evolution of the earth. As indicated by the many contradictory opinions and hypotheses on the subject its solution lies in the future. It is, however, from general considerations clear that any cause can always be classified as either external or internal with respect to the geological system in which rhythms are being formed (Lukyanov, 1972). If these external effects are periodic, a periodicity will arise in the evolution of the system itself; however, such a periodicity is superposed, having its origin outside the system and not related to its evolution. This can be described as a forced oscillation of a system. Internal causes give rise to oscillations directly related to evolution of the system under study, governed by its particular constitution and generated within itself, thus being self-excited oscillations (Andronov *et al.*, 1981). We will only consider internal causes since we believe them to be especially important for the understanding of many non-uniformities and 'phases' in the evolution of the earth and individual regions, since they seem to be capable of explaining their origins and specific features without resorting to debatable extraterrestrial effects. Also, being of terrestrial origin, they are accessible to investigation, and we would like to emphasize that the best method of investigation should be that which provides for a close mutual control between simulation and the global correlation of geological phenomena.

There are many geological systems capable of generating oscillations. We are interested in systems of geological importance: first, those with oscillation periods measured in geological time, that is, many thousands and millions of years; second, large-scale systems; third, those capable of effecting such processes as magmatism, sedimentation, rock deformation and so on. The study of such self-excited systems is still in its infancy and it is for this reason that the potential and the results of simulation are so interesting.

SELF-EXCITED SYSTEMS ARISING IN THE MAGMATIC PROCESS

The simulated models of self-excited systems in the magmatic process are concerned with heat transfer through a rock mass that undergoes melting and crystallization. Included are the constant heat flow, latent heats of melting and crystallization, specific heat and thermal conductivity depending on the state of the material, and the boundary conditions. Deformations and mechanical penetration of magma are not considered. The system does not generate oscillations in most of the cases. Under certain conditions, however, thermal oscillations are produced, causing pulses of heat flow and of melting (crystallization) of the rock matter. We will now discuss models related to lithospheric melting and crystallization of magma chambers.

Melting of the Lithosphere, the Homogeneous Model

A. N. Tikhonov *et al*. (1969, 1972) were the first to obtain seemingly paradoxical results about a pulsating melting of the lithosphere, if we disregard the 'thermal cycles' hypothesis of Joly (1930) and his followers as unsupported by calculations. The calculation of a theoretically constructed model has been carried out with the aid of a computer. The model describes an earth heated by radioactive decay with a resultant constant temperature at the surface. The values of thermal conductivity and capacity, melting temperature and latent heat values for rocks at different depths have been taken in accordance with the current concepts about the values for these quantities. In simulating the appearance of convection in the melt the effective heat transfer in the melt, λ_{eff}, was assumed to exceed the thermal conductivity for solid rock, λ. Calculations have shown that in this model a melt appears at depths 400–500 km from the earth's surface, and rapid melting occurs at the top of the molten layer. As a result, the upper boundary of the melt begins to rise (is displaced

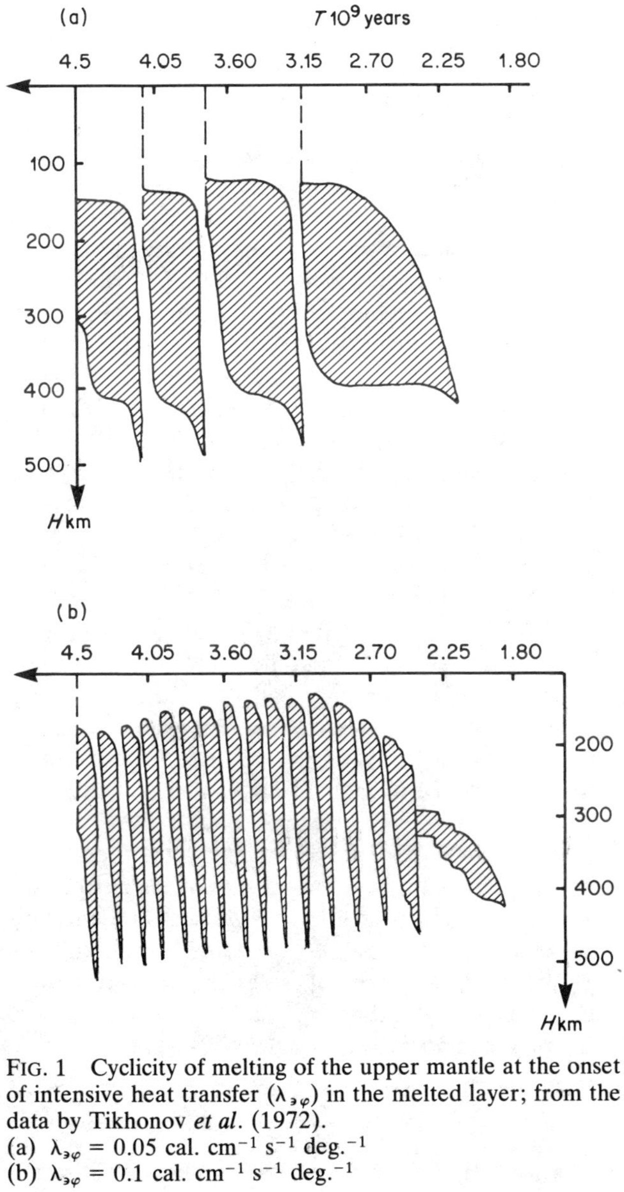

FIG. 1 Cyclicity of melting of the upper mantle at the onset of intensive heat transfer ($\lambda_{\text{э}\varphi}$) in the melted layer; from the data by Tikhonov *et al.* (1972).
(a) $\lambda_{\text{э}\varphi} = 0.05$ cal. cm^{-1} s^{-1} deg.$^{-1}$
(b) $\lambda_{\text{э}\varphi} = 0.1$ cal. cm^{-1} s^{-1} deg.$^{-1}$

upwards by melting). After some time, crystallization sets in at the lower boundary of the melt, causing this boundary to rise up.

In this way the giant molten layer, which may be as thick as 200–300 km, slowly moves up due to melting at its top and crystallization at the bottom; thus, a 'zone melting' is generated (Vinogradov, 1953; Shimazu, 1959, 1961; Magnitskiy, 1964). At first melting at the top is faster than crystallization at the bottom, thus increasing the thickness of the molten layer. Further up, however, the rate of melting at the top decreases, while the rate of crystallization at the bottom increases. After some time the bottom begins to move up faster than the top, so that the molten layer decreases in thickness, slowly at first, then at an increasing rate. The earth's surface is still not reached as the bottom overtakes the top, the melt vanishes completely. In various models this happens at depths of 150–30 km. Some time after the melt has vanished melting recurs at a depth of 400–500 km, and the process repeats itself. The rate of this process, and the period of a complete cycle, are dependent on the value of effective heat transfer in the melt (λ_{eff}), that varies in different versions of the model. This variation results in cycles with periods of between 140 and 500 m.y. (Fig. 1). Quoting the authors: 'We thus see that a cyclic character of the upper mantle melting found out by calculations alone, without invoking any extra hypotheses, is due solely to thermal conductivity switching over in the molten zone. We can make the cycle shorter by increasing the λ_{eff}. Again, short and long cycles may be made to alternate if λ_{eff} is varied repeatedly. Hence the ratio λ_{eff}/λ is the parameter that controls melting in the upper mantle heated by the energy of radioactive decay' (Tikhonov *et al*., 1972, p. 35).

We have studied a similar model by means of the hydrointegrator IGL*. We will quote the specific conditions of the experiment, since these were similar to those described above. The conclusions of Tikhonov *et al*. have on the whole been confirmed, though some details were different in our model. The most important correction to the scheme of the authors cited above has proved to be that the solution is clearly dependent upon the instant in time when convection sets in the melt, a fact which was overlooked by them. Cycles arise if a sufficiently thick molten layer is available at the instant when convection sets in (and so λ 'switches' to λ_{eff}), the thicker the layer the longer are the periods of the cycle (Fig. 2); while if higher heat transfer exists in small portions in the melt (for example, when its viscosity is small) then instead of a powerful pulsation of large molten masses, a zone of 'half-molten' rock appears whose mean heat transfer is slightly higher owing to numerous small molten portions rising in the same manner: 'melting at the front—crystallization at the rear'. These results show that the generation of thermal oscillations in the system and their frequency depend not only upon the ratio λ_{eff}/λ, but also upon the instant when λ_{eff} first appears. Both are governed by the mantle conditions, and these can be included in the model by making assumptions about the dependence of higher heat transfer on the thickness of the molten layer.

The above results of simulated lithospheric melting show that, proceeding from current notions about the internal composition and properties of the earth, we can expect a molten layer to appear in the upper mantle which, letting the heat flow through itself, will begin to move*, and will convert the upper mantle into a system generating thermal oscillations. The essential principle by which this system operates is as follows. The molten layer is a condenser of thermal energy accumulating in it, since melting proceeds through latent heat capacity. The amount of accumulated energy depends upon the thickness of the molten

* For the principles and methods of study by Lukyanov and IGL hydrointegrators the reader is referred to Lukyanov (1937, 1959) and Veviorovskaya *et al*. (1962).

* Again emphasizing that the molten layer does not move bodily, but through the melting-crystallization process.

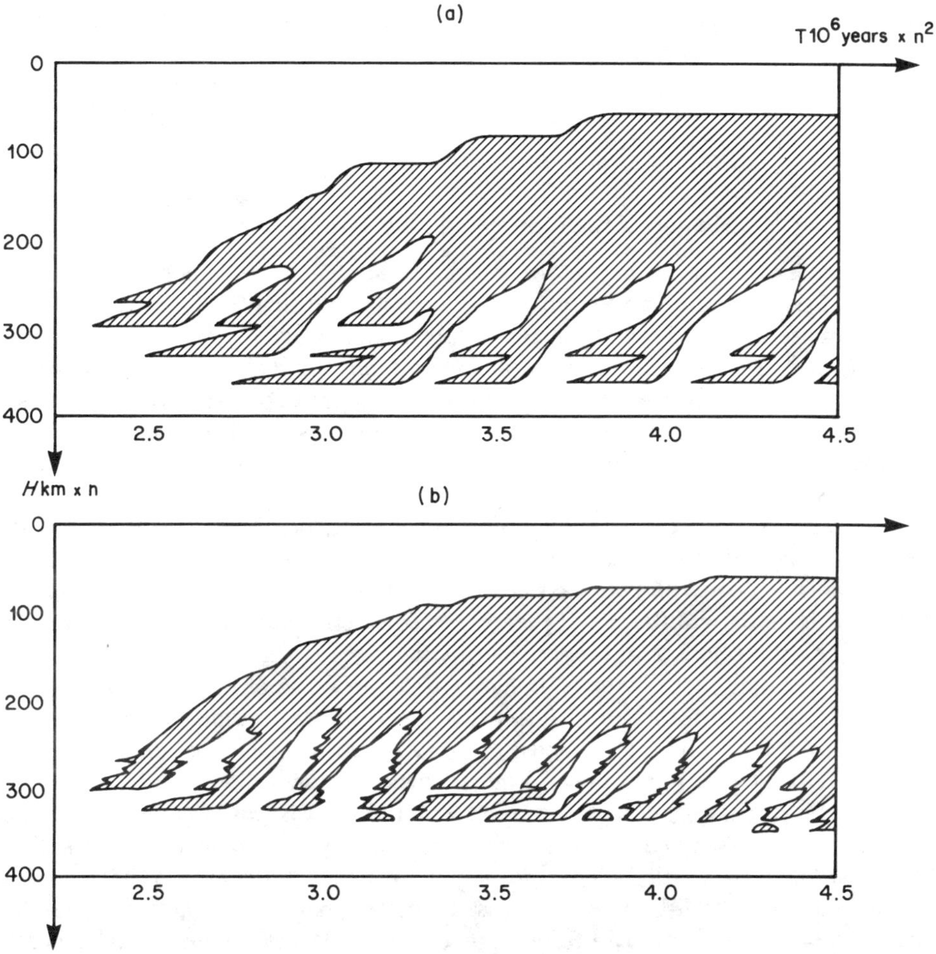

FIG. 2 Dependence of melting periods of homogeneous lithosphere on the moment of appearance of convection (onset $\lambda_{э\varphi}$). The thickness of melt at the moment of onset $\lambda_{э\varphi}$ reaches (a) 30 km, (b) 10 km.

layer. This layer also has a variable thermal resistance which drops sharply when convection sets in. The amount of energy accumulated in the molten layer, that is the 'thermal condenser capacity', is dependent upon the moment when convection is 'switched on'—the later it is the greater is the energy accumulated in the layer. After the appearance of convection and hence a sharp drop in the thermal resistance the accumulated energy rapidly begins to rise, thus melting the top and pushing the melt upwards towards the earth's surface. At this stage no accumulation of energy takes place and the upward velocity* of the melt is controlled by the value of λ_{eff}. As the melt approaches the earth's surface, the heat flow through a thin envelope that covers it increases as the envelope gets progressively thinner. The increasing heat outflow away from the melt uses energy accumulated in it, until it is completely exhausted† and the melt is frozen at some distance from the surface. After

* See the previous footnote.

† In many of our models, the melt did not freeze completely. A long-lasting molten layer of variable thickness was sandwiched between a solid layer near the earth's surface and a zone of pulsating melting at a great depth.

this the process will begin anew. The period of a cycle is thus a sum of the time of energy accumulation in the molten layer (depending on the instant when convection sets in), the melt rising time (depending on λ_{eff}), and the crystallization time (depending upon both), while the whole system functions like an RC-generator and is a relaxation self-excited system.

Melting of the Lithosphere, the Two-layer Model

A self-excited system occurring in the two-layer model of the lithosphere is still more interesting. This system is considerably more complicated than that analysed earlier. The processes discussed in the system originated during the passage of heat flow through the two layers composed of materials with different thermal-physical properties. When melted they mix to form a melt with variable composition. The most important processes took place near the boundary between these two media. Similar processes were expected to occur near the Moho discontinuity; the simulation was carried out by the IGL hydrointegrator of the Lukyanov system.

The changes in the heat transfer value of melt, depending on the distribution of concentrations of initial materials in it, proved essential for the model. Heat transfer in liquids is known to occur in two ways: with or without convection. Heat transfer value in both cases is essentially different. If there is no convection, the liquid phase transfers less heat than the solid phase. Convection sharply increases heat transfer in the liquid and it becomes greater than that in the solid phase. In the presence of heat flow passing upwards convection in the homogeneous melt appears if the thermal gradient is greater than the adiabatic. In the lithosphere this almost always happens because the adiabatic gradient is very small (~0.3 deg/km). It is by an order of magnitude less than the gradient of melting temperatures of rocks (~3 deg/km) (Magnitskiy, 1964). It is this circumstance that makes the molten mass convect and move *in the direction* of heat flow, namely upwards (Shimazu, 1959, 1961; Magnitskiy, 1964). This process was regarded as decisive in the analysis of the one-layer model of the lithosphere discussed earlier. This is valid, however, only for the homogeneous melt. An entirely different process takes place when the melt is generated on the boundary of two media and when its composition is heterogeneous. Concentration gradient can affect the density of the melt more than temperature. Therefore, its density can increase from the top down even in the case of considerable temperature gradients. Convection is not generated under these conditions and heat transfer in the melt remains low. Moreover, the gradient of melting temperature can be lower than the temperature gradient. As a result, the melt will be overheated at the lower boundary and overcooled at the upper boundary. This will inevitably result in melting at the lower boundary, crystallization at the upper boundary and a corresponding migration of the melt downwards, that is to say, 'zone melting' takes place with the movement of melt *against* heat flow. This process was discovered some time ago and experimentally reproduced (Whitman, 1926) in studying desalination of sea ice. It is now well known (Savel'yev, 1963; Pounder, 1965; Tsurikov, 1976). In our model the change of two types of heat transfer results in the origin of the self-excited system. At the first stage the melt migrates downwards, and at the second stage it moves upwards. In accordance with the direction of its movement the tendency of evolution of its composition is changed and some tectonically important consequences become apparent.

By using one of the calculated models as an example we shall show the development of the process. The model has two layers with a horizontal boundary between them. The melting temperature of the upper layer (A) is lower than that of the lower layer (B). It is assumed that the material (B) of the lower layer may dissolve in the melted material (A) of the upper layer, and its solubility depends on temperature according to the given eutectic diagram; the density of the material (A) is lower than that of the material (B), and the density of a melt of mixed composition depends on the material (B) concentration in it. The other conditions are not essentially different from those for a homogeneous lithospheric model: we assume a constant temperature at the earth's surface, radioactive sources of heat within layers, a heat flow rising from the deeper earth and thermal characteristics in accordance with current notions. The problem was solved in a one-dimensional version. The interface between the (A) and (B) layers may be thought of as simulating the Moho.

The initial distribution of temperature was taken to be such that both layers were not molten, and their temperature was still transient, increasing gradually with time and approaching the steady state distribution corresponding to given boundary conditions. Interesting results are obtained for the case where the thickness of the layer (A) is large enough for the temperature at its base to exceed the melting temperature for the material (A). Looking ahead, I will note that no steady state solution then exists and the system generates thermal oscillations. Because of this, the initial conditions do not affect the solution significantly, since the process becomes stable after only a few cycles and there is a specific temperature distribution for each phase of a cycle. For this reason it is advisable to take as initial conditions the distributions of material and temperature corresponding to any phase of a cycle.

Melting begins in the layer (A), near to its lower boundary. Complicated processes of dissolving, crystallization and melt migration immediately begin in the zone of melting. The layer of molten material (A) forms and increases as the temperature rises. Simultaneously material (B) is dissolved in it and the melt thus becomes mixed in composition. Material (B) dissolves at the lower boundary of the molten layer and diffuses upwards, thus creating a gradient of concentration in the melt. Under these conditions* the arguments contained in Shimazu (1959, 1961) and Magnitskiy (1964), aimed at demonstrating an inevitable occurrence of convection and upward migration of the melt, are inapplicable. The melt is a solution of heavy (B) in light (A), the concentration of (B) increasing downwards. Consequently, if the concentration gradient is sufficiently great, the melt density must increase downwards too, though the rate of increase in temperature with depth considerably exceeds the adiabatic gradient. As a result, no convection occurs, and the temperature gradient in the melt remains large.† Indeed it turns out to be greater than the gradient of crystallization temperature for material (B) in the melt, calculated with the concentration gradient taken into account. This leads to the crystallization of material (B) at the upper boundary of the melt. The melt is now included within material (B). In this process, material (B) continues to dissolve at the lower boundary, to diffuse upwards and to crystallize at the upper boundary. Thus a closed volume of a solvent, material (A), moves in a direction opposite to the heat flow, dissolving material (B) at the movement front, letting it pass through and crystallizing it in the rear. As the melt (or, better, solution) progresses towards higher temperatures its (B) content increases and so does the volume of the molten mass. The process just described has been well studied, both theoretically and experimentally, in investigations of the behaviour of a liquid phase in sea ice (Savel'yev, 1963; Pounder, 1965; Tsurikov, 1976). The movement against the heat flow of a drop of solution through ice,

* That is, where there is a gradient of concentration and so a density gradient of the melt.

† The temperature gradient is greater in the melt than in a solid phase, since thermal conductivity in a liquid phase with no convection falls below that for a solid phase.

independently of the direction of gravity, was experimentally reproduced by Whitman as early as 1926 (Whitman, 1926).

In our model, particles of molten material (A) are created repeatedly near to the interface and, dissolving (B), move away from the interface into the (B) layer against the heat flow. As a result, the lower boundary of the (A) layer slowly retreats upwards until it reaches a colder zone and reaches a position where material (A) is no longer melting. A 'half-molten' zone is formed below this boundary, consisting of material (B) saturated with melt inclusions of mixed composition migrating downwards against the heat flow. These inclusions grow during the downward movement and coalesce to form a melt zone. When the lower boundary layer (A) has risen to the point where melting ceases, the (A) layer stops generating molten inclusions, and the 'half-molten' zone begins to free itself of them. Gradually, all molten inclusions get out of it by descending into the melt zone. The model acquires a four-layered structure: (1) solid (A) layer (the thickness of which is considerably diminished); (2) a layer of newly formed material (B); (3) a melt zone of mixed composition; (4) a lower part of the layer (B) that had not melted. When in this state, the model stops evolving and is preserved almost without a change for a long time. It is only the lower boundary of the melt zone that continues to descent (by dissolving) still more slowly, so that it is barely noticeable. Its upper boundary continues to descend (owing to crystallization), and the concentration gradient in the melt slowly decreases. The last circumstance proved to be decisive: a thermal reversal of density is formed in the melt, and convection sets in.

With the commencement of convection heat transfer in the melt increases sharply (λ_{eff} is 'switched on') and the evolution of the model becomes very rapid. The process proceeds in a reversed sense. Material (B) crystallizes in the melt zone, not at the upper, but at the lower boundary, and dissolves at the upper boundary. The melt zone rapidly migrates upwards; simultaneously the concentration of (B) in the melt decreases, as does the thickness of the zone itself. The heat flow in the upper part of the model increases considerably. The melt soon rises very high, is completely free of material (B), and solidifies into a thick (A) layer at the top of the model. The model has come round to a state very little different from its initial condition. From that time, a new cycle of evolution begins repeating the stages of the preceding cycle.

Figs 3 and 4 give a concrete example illustrating the development of the two-layer model. The vertical shows depth, the horizontal marks time. It should be noted that the values are given in conditional units (in depth and time factors) since the model can be interpreted for different time and space scales in accordance with Fourier similitude criterion. The change of the depth scale by n times causes the change of the time scale by n^2 times. Thus at $n = 1$ the numbers marked along the axes represent depth in kilometres and time in m.y., whereas at $n = 0.1$ they indicate hundreds of metres and tens of thousands of years, respectively. This means that in the first interpretation of the model at the initial moment the interface boundary of layers (A) and (B) lies at a depth of 100 km and the length of the cycle is about 2500 m.y. In the second interpretation the boundary is at a depth of 10 km and the cycle terminates in about 25 m.y. We may assume any other value of n for interpretation. This is important because it allows the model to be applied to different geological boundaries, as for example, both to the M or C discontinuity. Moreover, by using the similitude criteria*

* In this model the similitude criteria are as follows:

$$\frac{\tau.\lambda}{l^2 c} = inv; \quad \frac{T.c}{Q} = inv; \quad \frac{T.\lambda}{q.l} = inv;$$

where τ is time (s), l is length (cm), T is temperature (degree), λ is specific thermal conductivity (cal. cm^{-1} s^{-1} deg^{-1}), c is specific heat capacity (cal. cm^{-3} deg^{-1}), q is specific heat flow (cal. cm^{-2} s^{-1}), Q is specific heat (cal. cm^{-3}).

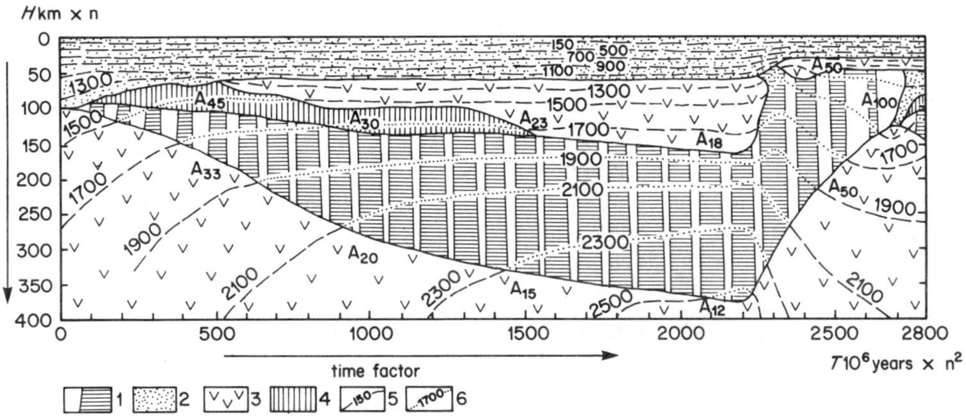

FIG. 3 Thermal cycle in the two-layer model of the lithosphere.
1 – liquid phase of composition A_nB_m ($m = 100 - n$); 2 – solid phase (A); 3 – solid phase (B); 4 – zone of partial melting; 5 – isotherms in solid phases; 6 – isotherms in melt and in zone of partial melting.

we can also vary the thermophysical properties of the model (temperature, thermal conductivity, heat capacity, and so on). The reader's attention is drawn to the similitude criteria in order to demonstrate the possibilities of interpreting the model as applied to a variety of geological conditions and to emphasize the relativity of the actual numerical values of thermophysical parameters, dimensions and time, used later.

If we assume $n = 1$ and the numerals on the isotherms in Fig. 3 show °C, then the following parameters are used. The thickness of the model is 400 km. The (A)/(B) interface boundary at the initial moment lies at a depth of 100 km. Specific heat capacities of (A) and (B) are 0.3 cal. g^{-1} deg^{-1}. Specific thermal conductivities of (A) and (B) in the solid state are equal and change at different depths depending on temperature in accordance with the diagrams taking into account different kinds of heat transfer (Lyubimova, 1968); thermal conductivity of the liquid phase (without convection) is half of the thermal conductivity of the solid phase, whereas effective heat transfer of the liquid phase with convection is three times as high as the thermal conductivity of the solid phase at the same temperature. The latent melting heat of (A) and (B) is 0.13 and 0.35 k. cal. cm^{-3}, respectively. Melting temperatures of materials (A) and (B) are 1100 and 1405°C, respectively, at the earth's surface; with depth they grow with gradient 3 deg. km^{-1}. Only material (B) crystallizes from the melt of mixed composition and crystallization temperature depends linearly on concentration. The easily fusible material (A) crystallizes only at 100 per cent concentration and can imitate eutectic mixture of material (B) with some other substance C (if we consider only one branch of their eutectic diagram, from A to B). The surface temperature was assumed to be constant and equal to 0°C (it proved more practical to assume 150°C as constant temperature at a depth of 5 km; this had little effect on the results). Heat flow from the interior through the lower boundary of the model was assumed to be $0.3 \,.\, 10^{-6}$ cal. cm^{-2} s^{-1}. The distribution of the inner sources of heat (radioactive decay etc.) was assumed to be irregular, the sources diminishing with depth, that is, every 10 km the heat flow is supplied from the inner sources with $0.6 \,.\, 10^{-6}$ cal. cm^{-2} s^{-1} in the depth range 0–30 km, with $0.15 \,.\, 10^{-10}$ cal. cm^{-2} s^{-1} in the range of 30–110 km, with $0.03 \,.\, 10^{-6}$ cal. cm^{-2} s^{-1} in the range 110–400 km. The distribution of material (A) concentration in the molten layer is assumed to be linear with the average concentration in the middle of the layer. The concentration gradient depends on the movement rate of the boundaries of the molten layer. The problem of the value of the gradient and of its dependence on the model development is rather complicated and is still insufficiently studied. In the different variants of the model we assumed slightly different dependences. which actually determined the difference in the variants shown in Figs 3 and 4. The concentration values of material (A) in the melt at its lower and upper boundaries in Fig. 3 are indicated by the subscripts to the letter A (for example, A_{12}, A_{18}) and by the thickness of the white unshaded band under the symbol 1. The instant of onset of convection was assumed at the moment when the concentration gradient became so small* that a sharply marked density inversion appeared in the molten layer. Since this moment is determined by the movement rate of the melt's boundaries and normally corresponds to the penetration of melt to a definite depth, the moment of onset of convection is in practice determined from these indicators. At this moment the effective heat transfer sets in and is six times as large as the thermal conductivity of the melt at the given temperature.

Fig. 3 shows the development of one cycle. At the initial moment (zero on the horizontal axis of the time factor) the model is composed of two layers of solid materials (A) and (B) (symbols 2 and 3). The temperature nowhere reaches the melting point of these materials, but it rises in the course of time (see the tilt of isotherms, symbols 5,6) and with the time factor reaching 100 layer (A) begins to melt and material (B) (symbol 3) dissolves in the melt. Concentration of material (A) and its concentration gradient are high (see the thickness of the white band under symbol 1). When the time factor reaches about 150, the crystallization of material (B) sets in the upper part of the melt and the zone saturated with molten inclusions moving downward is created (symbol 4). This zone grows upwards until the time factor reaches 500 due to the melting of layer (A) whose thickness visibly diminishes. When the time factor reaches 500, the melting of layer (A) ceases and the upper part of the layer saturated with melt inclusions is gradually freed of them. Here the pure solid phase (B) (symbol 3) appears. The base of this newly formed upper lens of material (B) still contains (until the time reaches 1500) the zone saturated with molten inclusions. By that time the melt becomes thicker (depth factor 200) and has low concentration of material (A) (A_{15}–A_{23}). During this period the temperatures persistently rise, as shown by the tilt of the isotherms: Moreover, below the molten layer the temperature rises quickly, while in the melt the rise is slow, and above the melt the temperature is practically constant. The heat supplied into the system accumulates in the melted layer as latent melting heat. Within the time interval 1500–2200 the thickness of the molten layer remains almost unchanged and its boundaries slowly sink, whereas the concentration of material (A) and the concentration gradient become still less (A_{12}–A_{18}). Convection begins at the moment 2200 and a higher effective heat transfer takes place in the molten layer. At the same time the concentration gradient disappears, due to mixing. Heat release from the lower boundary of the melt exceeds the latent heat escape of crystallization, and the temperature in the layer under the melt falls as is indicated by the tilt of the isotherms. A powerful heat flow runs into the upper part of the molten layer, overheats it and quickly melts the overlying layer. However, crystallization in the lower part of the molten layer occurs at a still faster rate. Consequently the thickness of the melt diminishes and the concentration of material (A) rises respectively (see the thickness of the white bands under symbol 1). By time 2700, the melt completely loses dissolved material (B) which crystallizes above and below. The temperature in the melt persistently falls to the point when the pure material (A) also solidifies. As soon as the convecting melt vanishes the increased heat transfer also disappears. The cycle is com-

* Less than the given value.

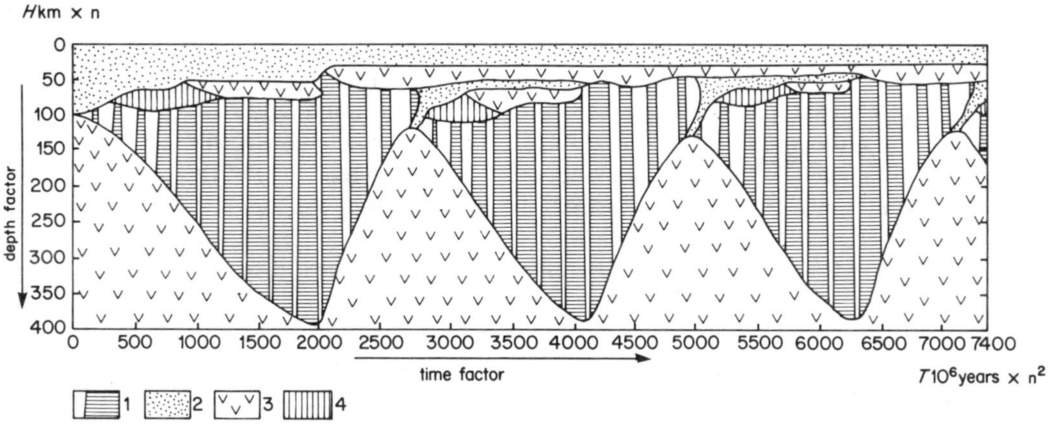

FIG. 4 Melting cycles in the two-layer model of the lithosphere. For key see Fig. 3.

pleted. The model differs from the initial state only by a thin interlayer of material (B) in layer (A), by a slightly lower position of its lower boundary, and by somewhat higher temperatures. But at that moment the temperature in the middle part of the model rises and the melting of the lower part of layer (A) sets in, thus starting a new cycle.

Fig. 4 shows a longer period of development of the model consisting of three full cycles. The scale on the time factor axis is reduced by more than 2.5 times. The figure is therefore compressed horizontally. Several parameters are changed in this variant of the model and consequently the molten zone is thicker. The general nature of the process, however, is the same. It is clear (Fig. 4) that an oscillatory regime quickly sets in and that the second and third cycles are almost identical. The newly formed interlayer of material (B) is reliably maintained above the molten zone. Under the intercalation the solid material (A) appears and disappears. The molten zone pulsates since the thickness and composition of the melt varies. The heat flow also pulsates, as shown in Fig. 3 (Fig. 4 has no isotherms). The period and amplitude of oscillations depend on the parameters of the model.

Using similitude criteria, we can demonstrate the dependence of the oscillation period on the model's parameters. With the nominal values of parameters given above the model represents a layer 400 km thick, which may correspond to the earth's crust and upper mantle. In this interpretation the upper layer of material (A), 30 km thick, simulates the earth's crust, the underlying solid layer, 30–100 km thick and composed mostly of material (B), simulates the mantle part of the lithosphere, whereas the asthenosphere is imitated by the molten layer 200–300 km thick. The temperature of the asthenosphere reaches 1900–2500°C. The period of the cycle is 2200–2700 m.y. Some researchers believe, however, that the asthenosphere is composed of partially molten material, that is, its refractory part forms a solid frame which is not involved in melting, whereas the easily fusible material fills the pores in that frame. Our model also allows an interpretation for this situation. In this case the linear dimensions of the model remain unchanged, but the value of specific heat is reduced because latent heat is absorbed or released during melting and crystallization not by the entire bulk of the material, but only by that part of it which experiences melting. Let us assume that the nominal values of parameters are equal to 1, and determine by similitude criteria how to change them for the new interpretation. We assume that 48 per cent of the asthenospheric volume melts. Then $Q' = 0.48$; $l' = 1$. The remaining parameters shall be: $T' = 0.8$; $C = 0.6$; $q' = 0.8$; $\tau' = 0.6$; $\lambda' = 1$. In this interpretation the size of the layers is the same; heat capacity and the flows are somewhat changed, the temperature in the asthenosphere reaching 1500–2000°C and the period of cycles is 1320–1620 m.y. The model can be also applied to the processes operating in the earth's crust, for example, on the boundary between the granite and basalt layers. In this case layer (A) is the granite layer and layer (B) is basalt; the melt is the source of magma of silicic or middle composition. Let us assume that the initial thickness of the granite layer is 20 km, the temperatures are half of the nominal temperatures, the content of the melt in the source is 50 per cent, i.e. $l'' = 0.2$; $T'' = 0.5$; $Q'' = 0.5$. Then the other parameters shall be: $C'' = 1$; $\lambda'' = 1$; $q'' = 2.5$; $\tau'' = 0.04$. Therefore the temperatures in the deep source reach 950–1250°C and the cycle is 90–110 m.y.

The analysed variants of interpretation demonstrate that self-excited oscillatory systems in the two-layer model of the lithosphere may have different periods of oscillation depending on the conditions of their excitation.

The performed analysis of models for lithospheric melting leaves no doubt about a possible internal source of pulsations in the crust and mantle of the earth, with a period of many millions of years. This source may extend over large areas and affect the character of the heat flow, magmatism and tectonic movements. The mutual correlation between these events, outlining their areas of occurrence, can help reveal the foci of such self-excited systems. It is important to study the chronological details of the process. The shape of the oscillations, that is, duration of individual stages of the cycle, depends on the thermal-physical characteristics of the system, large variations of these are allowed in the model (through lack of knowledge), resulting in a great variety of shapes and periods. The study of natural non-uniformity in the temporal development of the tectonomagmatic process (slow movements replaced by rapid ones, amagmatic by volcanic development) gives us invaluable phenomenological material for a more accurate interpretation of the model and in determining the thermodynamic characteristics of material at great depths.

Crystallization in Magma Chambers

The question of why pulses appear during magma crystallization arises in the first place in relation to the study of layered intrusions. These have been known for a long time. They have mafic composition, thin-layered structure and a chemical composition that changes regularly across the section. There is a vast literature devoted to these, for example a comprehensive review by Wager and Brown (1970) and a summary by Sharkov (1980) offer a good introduction to the problem. In spite of several existing hypotheses, the origin of such layering is still open to question. From the point of view of correlation between geological phenomena, it is very important to know whether there is any connection between the formation of 'layers' in an intrusion and phases of tectonic movements, magma penetration and similar manifestations of external activities. The available hypotheses entail different opinions on this point. Our own investigations support the view that the rhythmicity is a purely internal characteristic of a system, unrelated to any periodic external activities.

The modelling was carried out on an IGL hydrointegrator. The model is rather complicated, since we are modelling the crystallization of non-eutectic melt taking into account diffusion, heat flow and convection which appears and disappears depending on the distribution of temperature and concentrations. In the simplest case of a two-component material a self-excited system appears with a convection such that heat transfer in the chamber appears more rapidly than does diffusion in the zone of crystallization. A rapid heat escape gives rise to crystallization at the bottom of the chamber, first of the more refractory component and then of the eutectic, when the temperature lowers as required. Simultaneously, an intense heat flow towards the top of the chamber maintains a high temperature there, sometimes even melting the top a little. This results in a lowering of the temperature difference between top and bottom, thus stopping convection and causing a sharp drop in the heat transfer. This results in the top beginning to cool, accompanied by crystallization of the refractory component and afterwards of the eutectic near to the upper boundary of the chamber. In contrast to this, the temperature is rising, and crystallization ceases at the bottom of the chamber, while diffusion is gradually levelling the concentrations in a near-bottom layer. When the gradient of temperature has sufficiently increased, convection sets in again, heat transfer sharply increases in the chamber and the process repeats itself, forming the next asymmetrical layer. The modelling takes into account that some overcooling of the melt is needed to initiate crystallization. Active self-excited systems occur in more complicated models as well, but these will not be discussed. It should be stressed that they arise under rather restricted temperature conditions, and only when a heat flow impinges against a chamber containing a melt of non-eutectic composition, in which heat transfer is a non-linear function of the temperature gradient, the crystallization of the refractory component begins under overcooling conditions and diffusion does not have enough time to be able to level the concentration in the zone of crystallization.

This example shows that due caution is required in chronological comparisons between rhythmically built complexes, since different self-excited systems may be tuned to a particular frequency because of their special internal structure rather than as the reflection of a periodicity imposed by external activities.

Differentiation of Matter in the Deep Interior

An interesting self-excited system has been recently demonstrated in the course of simulation of the process of gravitational differentiation of material in the lower and middle mantle of the earth (Keondyan and Monin, 1980; Keondyan, 1980). The self-excited systems of the models analysed earlier were based on thermal and diffusion processes, whereas Keondyan and Monin's model considers hydrodynamic processes occurring during differentiation of two-component Newtonian liquid in the gravity field. The model was constructed for the purposes of plate tectonics; it simulates the supposed process of formation of the heavy (iron or iron-nickel) core from the mantle of mixed (iron-silicate) composition. Therefore the basic variant of the model was calculated for a spherical layer 3.000 km thick ($3.500 \leqslant r \leqslant 6.500$ km) imitating the earth's mantle. Gravitational differentiation in the layer was studied as the progress from 'the proto-earth, a celestial body practically homogeneous in composition, its mass equal to the mass of the modern earth, to its modern layered state' (Keondyan, 1980, p. 3). Without debating the geological argument and the significance of the model, we shall only emphasize its special importance to us since at certain parameters the development of the model pulsates.

The model assumes that the entire space of the layer is filled with two components chemically inert to one another; each of the components is incompressible Newtonian liquid with certain viscosity and density; one of the liquids is heavier than the other. Both components are represented by particles of very small size as compared to the dimensions of the layer. Without gravity the particles would have been evenly distributed in space and their mixture would have had universal concentration corresponding to average concentration. But in the gravity field (gravity force is directed along the radius) the heavy particles sink while the lighter particles rise, thus inducing their concentration (and, consequently, the density of the mixture) to vary in different places. Sedimentation rate depends on viscosity and the size of particles. The heavy particles easily cross the lower boundary, join the core and increase it. In the process of sedimentation, that is gravitational differentiation, the potential energy of the mixture decreases and the corresponding excess energy escapes into space. If in the process of sedimentation the inversion of densities occurs in large volumes of the mixture, then motion starts and convective flows originate. This is the qualitative specification of the model. The authors accomplished its accurate mathematical description, assumed definite values of parameters and carried out a numerical simulation.

In the course of solution of the problem the authors varied the values of some of the parameters having the greatest effect on the development of the process. Such parameters of the model are the nature of viscosity variations with depth, 'separation rate of components' (depending on viscosity and size of particles) and the geometry of the region. The values of these three parameters, especially of the first two, mainly determine the behaviour of the model.

The adopted model was axially symmetric which reduced it to a two-dimensional model. One of the discussed variants represented a square region with each side 3.000 km long, another variant was a section of the spherical layer 3.000 km thick ($3.500 \leq r \leq 6.500$). The separation rate (a) in the basic variant is about $10^{-5} \leq \alpha(\nu) \leq 10^{-1}$ cm/y (i.e., the size of the heavy particle is 100 m at assumed viscosity values). The kinematic viscosity (ν) in the major part of the layer is 10^{23} cm^2/s, whereas in the lowest layer 300 km thick it linearily decreases from 10^{23} at its upper boundary to 10^{19} cm^2/s at the boundary with the core. This reduction of viscosity is of special importance since it actually deprives the lower layer of the heavy component, thus inducing density convection. The density of components is 5.75 (heavy) and 3.42 g/cm^3 (light). At the initial moment the concentration of the heavy component (and consequently the density of the mixture) slightly increases with depth but remains unchanged in the horizontal layer. 'Initial disturbance' of density was introduced in the lower layer. The mean concentration of the heavy component was 0.1.

The numerical experiment revealed that the model could develop in two different ways. First, simple sedimentation of the heavy component occurs without provoking convective flow in the mixture. This process takes place either at very high ($\alpha > 10$ cm/y) or very low ($\alpha < 10^{-5}$ cm/y) rate of separation of components. Second, in the process of sedimentation density inversions may appear in the mixture and form convection flows grouped into one or several large convection cells. This case was analysed in the basic variants of the experiment. In the principal part of the model sedimentation of the material caused the concentration of the heavy component to grow in the lower layers, thus impeding the initiation of convection. On the contrary, in the lowest layer, with lower viscosity and free admission of the heavy component into the core through the lower boundary, rapid sedimentation of the heavy component caused a reduction of its concentration and formation of an anomalously light mixture rising to the upper part of the model.

The convection thus excited involved the entire thickness of the layer. Convection started with the formation of two large opposite circulations with the rising flow in the equatorial part of the model. But soon this pattern of circulation became unstable and divided into several (4–5) smaller cells. This division was also unstable. Two cells grew again and dominated their neighbours. The two-cell convection was resumed, which later again divided into 4–5 cells, and so on. As the result of these processes, 'plumes' of relatively light matter appeared over the rising flows, while heavy lenses accumulated in the lower parts of the convection cells.

The energy flow emitted by the model waxed and waned. The same happened to the supply of heavy matter into the core. Oscillations occurred in the background of gradual attenuation of the process, because the source of the heavy component in the layer

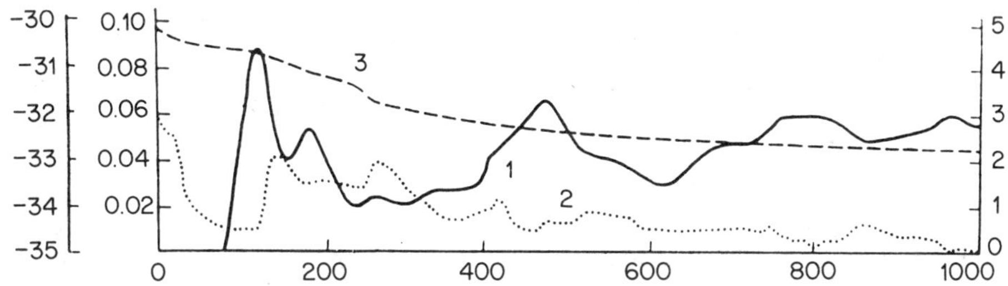

FIG. 5 Change in time of integral convection characteristics in the spherical layer from Keondjan's (1980) data.

1 – natural logarithm of average density of kinetic energy; 2 – flow of heavy component mass through the core-mantle boundary (in per cent of the initial mass for 10 m.y.); 3 – average concentration of the heavy component in the mantle.

expended with time. Keondyan (1980, p. 18) writes: 'The fundamental peculiarity of the analysed process is its pre-eminent non-stationary nature realised as the oscillatory regime of its basic integral characteristics against the background of the slow trend of potential energy. The typical dependences of integral characteristics, i.e., the average density of kinetic energy, the flow of mass of the heavy component into the core, and its mean concentration in the layer, are given in Fig. 5. It is obvious that the mean convection intensity and the rate of its mass exchange with the core oscillate with a specific period of $\approx 10^8$ years. Moreover, the minimal mass flow into the core (or its growth rate) is observed during the phase of maximal intensity. The amplitude of oscillations of these characteristics diminishes with time.'

The model, suggested by Keondyan and Monin (1980), demonstrates the possibility of the appearance of a self-excited oscillatory system in the deep interior of the planet. No doubt, this system should essentially affect the course of its tectonic development. Separate phases of movements have planetary importance, but in particular regions they are apparent in various forms.

SELF-EXCITED SYSTEMS ARISING IN SEDIMENTATION

The models of self-excited systems related to sedimentation consider material transfer by various agents whose activity depends on features typical of the basin of sedimentation and on the structure of the sediment itself. One model only will be considered, applicable to different types of subaerial and subaqueous sedimentation, but which is of special importance for understanding rhythmically built masses of marine sedimentary rocks (among these are apparently the flysch, olistostromes and so on). The many hypotheses explaining the origin of rhythmic stratification—from the oscillatory hypothesis to that invoking turbidity currents—are too familiar in geological circles to need further discussion (Bobrov and Romanovskiy, 1970; Vassoyevich, 1940, 1948, 1951, 1960; Keller, 1949; Leonov, 1972a,b; Romanovskiy, 1976; Strakhov, 1962; Khvorova, 1958; Kuenen, 1958, 1967). The only problem to be considered is that of the role of external activities in the formation of rhythms. Unfortunately many, if not most, geologists still tend to believe that the rhythms reflect a periodicity of external forces (tectonic, seismic forces and so on). An example is the point of view of such well-known geologists as Khain and Lomidze (1961). We consider the views of Romanovskiy (1976) and Savel'yev (1977), in many respects corresponding to the model given below, to be more correct.

Modelling is carried out on an IGL hydrointegrator. Since problems of this type have not yet been tried on an IGL, we should here give a justification for this attempt. The anolog IGL hydrointegrator can model processes described by partial differential equations of the parabolic type, that is ones having the form $\partial u/\partial t = a^2\nabla^2 u + \phi$ (problems involving diffusion, heat condition, filtering and so on).

The transfer and accumulation of sediments in a basin is described by a similar equation if the displacement of sediments over the basin bottom area is regarded as being caused by unordered wave motion. It is assumed that waves give rise to irregular particle movements in a near-surface 'mobile sediment layer'. The mobility of a sediment is measured by a 'mobility constant' relating the magnitude of 'sediment flow' to the bottom slope value. This constant is dependent on the nature of the basin, its depth and the character of the sediment. The problem of sediment transfer thus formulated is described by the equation

$$\frac{\partial H}{\partial t} = k\left(\frac{\partial^2 H}{\partial x^2} + \frac{\partial^2 H}{\partial y^2}\right) + \eta,$$

where H is the height of sediment surface at the bottom, k is the sediment mobility constant, η is the rate of influx of sedimentary material. This process is modelled on the integrator.*

Modelling requires that the sediment mobility constant as a function of the basin depth, $k = k(H)$, should be given. This functional relationship is not yet known exactly. The relationship $k = k(H)$ may certainly vary for different basins, even for different portions of the same basin. The sediment mobility is, however, known to be invariably decreasing with depth at a fast rate (non-linearly). This is sufficient to build a model that would represent the natural process with some accuracy.

Model studies with the above considerations taken into account show that the bottom relief of a basin will tend, as sedimentation goes on, to a certain curve determined by the relationship $k = k(H)$, and independent of the initial relief. This curve is convex, that is the bottom slope increases with depth. The rate of the process depends on the influx rate of the sedimental material.

It follows from the above that at some depth in any sufficiently deep basin, the increasing slope of the sediment surface must reach a critical value at which the slope would become unstable and the sediment would slump down. If the sediment is thixotropic,* that is if its

* It is assumed in our formulation that waves in the basin set sediment particles into motion only in a near-surface layer. Below this (the 'mobile layer'), the particles do not move, they are in a buried state. A certain average number of sediment particles pass through a unit section area of the mobile layer per unit time, to be characterized by a mobility constant. The mobility constant of a sediment is dependent on the character of the sediment itself and the character of mobility in the medium, that is to say, its values may vary at different points of the bottom. The number of particles that pass through the unit section at a horizontal bottom in forward and backward directions is the same on an average (because of wave motion irregularity). For this reason, if particles passing in one direction were counted with a plus sign ascribed to them and those coming in the opposite direction as having a minus sign, the sum total would be zero. However, the number of particles passing through the section down a sloping bottom is greater than that coming in the opposite direction, and the sum total would not be zero. When the slope is small it is proportional to the gradient of the bottom level, $\mathrm{d}H/\mathrm{d}x$. If the sediment flow (q) is defined as the net amount of sedimental material passing through unit section per unit time for a given bottom slope, we may then define the sediment mobility constant as the sediment flow divided by the bottom slope: $k = -q/(\mathrm{d}H/\mathrm{d}x)$ (the minus sign indicates the flow to be down the slope, that is towards decreasing H). The expression $q = -k(\mathrm{d}H/\mathrm{d}x)$ can be used to derive the diffusion equation by a normal argument.

* The term 'thixotropic' is derived from the Greek ϑιξισ (touch) and τροπη (turn, change), and so means 'changing its properties from having been touched, shaken'. In thixotropic substances under moderate stresses, some bonds between the particles are broken, resulting in a sharp drop in strength and viscosity. The bonds, and the mechanical properties, are re-established under quiet unstressed conditions. Thixotropic substances

deformational properties change on shaking so that it acquires greater mobility, then a considerable mass of buried sediment must be involved in the slumping process. The slope angle will in this case be considerably diminished after the slipping.

If we include two angle quantities in the model (the maximum slope angle α_{max} at which slumping begins, and the minimum slope angle α_{min} at which no slumping occurs even in a sediment weakened by shaking) we obtain a self-excited system whose frequency of oscillation depends on its own parameters, but not on any external action.* These parameters are the influx rate of sedimentary material, η, the sediment mobility constant, $k = k(H)$, the maximum (α_{max}) and minimum (α_{min}) slope angles. For indeed the relationship $k = k(H)$ determines the depth where the critical slope angle of the sediment arises, α_{max}; the difference $\Delta\alpha = \alpha_{max} - \alpha_{min}$ determines the thixotropy of the sediment, that is, the size of the sediment lens set into motion in the slumping; the rate of influx of the sedimental material, η, determines the time for the lens to accumulate, that is, the period of oscillation.

The results of modelling with the maximum and minimum slope angles taken into account confirm these considerations; they can be summed up as follows. The surface of accumulating sediment in the basin to be simulated acquires a slope during sedimentation that depends upon depth. The slope angle reaches the value α_{max} at a certain depth. This depth proves to be critical for a basin, since below it very little material accumulates in the basin in the manner discussed above. On reaching α_{max} the sediment becomes unstable at the critical depth; as a result, it is set into motion, is destroyed, and slumps down, probably by landslide or turbidity current. Destruction and slumping affect a more or less considerable lens of sediment accumulated during a long time. The lower boundary of the lens is determined by the slope angle α_{min}, because for gentler slopes the sediment remains stable even when violently shaken. Sedimentation continues on the slope left after the landslide (with an angle equal to α_{min}) until α_{max} has again been built up at a critical depth. The slumping process then repeats itself. No slumping occurs when the basin depth is less than the critical value. If, on the other hand, the basin depth exceeds the critical value then the basin

* External action may of course disturb the period of self-excited oscillation of the system, or even produce forced oscillations, but is not at all necessary.

---

are thus either in a 'solid-like' or in a 'liquid-like' state, and can change from one state to the other under certain mechanical actions. The term is in general use in relation to gel-like, colloidal substances in which bonds are broken and re-established at a molecular level (Freindlich, 1939), that is for substances that are tixotropic at a microlevel. I am here using it in a broader sense, imposing no restriction about the molecular level on the size of particles, bonds between which are broken and re-established depending on the character of stress and strain. There are a great number of composite materials containing microscopic particles and particles observable with the naked eye; between these bonds can arise and be broken bringing about sharp changes in the mechanical properties of the material. A geological example of such materials may be quicksands whose particles, when in a quiet state, form a frame of some strength, but which breaks down when shaken, after which the quicksand is like a viscous fluid. It is consolidated again in a quiet state. This is a substance that is thixotropic at a minilevel. In some substances and under some conditions (especially in geological objects) bonds between large blocks are broken; as a result, enormous masses acquire higher mobility ('block flow') with subsequent consolidation under quiet conditions. An example is pulsating glaciers, to be discussed below. Thixotropy here manifests itself at a macrolevel.

The thixotropy phenomenon is extremely important for the production of self-excited oscillations (pulses) in deformable mechanical systems. Thixotropy has common features (sharp changes in the properties produced by mechanical action) at all levels; it is precisely those features which determine the appearance of pulses. The individuality of thixotropy at each level is of course retained; this is expressed in specific features of the physico-chemical and mechanical processes that cause inter-particle bonds to be established and broken. Distinctions should be kept between micro-, mini- and macro thixotropic media and substances. This corresponds to characteristic deformation shapes at micro-, mini- and macrolevels observed in structural geology (Lukyanov, 1980).

is clearly divided into two zones that are different as regards the manner in which transfer and accumulation of sediment occur: a pre-critical and trans-critical zone. In the trans-critical zone, where depths exceed critical values, material is mostly transferred and accumulated in the second (slumping) manner, that is, not continuously but in portions, by landslides (or turbidity currents) whose formation occurs at critical depths. The geological importance of the model lies exactly in this division of a basin into two zones, at the boundary of which the self-excited system is functioning and an essential change in the style of sedimentation occurs. Terrigenous material comes into the deeper zone mostly by means of rhythmically occurring slumpings.

The rhythm is controlled by the basin parameters. The self-excited system formed in the basin produces either rare and powerful or frequent and weak pulses, and functions either powerfully or sluggishly. The strength of the pulses depends on the character of the sediment accumulated. If its properties admit of large values of $\Delta\alpha$ then occasional powerful pulses occur and, correspondingly thick sedimental layers (rhythms) are formed in the deep part of the basin. With small $\Delta\alpha$, frequent weak pulses occur with thin rhythms of accumulation. A system's period of oscillation is also dependent upon the rate of influx of sedimental material: with higher rates the system functions energetically, with low rates, weakly—that is to say, although the rhythms are as powerful (for the same value of $\Delta\alpha$), they occur much less frequently, since a low rate will require a longer time to accumulate an equal lens of sediment. Therefore the rhythmic character of a rock mass may change as a consequence of slight changes in the composition of the inflowing material (which will influence $\Delta\alpha$). At the same time, similar looking rhythms may be accumulated rapidly or slowly. These results make one cautious about attempts at a stratigraphic correlation of distant regions based upon the character of rhythms.

The above modelling results seem to reflect natural processes in basins correctly enough. The investigators of flysch, in a further development of Kühnen's ideas, have repeatedly emphasized that there are three zones where flysch is formed: (1) a shallow-water zone where the material undergoes transfer, grinding and mixing; (2) 'a kind of trap where all the inflowing material is retained for some time by sedimentological barriers, expressed morphologically either as island chains or as internal elevations like submarine ridges.* The most important task of this zone may be considered material accumulation by separate portions and its subsequent unloading into a third accumulative zone where flysch sedimentation proper is realized.' (Romanovskiy, 1976, pp. 150–1). Material can be transported over great distances: 'When the bottom surface is more or less smooth the current [turbidity current] can move through hundreds of kilometres with no apparent loss of velocity. A full unloading occurs only when the current has reached the abyssal plain' (ibid., p. 41).

The same three zones were distinguished in the process of simulation. We shall describe one of the studied models for illustration (see Figs 6–10). The model simulated the flow of terrigenous material, continuously put into the basin, and the parameters of the sediment and basin, that is the mobility coefficient of the sediment, the regularity of its change according to depth (it diminishes in reverse proportion to the square of depth), and three critical angles showing the sediment's properties in undisturbed, disturbed and turbid states. Angle α_1 indicated the maximal slope angle of undisturbed sediment and corresponded to α_{max}; angle α_2 indicated the slope angle left after slumping and corresponded to α_{min}; angle α_3 showed the tilt at which slumping of the turbid material stops. The relation of angles was

*The results of our model studies show that ridges and submarine ranges are not really necessary. The role of a 'barier' is played by the inevitable variation according to depth of the sediment mobility constant, while a 'trap' may be merely a portion of a slope, not necessarily a depression.

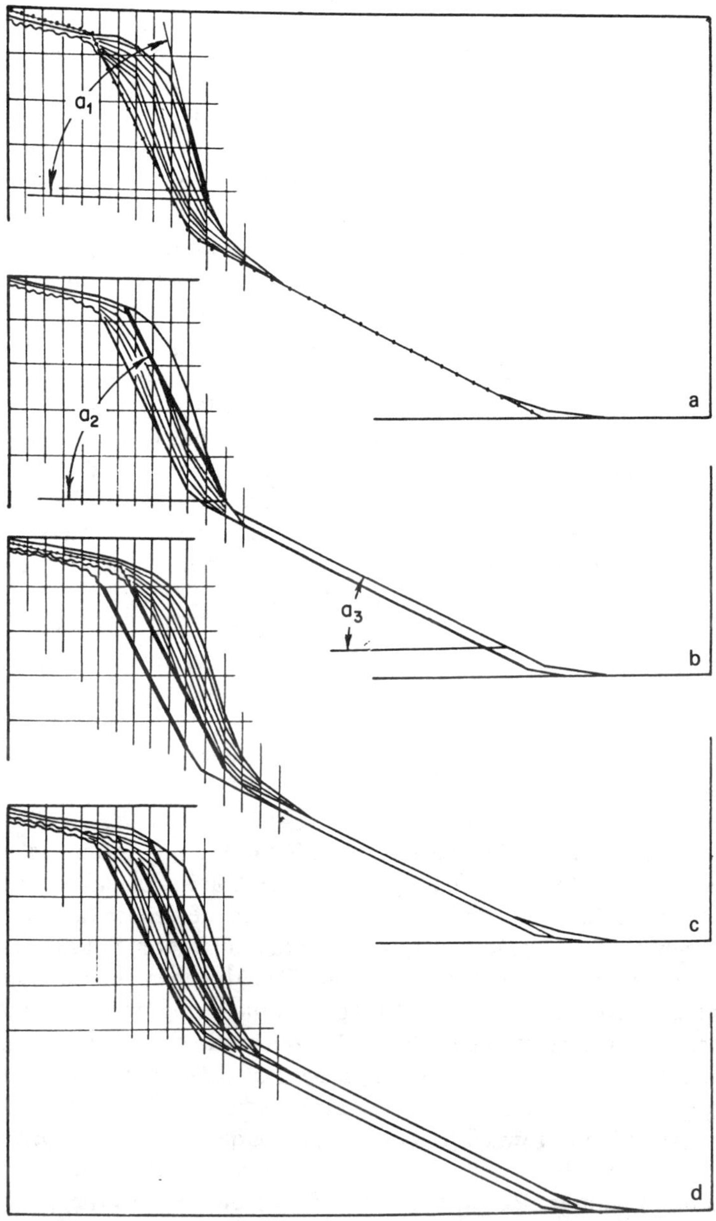

FIG. 6 Development of sedimentation cycles during settled self-excited oscillatory regime. See explanation in the text.

assumed to be $\alpha_1 : \alpha_2 : \alpha_3 = 8{:}4{:}1$. The depth of the basin and bottom topography were the initial conditions which varied.

Fig. 6 shows the development of rhythms in the settled self-excited oscillatory regime. The topography in the basin as it was immediately after one of the slumps is regarded as initial conditions. The section represents it by the line marked with dots and having the shape of a tilted step (see Fig. 6a). At the initial moment the water in the basin is to the right and above that line, whereas rocks and bottom sediments are to the left and below the line.

During further development of the model, the upper part of the step becomes slightly eroded (see undulating line); after that moment the accumulation of the sediment takes place universally. However, the intensity of sediment accumulation is variable: in the upper part it is low, on the steep slope it is very high, and on the larger part of the lower gentle slope it is practically nil. On the steep slope a thick lens of sediment piles up, the thin lines showing separate stages of this accumulation. The highest rate of piling occurs in the upper half of the slope, thereby gradually increasing its steepness. Eventually, the tilt of the surface in the lower part of the lens reaches α_1, and at that moment a part of the lens of the accumulated sediment slumps down. In Fig. 6b the slumped part of the lens is unshaded. The slope left after the slump is tilted at an angle of α_2. The transported turbid mass is deposited on the lower gentle slope, its surface is tilted at an angle of α_3 and it forms the thin extended layer shown in Fig. 6b. The cycle is thus completed. The newly settled surface is the initial condition for the next cycle. Figs 6c and 6d largely repeat Figs 6a and 6b, demonstrating the next cycle and interrelation of cycles, that is, the position of erosions (undulating lines), strippings (thick lines), stripped layering, wedging and intertransitions of various lithological types. Theoretically both cycles are expected to be identical. In the figure they are slightly different because it shows the results of concrete simulation, when the introduced values contain random deviations from the theoretical ones. In spite of these deviations the solution remains stable. But how do initial conditions effect the solution? In the given example the initial conditions correspond exactly to the topography settled after the cycle has been completed. What will happen if the initial conditions are different?

Figs 7 and 8 show the models simulating the effect of initial conditions on solution. Fig. 7 demonstrates the slightly tilted bottom of the basin ($\alpha_3 < \alpha < \alpha_2$) as the initial condition. In this case, at the start the sedimentation will be completely different from the one described earlier. Over a considerable period of time a thick lens of sediment is accumulated without strippings because its surface is universally tilted at less than α_1 (see Fig. 7a). Finally, however, the tilt reaches α_1 (the line marked by circles in Fig. 7a). After that the first slump occurs along the surface tilted at angle α_2 (thick broken line in Fig. 7a) and the transported masses are deposited at the base of the slope as a lens with the surface tilted at angle α_3 (thin broken line). The cyclic development begins from that moment on, but the duration of each cycle is not yet the same. As the initial condition, Fig. 7b shows the topography after the first slump demonstrated in Fig. 7a. After four cycles, the topography of the basin becomes reduced to the profile corresponding to the settled self-excited oscillatory regime (Fig. 7b). During this process, at the base of the slope, a layer of flat-parallel strata of the slumped material accumulates and its surface is tilted at α_3, whereas the steep slope tilted at α_2 slightly advances. The bend from α_2 to α_3 appears at a quite definite critical depth.

In Fig. 8 the steep slope ($\alpha_2 < \alpha < \alpha_1$) is assumed as the initial condition. In this case the onset of pulses is immediate. But at the outset slumping involves the entire bulk of the accumulated material and bedrock is exposed after the landslide. Again exactly the same amount of sediment accumulates to slide down again in a single slump (dotted line in Figs 8a and 8b). Therefore the upper part of the slope does not retain any sediment. At the base of the slope, however, the slumped material forms a thick, rhythmically structured layer whose surface is tilted at α_3. The upper edge of that layer persistently rises and approaches critical depths. In the shallow basin this process takes less time (Fig. 8b) than in the deep basin (Fig. 8a). But finally the layer starts to push the temporarily accumulating and slumping lenses from below. Eventually the moment will come when the sedimentary lens ready to slump is slightly larger than the preceding ones (the lines with circles in Fig. 8b). From that moment on (Fig. 8c) the slumping lenses advance progressively and each time a greater part of the lens remains on the bedrock slope uninvolved in slumping. At last after one of the

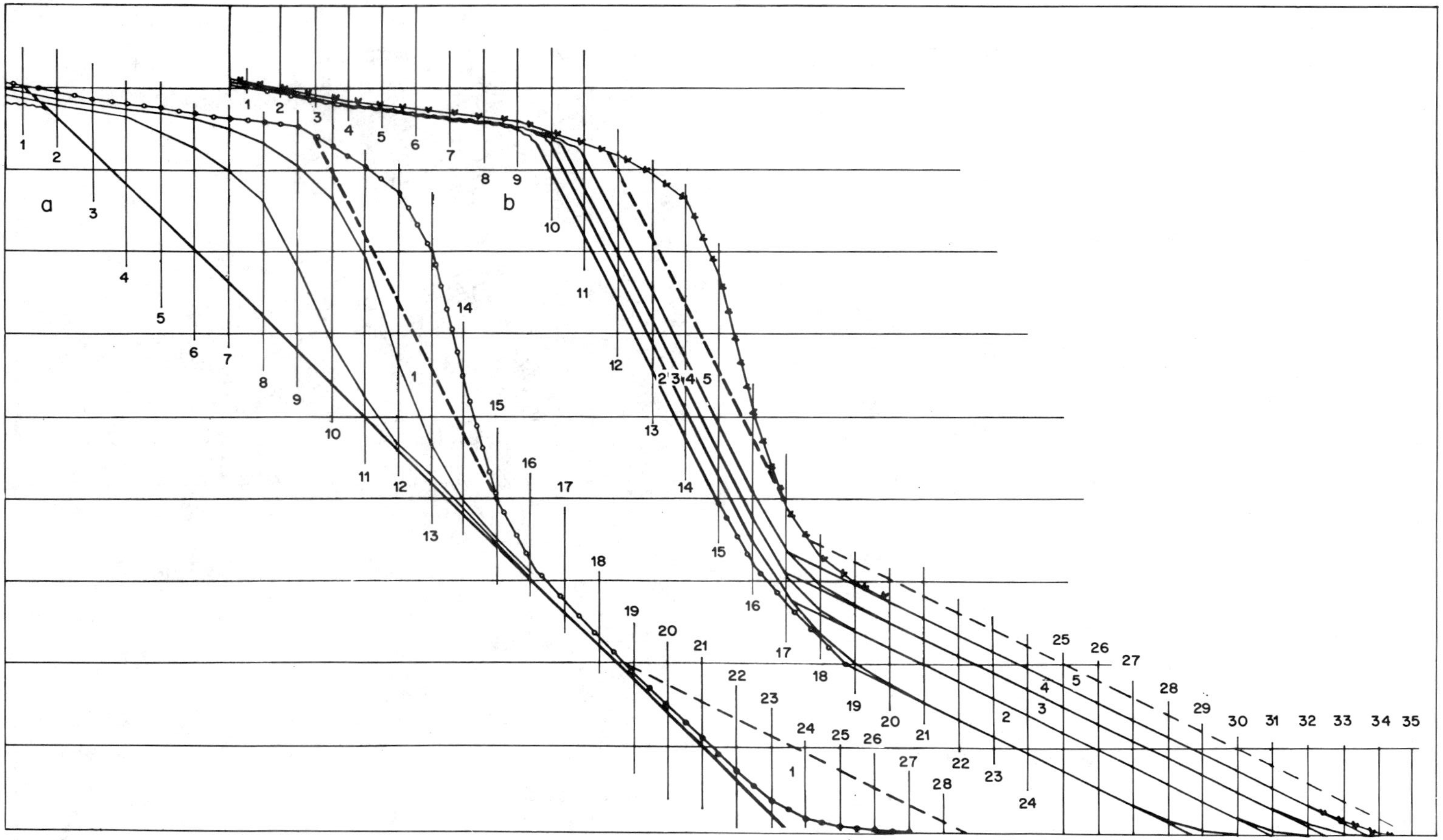

FIG. 7 The effect of initial conditions on the development of a self-excited oscillatory system in a basin with a gently tilted slope. See explanation in the text.

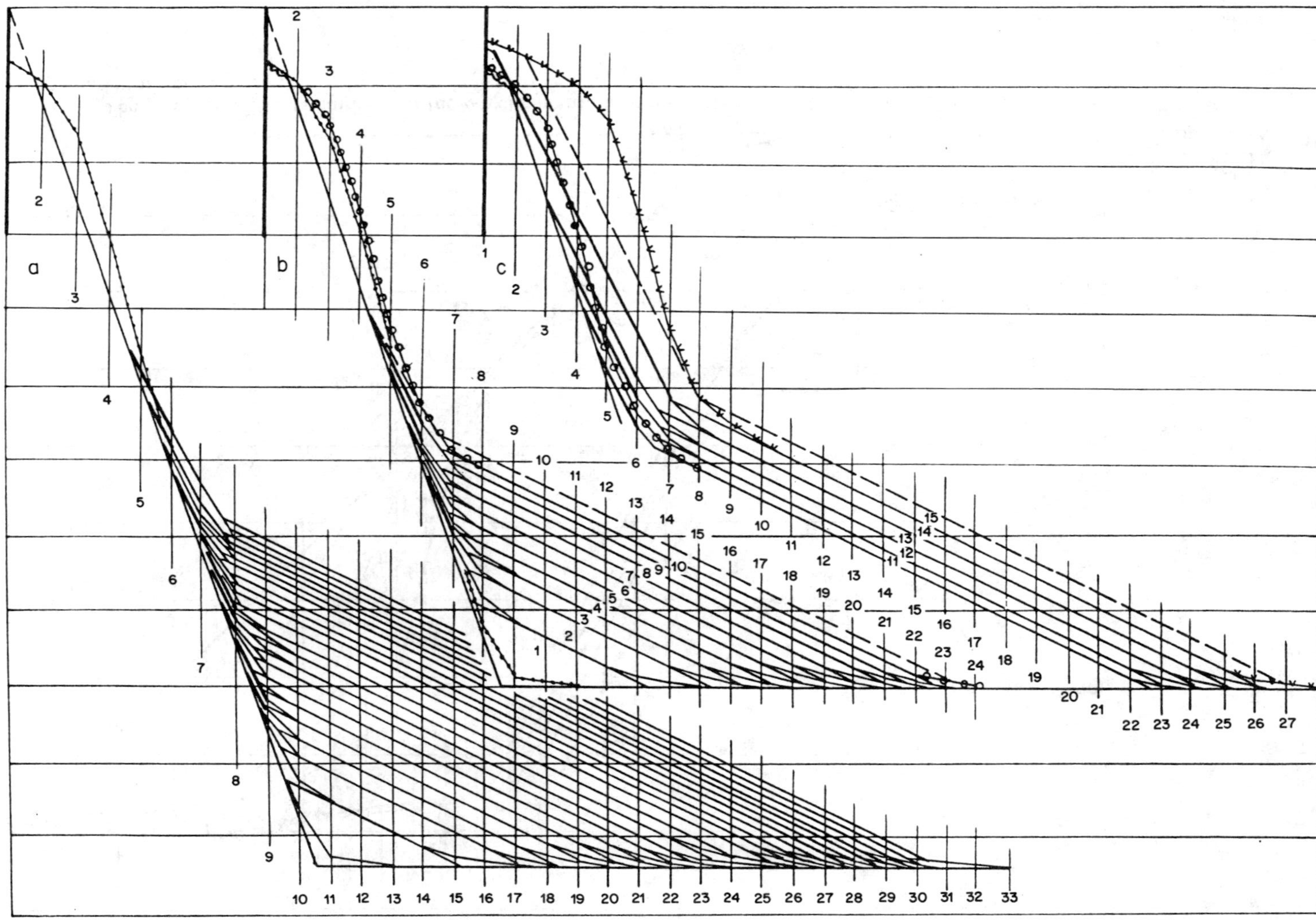

FIG. 8 The effect of initial conditions on the development of a self-excited oscillatory system in a basin with a steep slope. See explanation in the text.

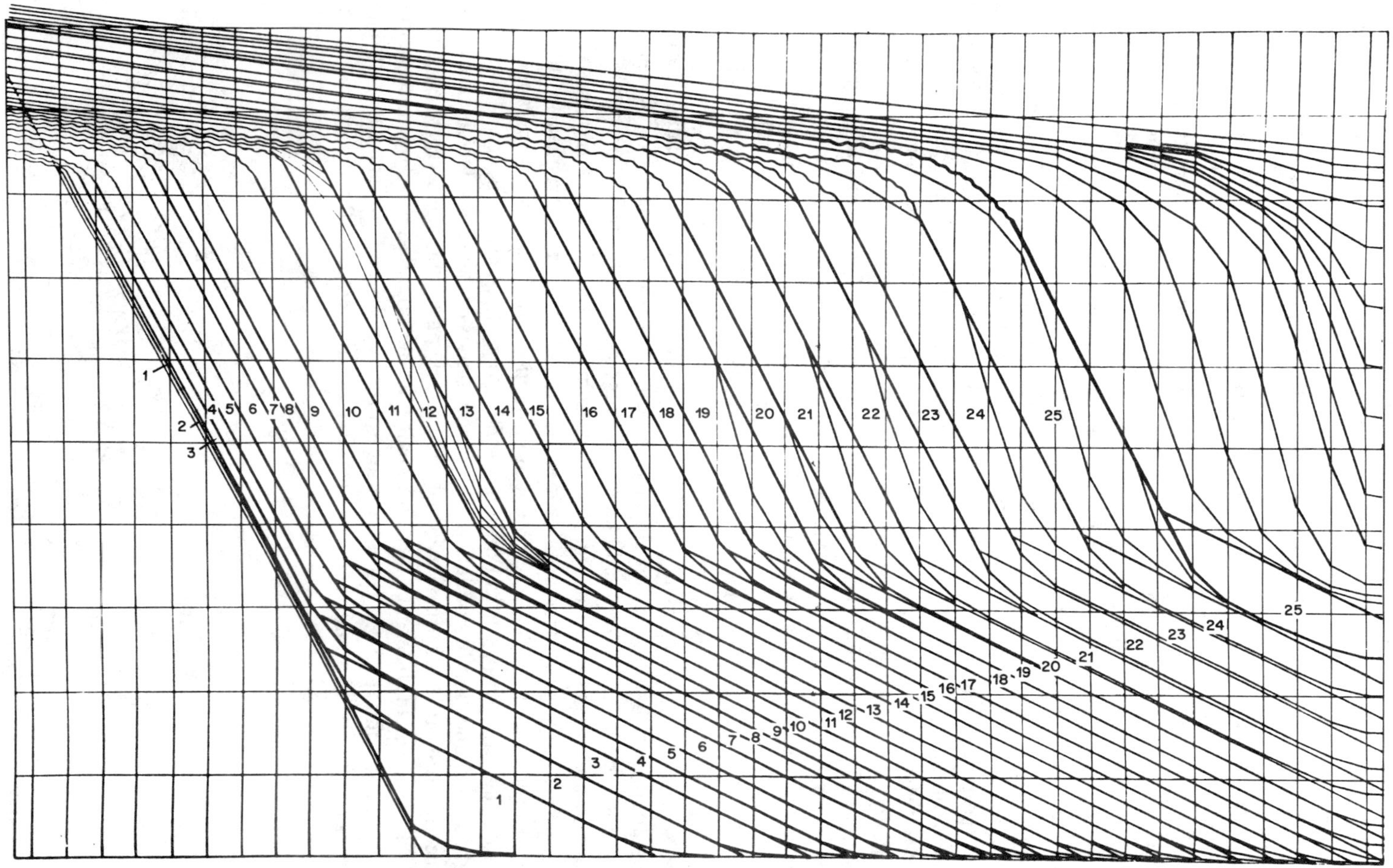

FIG. 9 Long-term development of a self-excited system in the basin. See explanation in the text.

slumps the surface of the slip (angle α_2) fails to reach the bedrock slope (cycle N 14 in Fig. 8c). This means that the profile now corresponds to that of the settled regime and the bend from α_2 to α_3 appears at the critical depth. Obviously, the initial conditions affect sediment accumulation only at the onset and from then on their effect is gradually mitigated. The development of the system in the settled regime is independent of the initial conditions.

Fig. 9 represents the development of the model over a long period of time. The bank of the basin is on the left side and the right-hand edge of the figure is the basin's axis of symmetry. The slope with α_2 angle is assumed to be the initial condition (the inclined thin line on the left). The depth of the basin was greater than critical. Pulse development set in immediately but in the first eight cycles the effect of the initial conditions was still felt. Cycles 9–20 developed in the settled regime of the self-excited oscillatory system for this depth of the basin. Beginning with cycle 21, the nearness of the axial zone became apparent, that is, the slumped material joined the material supplied from the opposite bank and consequently the depth of the basin was reduced after each slump. As the depth of the basin reduced, the slumps occurred less often and the lenses became thicker. Cycle 25 was the last. The depth of the basin became less than critical and the pulses stopped. They ceased because the shallowness of the basin prevented the onset of the first critical angle α_1 and consequently of the slumps carrying the material away from the zone of intermediate accumulation. The system lost the negative factor and ceased functioning in the self-excited oscillatory regime. This is reasonable because 'the properties of periodic movements, depicted by ultimate cycles with negative characteristic indices, i.e., (a) stability as regards small changes in the system itself; (b) independence (in the indicated sense) of the period and "amplitude" from initial conditions, comprise a specific feature of the real self-excited oscillatory processes' (Andronov *et al.*, 1981, p. 327).

If we scrutinize Fig. 9, we shall see that our model shows several peculiar features of sedimentation, which allow us to distinguish a number of situations or zones in the sedimentation basin and a few sedimentation complexes with specific properties. These peculiarities are summarized in Fig. 10, where symbols show seven series of sediments which are obviously petrostructurally different. These series are arranged in horizontal bands corresponding to definite sedimentation depths. They are not stratigraphical units (synchronic units), though vertical drilling would reveal the same sequence of these series throughout. Stratification lies obliquely to the boundaries of the series, which means that if we trace one stratigraphic horizon we shall successively pass from one series to another and if we drill the layer in different places we shall find the boundaries between the series at different stratigraphical levels. But according to the type of bedding and other structural features, each series is a unity. The series can be divided into three large lithological

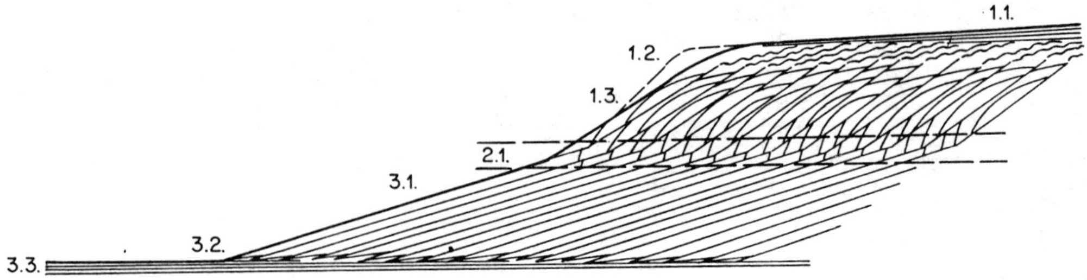

FIG. 10 Scheme of facial zonality in the basin developing a self-excited oscillatory regime. See explanation in the text.

complexes as regards accumulation process. These complexes and series are indicated in Fig. 10 by numerals.

1. The complex of remnant sediments. This group consists of remnants of sedimentary material temporarily accumulated in this zone of the basin before moving later into the next zone. In this connection the three series comprising the complex have specific layering: (1.1) the series of universal interbedding of rocks. It consists of the most coarse-grained sediments; it is thin and was formed in the subzone through which most of the material was transferred; (1.2) the series with eroded bedding due to periodic erosions after every slump; (1.3) the series with stripped incline-bedding. This is composed of the remnants of layers accumulated into lenses which later slumped. The series is therefore characterized by oblique layering stripped in the upper part by the slumping surface.

2. The complex of mixed sediments. This is represented by one series (2.1) of complicated layering. In this series the layers of remnant sediments alternate with the layers of sediment displaced during slumps.

3. The complex of transported and intact sediments. This is composed of three series, the second of which is transitional from the first to the third series: (3.1) the series with gradation layering. All or most of the series consists of transported material supplied to this zone of the basin during periodic slumps. This material is expected to contain layers corresponding to every slump and having parallel boundaries, traces of erosion on the upper surface, refinement of material from below upwards, traces of the flow of sediment and other features of turbidites; (3.2) the series with wedging gradation layering; (3.3) the series with horizontal layering without material displaced during slumps.

Zones and subzones of sedimentation correspond to the complexes and series of sediments: 1 – the zone of pre-critical depths, or the zone of unstable accumulation of sediments. This can be divided into subzones: (1.1) transportation; (1.2) periodic erosions; (1.3) periodic strippings of sediment. 2 – The zone of critical depths, also the subzone (2.1) of mixed accumulation of sediments. 3 – The zone of post-critical depths, or the zone of stable accumulation of sediments. This has subzones of (3.1) portioned supply of material, (3.2) degradation of slope effect and (3.3) deep accumulation of sediments.

The following similitude criteria should be used in the interpretation of the model described:

$$\frac{\tau.K}{l^2} = inv; \quad \frac{H.K}{l.q} = inv; \quad \frac{H}{l.tg\alpha_i} = inv,$$

where τ is time (years); l is length (horizontal) (m); H is depth (vertical) (m); α_i are critical angles (degrees); q is the specific flow of terrigenous material ($m^2.y^{-1}$); K is the sediment mobility coefficient ($m^2.y^{-1}$).

The following cases can serve as examples of interpretation.

(1) Continental slope with temperate supply of terrigenous material: depth $H = 1650$ m; critical angles $\alpha_1 = 8°, \alpha_2 = 4°, \alpha_3 = 1°$; length of model $l = 68650$ m; specific flow of terrigenous material $q = 3\ m^2\ y^{-1}$; sediment mobility coefficient in the lower part of the model $K = 0.7\ m^2\ y^{-1}$ (it gradually grows in reverse proportion to the square of depth 100 times); accumulation time of 19 cycles $\tau_{19} = 21 \times 10^6 y$.

(2) Continental slope with considerable supply of terrigenous material: $H = 1650$ m; $\alpha_1 = 8°$, $\alpha_2 = 4°$; $\alpha_3 = 1°$; $l = 68{,}650$ m; $q = 7\ m^2\ y^{-1}$; $K = 1.63\ m^2\ y^{-1}$; $\tau_{19} = 9 \times 10^6 y$.

(3) The slope of a smaller and more shallow basin with considerable supply of material: $H = 550$ m, $\alpha_1 = 8°$, $\alpha_2 = 4°$, $\alpha_3 = 1°$, $l = 22{,}840$ m, $q = 7\ m^2\ y^{-1}$, $K = 1.63\ m^2\ y^{-1}$, $\tau_{19} = 10^6 y$.

All the above cases of interpretation refer to the model shown in Fig. 9. The critical depth is in the middle part of the model and the thickness of separate cycles is therefore great. If an analogous model with considerably larger thickness of the post-critical zone is constructed the thickness of cycles will diminish in reverse proportion to the thickness of the post-critical zone (compare Figs 8a and 8b). The time of accumulation of cycles will then slightly diminish (the order of values remaining the same). Thus the model can represent sedimentation basins within a wide range of natural conditions.

The obtained results of model studies show that the rhythmic structure of sedimentary layers is insufficient to prove rhythmic external influence on the sedimentation basin. This structure can be determined by the internal properties of the basin, or a part of the basin, in other words, it might be locally important. Moreover, the rhythmicity of the basin's inherent oscillations can also be superposed by forced oscillations caused by external factors. Most of the real rhythmically structured layers apparently contain information about the self-excited oscillatory system and about the external effects on it. It is not easy to distinguish between these two causes. But further study and simulation of rhythmicity in sedimentary layers should provide a solution to the problem.

SELF-EXCITED SYSTEMS ARISING FROM ROCK DEFORMATION

The self-excited oscillations discussed above in systems related to sedimentation are similar to the oscillations of tectonic systems. This can be easily seen as follows. We have been simulating sediment accumulation accompanied by increase in the slope angle, that is, in the internal stress of the sediment. When the stress exceeds the strength of the sediment, it undergoes a failure, acquires higher mobility and is transported from the zone of intermediate accumulation into that of deposition. After that, the place formerly occupied by the sediment is gradually filled by new material until the occurrence of another failure in the sediment. The chief factor responsible for producting of a self-excited system is a sharp fall in strength of the sediment after failure, which occurs upon a background of increasing stress. As a result, the stress falls considerably below the strength of an undisturbed sediment, so that it takes some time to rise to this level to produce another failure. In principle, the same pattern is observed in the deformation of rocks.

When we observe the boudinage of thick and thin competent layers in a yielding plastic mass, we can easily see that the boudin size is proportional to the strength of the relevant layers. This is a good example of the fact that rock deforms non-uniformly under increasing stress, the periodicity of deformation depending on the parameters of the deformed system. Extending this observation to large units, we can presumably expect phases of tectonic activity to be related to the evolution of local systems. Therefore one of the tasks prior to carrying out a correlation study of tectonic events is to find those local systems and determine their structure. Equally important is separation of the phases of tectonic activity related to local causes from the phases of global scale.

Models of local oscillatory systems due to rock deformation could be pulsating glaciers, which have been studied by glaciologists in great detail. Three glacier classes are known: normal, rapid and 'flood' glaciers pulsating glaciers (Krenke, 1974). Pulsating glaciers are characterized by a varying mode of movement corresponding to relaxation self-excited oscillations of a system (Shumsky, 1973). Only a small fraction of glaciers are pulsating but they are still numerous: 204 in North America, about 40 in the USSR, a considerable number in Central Asia, in the Alps, in South America, New Zealand, Iceland, Spitzbergen and elsewhere (Dolgushin and Osipova, 1971). Mayer and Post summarize the characteristic features of pulsating glaciers as follows:

(1) all pulsating glaciers pulsate repeatedly, some of them 10 times or more; (2) most of the pulsations for which data are available concerning their temporal behaviour are uniformly periodic, and are not due to any of the external factors (earthquakes, climatic changes); (3) the active stage of a pulsation is relatively short (from less than one year to six years), and is followed by a long quiet stage (15–100 years or more); (4) the time of a complete cycle, or of an active stage, has no immediate relation to glacier length and area (the pulsation period is probably a complex function of the total ice displacement during a pulsation and the value of mass balance during a quiet period); (5) the velocity of ice movement in the active phase is always much greater, at least by an order of magnitude, than that at the same places in the quiet period. (Dolgushin and Osipova, 1971, p. 212)

Dolgushin and Osipova think that

a complete pulsation cycle ... consists of three main phases which are periodically recurring: (1) the phase of active advancing movement in which ice velocities suddenly increase by one or two orders of magnitude, and the flood wave rushes down the glacier causing its end to move down the valley into the area of dead ice, moraine, and fluvioglacial deposits, left after the preceding displacement; (2) the phase of stagnation and degradation affecting the advanced part of the glacier, and (3) the restorative phase during which the glacier is gradually, beginning from above, restoring its longitudinal profile to get back to the initial position. Accordingly, the pulsating portion of a glacier (or the whole of it) may be subdivided into three zones, in which ice amount is diminished, increased, and through which ice merely passes. (ibid., p. 213)

A classical and well-studied case of a pulsating glacier is the Medvezhiy glacier in the Pamirs (Fig. 11). It pulsated in 1937, 1951, 1963 and 1973.

In 1963, from the middle of April to 24 June, the front of the Medvezhiy glacier, swollen to reach 120–150 m, had moved down the valley through 1600 m with an average rate of 40 m per day. ... As a result of this displacement, which began from the foot of the icefall, the glacier tongue lowered by 50–70 m on an average in its upper two-thirds, leaving ice blocks and strips of 'tills' over the valley slopes marking its former surface. During the displacement, the glacier had been cut by numerous intercrossing cracks into prismatic blocks like basaltic columns, and huge (several kilometres in length) longitudinal fractures had been formed between the axial glacier zone and the marginal zones of ice crumbling. The ice transported from the upper portion of the glacier tongue into the advanced portion amounted to 140 million m^3 in volume, while the mass of the moraine simultaneously transported was 3 million tonnes at the least.

Towards the autumn of 1963 the lower portion of the glacier tongue was entirely dead along a length of 5–6 km, and then began consolidation, melting, and surface polishing. ... A new kinematic wave was noted in 1965–66 with mean rates of ice movement 250–300 m/yr in the upper portion of the glacier tongue, under the icefall, close to the upper boundary of the ice removal zone. In the succeeding years 1967, 1968, 1969, the peak of that wave had been moving down the glacier with a velocity of 1200–1500 m/yr; behind it a tension zone was forming with gaping deep cracks, moving along the glacier behind the peak. It has been noted that when ice velocities exceed 70–100 cm per day the glacier body splits into prismatic blocks along a system of intercrossing transverse and longitudinal (with respect to ice movement) cracks, while marginal fractures are formed along the valley sides. In the activated part of the glacier tongue which is gradually consuming its lower dead part, the ice surface rises. The glacier tends to restore its longitudinal profile that had existed prior to the 1963 displacement, and been levelled

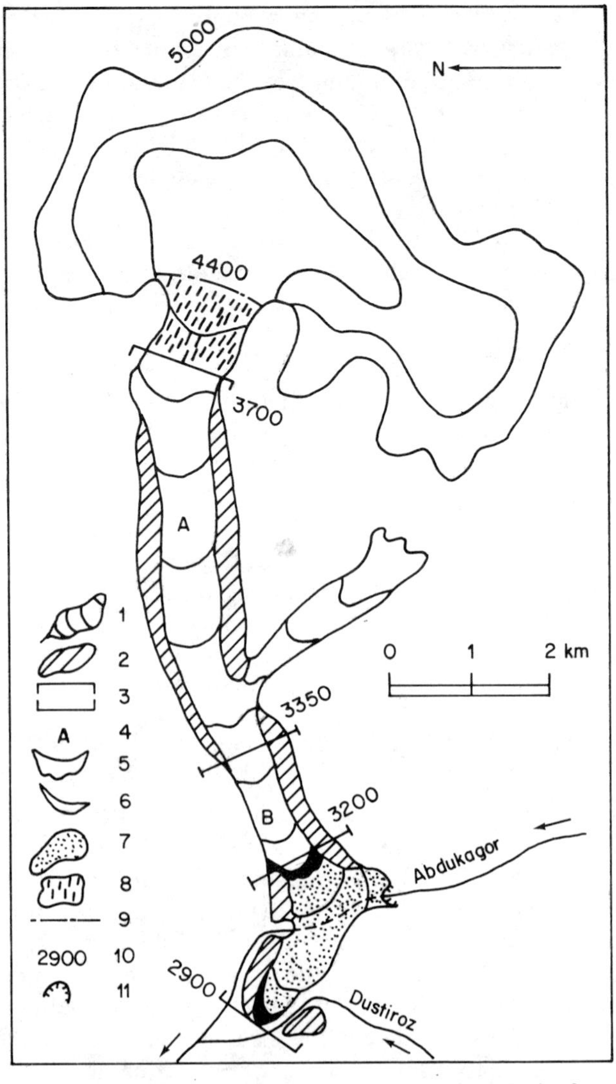

FIG. 11 Scheme of structure of the Medvezhiy glacier from the data of Dolgushin and Osipova (1971).

1 – glaciers; 2 – ancient moraines; 3 – boundaries of zones on the glacier in the active phase of pulses; 4 – zones: A – waste, B – supply, C – advance; 5 – position of tongue of glacier before movement in 1963; 6 – position of tongue of glacier after 1963; 7 – advanced part of glacier after movement in 1963; 8 – icefall; 9 – firn line; 10 – altitude indications; 11 – grottoes.

towards the end of the active pulsation phase. Towards the autumn of 1969, the peak of the frontal kinematic wave which was towering as a steep convex slope 40–50 m above the dead part of the glacier tongue, had passed the mouth of the right (and only) tributary, and was about 2 km from the place where the Medveshiy glacier had its end before the 1963 displacement. It advanced 1000–1200 m more in 1970. In about two or three years the wave will reach the glacier end, and after one or two years more the longitudinal profile will exceed the critical level, so that the glacier will be ripe for a next displacement, probably due to take place in 1974–75. (Dolgushin and Osipova, 1971, pp. 203–4)

The predicted glacier displacement actually occurred in 1973.

The glaciologists consider it as well established 'that glacial pulsations are not a direct consequence of fluctuations in the input-output balance of a glacier, since pulsations occur both during general advances and general retreats of glaciers. They represent a dynamic expression of liberation of stresses accumulated in glacial systems themselves (ibid., p. 213). The characteristic features in the structure of pulsating glaciers seem to be such that 'the amount of ice fed into the accumulation region of a glacier is during most of its life much greater than ice removed out of there bodily and, reversely, ice expenditure in the ablation area greatly exceeds ice income there. It is only during displacements that this picture is reversed, leading to a sharp lowering of glacier surface in the upper portion and a swelling of the lower part' (Krenke, 1974, p. 275). Ice crumbling is noted during rapid displacements, increasing its mobility. In some glaciers, it is true, cracks seem not to cut through the entire glacier body, and plastic flow of ice is preserved in the deeper portions. Even so, it is clear that glacier mobility increases sharply after breaking under critical stresses, and also that movement during a 'flood' occurs according to different laws compared with movement under 'normal', undisturbed conditions.

There are numerous hypotheses about those laws, taking into account both mechanical and thermal factors, but no consistent, exhaustive theory of glacial movement has yet appeared. Nor is it necessary for our purposes, since details of deformations and movements may be essentially different in different geological objects. The glacial model of a relaxation self-excited system proves the main thing: such systems can exist among geological objects, may be large, have long periods of oscillation and be the determining factor for tectonic movements within their range of action.

SELF-EXCITED SYSTEMS ARISING IN THE PROCESS OF CLIMATE FORMATION

The problem of climate pulsations has been chosen not so much because paleoclimatic research plays an important part in the correlation of geological events, but mainly because in recent years considerable progress was achieved in simulation of climate oscillations with results of universal theoretical value. During the past decade, V. Ya. and S. Ya. Sergin (1966, 1978) constructed the mathematical model of the glaciers-ocean-atmosphere (GOA) system, which takes into account a large number of various interrelations known to exist between the elements of the system. The model's behaviour was calculated and analysed by means of analog and digital computers.

Under certain conditions, the GOA system started generating pulsations and became a self-excited oscillatory system. Under other conditions, it ceased to be a self-excited oscillatory system. While in the self-excited regime, the system was highly sensitive to external disturbances resulting in modulations of oscillations by both frequency and amplitude. When generated locally, i.e., within certain regions of the world, the self-excited oscillatory system actively affected the neighbouring regions and produced planetary, global climate oscillations with peculiar phase shifts of oscillations. All this should be carefully analysed.

The GOA system model was created to find out the causes of pulsations of glaciation on our planet in the Pleistocene and to predict climatic changes. The problem of great glaciations, the repeated advances and retreats of ice, the change of the warm Miocene to the cold Pleistocene are longstanding subjects of study for geologists. Much has been done by many generations of geologists to solve the problem. The history of the ice age has been studied in detail. Numerous hypotheses have been suggested to explain the causes of glaciations and their periodicity. To this end, geologists have first of all tried to find the cause of changes and oscillations of climate in the external influences of cosmic and geological factors. External factors have served as the foundation of many still popular hypotheses. Such factors are numerous: changes in solar activity; the passage of the solar system through the clouds of cosmic dust; variations of parameters of the earth's orbit and of its rotation around its axis; orogenic and volcanic processes on the planet; the rise and subsidence of continents; drift of continents, and so on. The existence of some of these factors is problematic and they are assumed specifically to deal with glaciations; other factors actually exist and may cause changes and oscillations in climate. As much has been proved so far. But do they comprise the basic inner meaning of events? For more than a century now scientists have tried to use the peculiarities of the internal interrelations between the earth's surface and the atmosphere to solve the problem. And with time this trend seems increasingly promising.

In 1884 Vojejkov drew attention to interrelations between the earth's surface and the atmosphere joining them into one system. He emphasized that the ice cover causes cooling which encourages glaciation (i.e., it has a positive feedback with the atmosphere). He wrote: 'The snow cover has in itself the conditions for further existence' (Vojejkov, 1952, p. 268). 'The snow forms snow-flakes, then glaciers and ice covers, from which ice mountains break off and in their turn cool the seas of temperate latitudes and create favourable conditions for snow generation. Therefore, consequence affects the cause, and vice versa' (ibid., p. 290). Feedback relations are effected by changes in radiation and absorption of solar energy by the surface (the change of albedo), by the change in CO_2 and H_2O content in the atmosphere (hot-house effect), and so on. Some of the relations are negative. According to Vojejkov, proximity of the ocean is necessary for glaciation, since it produces appropriate humidity of the atmosphere. In 1926, Brooks pointed out that the ice cover affects the atmosphere in such a way that on reaching a certain volume the ice starts autogenic accumulation, cooling the atmosphere until ablation in its marginal zone is equal to the amount of precipitation (Brooks, 1952). In 1930, Gernet published a logically perfect theory of formation and succession of glaciations based on the analysis of interrelation between the atmosphere and the earth's surface operated by means of the changes in temperature and humidity (Gernet, 1930). He was not acquainted with the work of Vojejkov and Brooks, but his theory was close to theirs.

Gernet's arguments are based on the fact that the positive balance of ice accumulation, necessary for glaciation, is principally determined by atmospheric temperature and humidity. It is these two parameters that create such a state of the atmosphere ('snow-excessive layer') when snowfall during the winter is more than can thaw during the summer. Therefore the consequence (formation of glaciation) influences the cause (the state of the atmosphere) through the same two parameters, that is, it reduces both temperature and humidity. The fall of temperature is positive feedback; lower humidity is negative feedback. The snow, which does not melt during the summer on a flat upland, lowers the atmospheric temperature to a level where the formation of a glacier and its autogenic growth on islands and continents is inevitable. Moreover, Gernet arrives at the conclusion that 'with the humidity of climate and intensity of solar radiation, existing at the present moment, our planet can be in two entirely natural states: iceless and glaciated' (Gernet, 1930, p. 81). The positive feedback between glaciation and the atmosphere through the drop in temperature results in the 'self-generated growth of ice' and reduces the earth to an ice-covered state.

Reversed transition is impossible at the existing humidity of climate and solar radiation intensity. 'This transition is effected if either the solar rays become warmer, or the climate drier; practically the latter' (ibid., p. 82). Humidity is reduced by glaciation itself since it lowers the temperature and covers the ocean with ice. The combination of positive and negative feedbacks brings the system to dynamic equilibrium, which localizes the spreading of the earth's glaciation. Gernet distinguishes two types of localization: glaciation is either 'localized stationary' (Antarctica) or 'localized by pulsation' (the northern Arctic region).

The first type is due to the fact that the southern continent is surrounded by oceans. The ice, having completely covered the polar continent, starts covering the coastal waters of the ocean with icebergs, thus gradually reducing the humidity of the air and the amount of precipitation in the glacier-feeding area. When the waste of ice into the ocean balances the decreased feeding of glaciers, the area of glaciation becomes more or less permanent. The area of ice cover is confined by the ice-flow boundary around Antarctica.

The second type of localization is more interesting. The Arctic Ocean is surrounded by continents and this fact determines the type of localization. The Greenland glacier was the first to appear; it was abundantly supplied with snow from the polar ocean, produced a great amount of icebergs and caused 'reflected' glaciations in Europe and North America. These ice covers grew further into the continents and encircled the polar ocean, which fed them with water (snow), with an ice ring. Eventually they caused the polar ocean to freeze, thus cutting off their own supplies. Without sufficient feeding, continental glaciation lessened and finally disappeared. Only after this was the polar ocean also freed of ice. The intensive feeding of the Greenland glacier resumed and started another cycle of glaciation: icebergs; cooling of the ocean; reflected glaciation of continents; freezing of the polar ocean; cut-off of supplies, lessening, and so on. Pulses occur; the self-excited oscillatory system becomes active. The principle of its operation lies in the fact that specific contours of continents and of the ocean induce the feedbacks existing in the system (positive and negative) to switch on, not simultaneously but with delay. This means that whereas the positive feedback (through temperature) is active for a long time and creates vast continental glaciations, the negative feedback meanwhile remains practically inert (humidity remains almost constant). As soon as glaciation is reached and the ocean freezes, the negative feedback is quickly switched on (through humidity), while the positive feedback remains passive until the ice disappears and high humidity returns. Thus this system follows the principle of multivibrator or flip-flop generator. Though Gernet naturally did not use the terms 'feedback', 'self-excited oscillatory system', and so on, he in fact described this system exactly.

In 1955 Stokes proposed a theory almost identical to Gernet's theory (Stokes, 1955). He developed Brooks' ideas and was unaware of Gernet's work which had been published in Tokyo in Russian. In 1956 a very similar theory was published by Ewing and Donn (Ewing and Donn, 1956, 1958; Donn and Ewing, 1966) and caught the attention of many scientists. During those years a number of researchers also attempted to explain the oscillations of climate in the Pleistocene period by applying the idea of self-excited oscillations in the earth's surface-atmosphere system, but using other factors as the basic causes (for example, the amount of CO_2 in the atmosphere released by volcanoes and fixed by organic life). Various feedbacks were eventually revealed, but the relations determined by temperature and humidity apparently remain basic (Chizhov, 1976, 1981; V.Ya. and S.Ya. Sergin, 1978). 'It is obvious now that the presence of complex interrelations and feedbacks between natural phenomena is a typical and important property of geographical medium . . . Chizhov (1976), proceeding from a similar evaluation of hypotheses of the types discussed here, accomplished a careful check of these hypotheses on the basis of glacioclimatic and palaeogeographic data now available. He arrived at the conclusion that the successive states of climate and glaciations of the earth correspond to observation data, as suggested by the Gernet-Stokes hypothesis. He believes that this hypothesis provides a logical basis for apprehension of causes for climatic changes and glaciation of the earth in the Pleistocene' (V.Ya. and S.Ya. Sergin, 1978, p. 40).

As was noted earlier, in constructing their model V.Ya. and S.Ya. Sergin tried to take into account all appreciable interrelations between glaciers, the oceans and the atmosphere. They represented these interrelations in a form suitable for introduction into the mathematical model by means of corresponding parameters. In addition to initial and boundary conditions, the number of parameters was more than twenty (V.Ya. and S.Ya. Sergin, 1978, pp. 107–8). They need not be enumerated because not all of them are easily reduced to simple notions and the interested reader can study the model and the introduced parameters in the excellent monograph by V. Ya. and S. Ya. Sergin (1978). I shall only mention that the authors assumed for the parameters, as they believed, the most probable values ('nominal values') and introduced them into the model. They varied the most important parameters in the process of solution to investigate their effect on the behaviour of the model. The authors persistently complicated the model itself; at first the variants were solved disregarding external effects for the northern and southern hemispheres only, then for the whole globe, and finally for the whole globe taking account of external effects. Figs. 12–18 show some of their results. They are significant and completely substantiate the possibility of the appearance of self-excited oscillations in the GOA system. 'The results of simulation give grounds for regarding the self-excited oscillatory properties of the GOA system and the modulating effect of external disturbances as the cause for changes of climate and glaciation of the earth in the Pleistocene' (V.Ya. and S.Ya. Sergin, 1978, p. 197).

A separate analysis of the GOA system behaviour within the boundaries of the northern

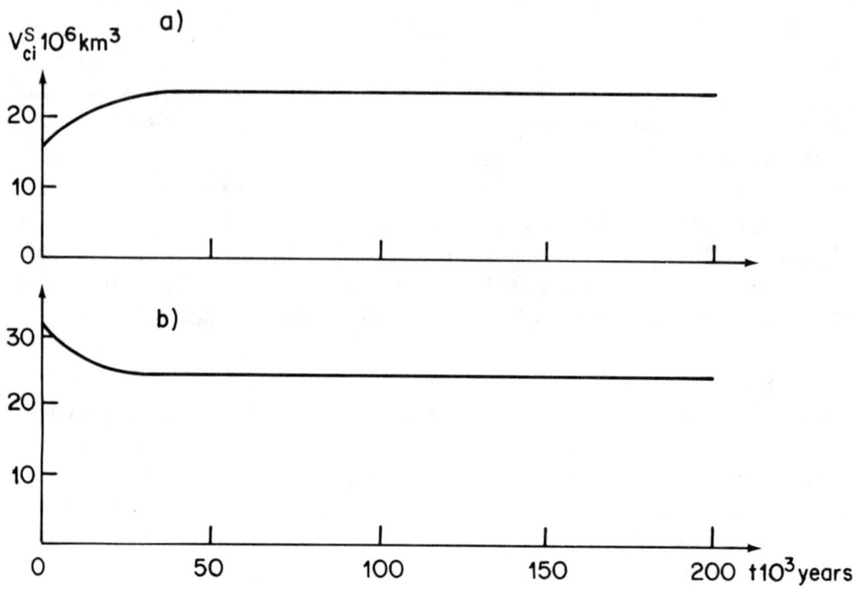

FIG. 12 Transition processes in the model of the southern hemisphere at different initial conditions $V^S_{ci}(o)$: (a) $V^S_{ci}(o) = 15.10^6$ km^3, (b) $V^S_{ci} = 30^6$ km^3. V^S_{ci} = volume of continental ice in the southern hemisphere.

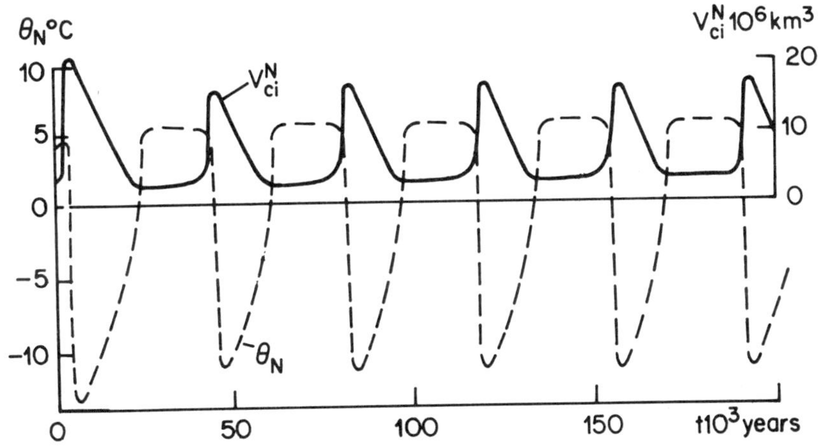

FIG. 13 Self-excited oscillations in the model of the northern hemisphere at nominal values of parameters by V. Ya. and S. Ya Sergin (1978). V_{ci}^{N} – volume of continental ice in the northern hemisphere. Θ_N – average temperature deviation of the earth's surface in the northern hemisphere.

and southern hemispheres was undertaken to reveal the role of geographical peculiarities in each hemisphere. The model imitated the absence of connections between the hemispheres by disregarding the energy flux across the equator. A stationary solution was obtained for the southern hemisphere (Fig. 12). Independently of the volume of glaciation assumed as the initial condition, the GOA system in the southern hemisphere achieved a stable balanced state after a certain period of time. An entirely different solution was obtained for the northern hemisphere (Fig. 13). The system did not reach the stationary state and started pulsating with a certain frequency. Variations of initial conditions showed that they had practically no influence on the solution, since every 2–3 periods the system started identical pulses under different initial conditions. The changes in introduced parameters appreciably affected the solution, causing changes in the period of oscillations, in their amplitude and the shape of curves. We can illustrate this by quoting solutions obtained by varying parameters which characterize the average depth of the ocean, H (Fig. 14-I), the time constant of the ice cover T_Λ (Fig. 14-II), and the hot-house effect of the atmospheric water K_Θ (Fig. 14-III). As seen in the figures, the change in the average depth of the ocean, H, has a great effect on the amplitude of oscillations and a smaller influence on their period; on the contrary, the change of the time constant of the ice cover T_Λ actively affects the frequency of oscillations; and, finally, a considerable growth of the hot-house effect K_Θ results in the disappearance of pulses, in other words, it destroys the self-excited oscillatory system. Other parameters were likewise analysed; they characterized the state of the atmosphere, of the ocean and of the earth's surface (cloudiness, albedo, summer temperature, thickness of the ice cover, and so on). This analysis revealed especial features of the self-excited system of the northern hemisphere.

The interrelation between the northern and southern hemispheres was expressed by a special coefficient K_Θ showing the energy transfer rate across the equator. By introducing this coefficient the authors of the GOA model studied the effect of the self-excited oscillatory system which appeared in the northern hemisphere, on the southern hemisphere. They found that the system not only induces oscillations in the southern hemisphere, but also experiences a reciprocal influence in the southern hemisphere which transforms the

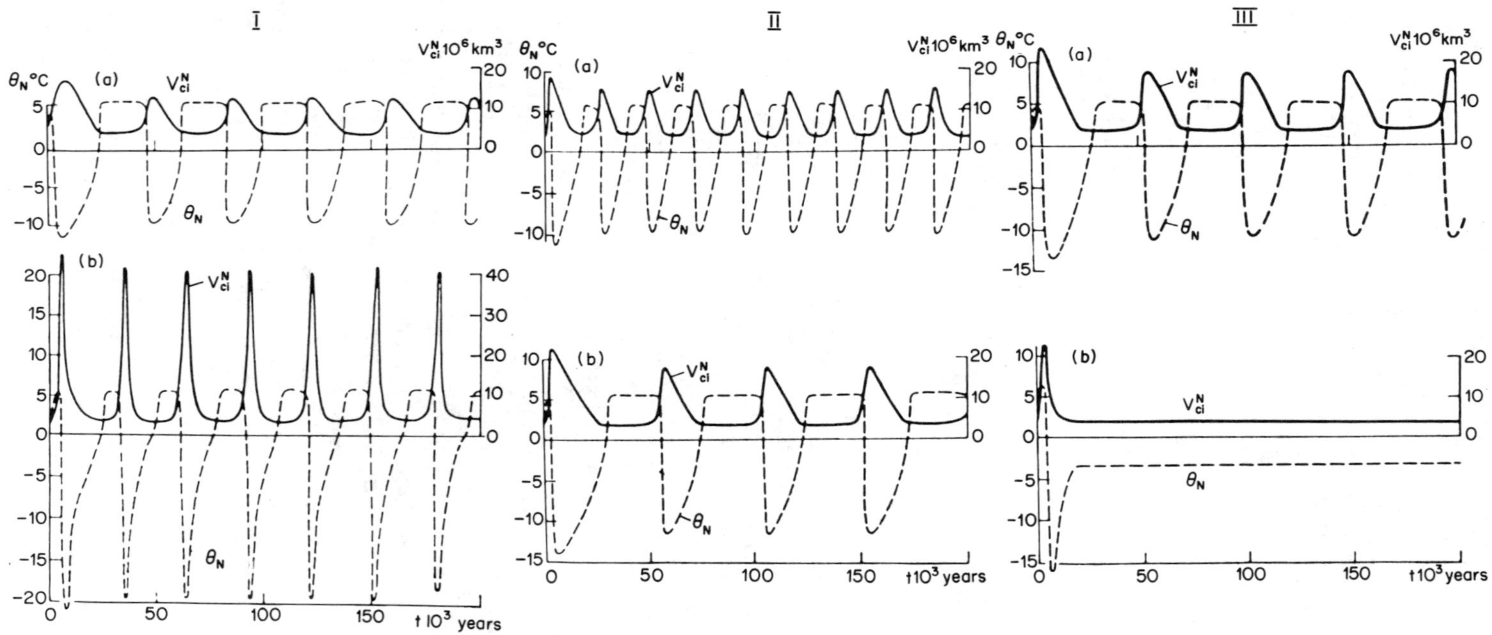

FIG. 14 Effect of the change in parameters on self-excited oscillations in models of the northern hemisphere by V. Ya and S. Ya. Sergin (1978). Self-excited oscillations in the model
I – at different average depth of the ocean (H):
(a) $H = 2$ km; (b) $H = 8$ km.
II – at different value of the time constant ice cover (T_{gl}):
(a) $T_{gl} = 5.10^3$ y, (b) $T_{gl} = 15.10^3$ y.
III – at different value of hot-house effect (K_Θ):
(a) $K_\Theta = 0.4$, (b) $K_\Theta = 0.8$.
For explanation of the symbols on the curves see Fig. 13.

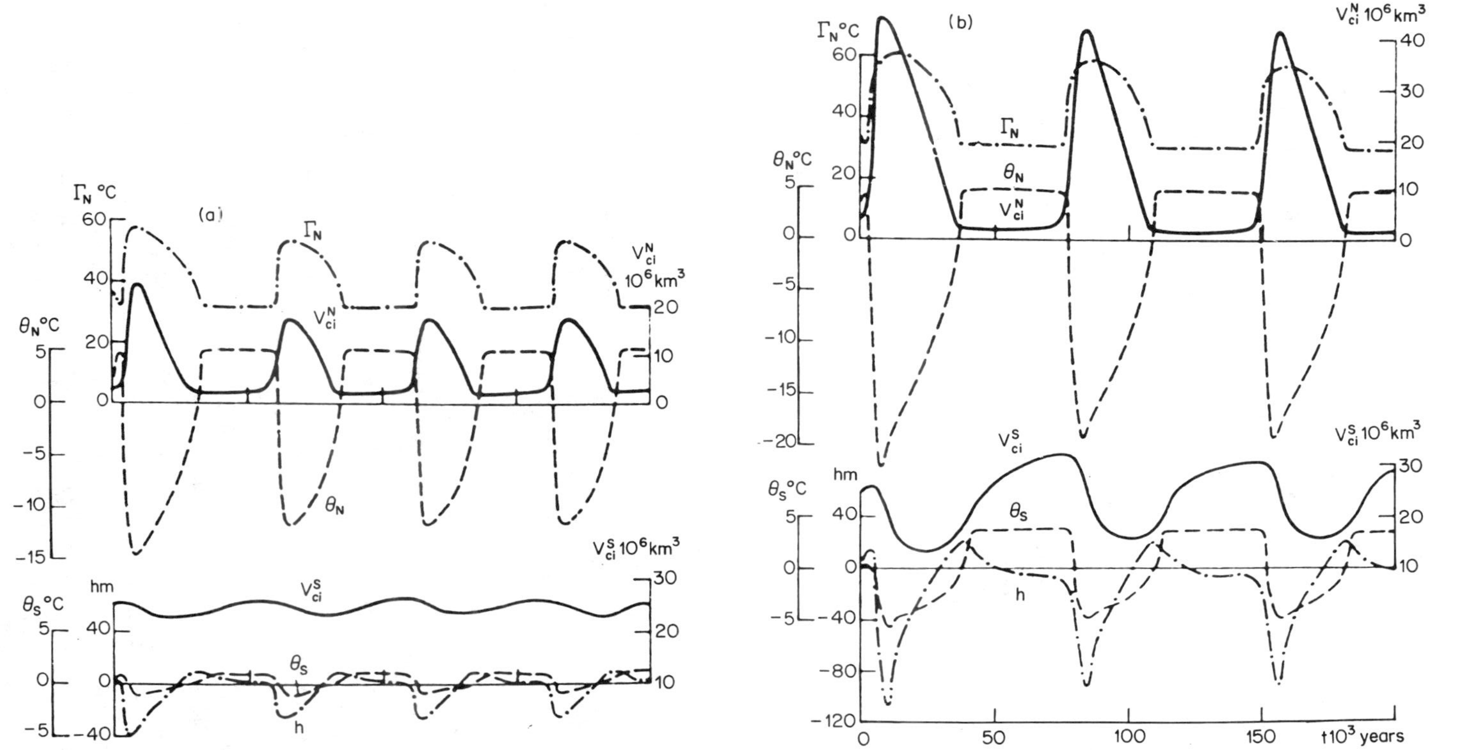

FIG. 15 Self-excited oscillations in the planetary GOA model at different values of coefficient of energy transfer across the equator ($K_э$) after V. Ya. and S. Ya. Sergin (1978): (a) $K_э = 0.1$; (b) $K_э = 0.5$; Θ_N – deviation (from modern) of the average temperature of the northern hemisphere; Γ_N – difference between temperatures on the equator and at the pole for the northern hemisphere; V_{ci}^N – volume of continental ice of the northern hemisphere; h – deviation of the world ocean level; Θ_S – deviation of the average temperature of the southern hemisphere; V_{ci}^S – volume of continental ice of the southern hemisphere.

frequency and amplitude of oscillations in the northern hemisphere. In other words, a single planetary GOA system is generated, which maintains its self-excited oscillatory nature. Fig. 15 represents the dynamics of development of the GOA planetary system at different energy transfer coefficients across the equator K_Θ Almost imperceptible changes in the volume of continental ice in the southern hemisphere at small K_Θ expand to rather significant changes at large K_Θ. The amplitude of oscillations of the whole system grows simultaneously. Moreover, the oscillations in the southern Hemisphere are phase-shifted in relation to the oscillations in the northern hemisphere, and the curves have different shapes.

Variations of values of separate parameters in the planetary GOA model caused changes in its development similar to the changes observed in the model of the northern hemisphere. The resulting change in several parameters simultaneously is seen in Fig. 16. Attention should here be paid to the phase shift of individual curves and to a clear dependence of the amplitude and frequency of oscillations on the value of the changed parameters.

After investigating the self-excited oscillations of the GOA planetary model, the authors made an attempt to take into account the effect of astronomical and geological factors on the model, that is, the effect of external factors. Astronomical factors (the effect of changes in the orbital parameters of the earth on the differences in the equator–pole temperatures and on specific ablation of glaciers) were mostly taken from the classical work of Milankovitch (1941). The reckoned tectonic influences were restricted to the progressive uprising, to a certain expansion of the area of the continents, and to the deepening of the ocean during the Pleistocene. The tectonic factors were, moreover, assumed as aperiodic, linearily growing, whereas the astronomical factors experienced complex oscillations, whose frequency did not coincide with the inherent frequency of the self-excited oscillatory system. As a result of the effect of external disturbances on the inherent oscillations of the system, complicated curves were obtained, modulated by both amplitude and frequency (Figs 17–18). V.Ya. and S.Ya. Sergin (1978, p. 175–9) write:

> The changed nature of oscillations and a considerable scatter of amplitudes of calculated variables are easily observed. Not apparent previously, long waves of climatic changes now appear. The results of calculations, presented in Fig. 18, show the effect of external disturbances on the model with a small period of inherent oscillations. Most interesting is the fact that here the average period of oscillations is changed, now lasting 40 th.y. (instead of 20 th.y. during the self-excited oscillatory regime). This fact demonstrates that the modulating influence of external disturbances, caused by the changes in the oribital parameters of the earth, grows with the reduction of the period of inherent oscillations.

By analysing a longer period of time the authors succeeded in explaining the transition from the steady warm climate in the Miocene to intensive oscillations of a cooler climate in the Pleistocene. The self-excited oscillatory system depended on the existing glaciers, but it deteriorated during the general warming of climate and disappearance of glaciers.

> A prerequisite for the inherent oscillations in the earth's surface-atmosphere system is the presence of at least two inertial links with large time constants. Such links are the ocean and continental ice. During the warm periods only one of the links is present, the ocean, owing to which the system is aperiodic and the changes in climate only occur under external effects. During the cold periods the second inertial link, continental glaciation, appears; the system becomes oscillatory, the climatic changes become very great. This observation allows understanding of both occurrence of large climatic oscillations in cold periods and their absence in warm periods (V.A. and S.Ya. Sergin, 1978, p. 199)

The GOA model, composed and analysed by V.Ya. and S.Ya. Sergin, provides a much

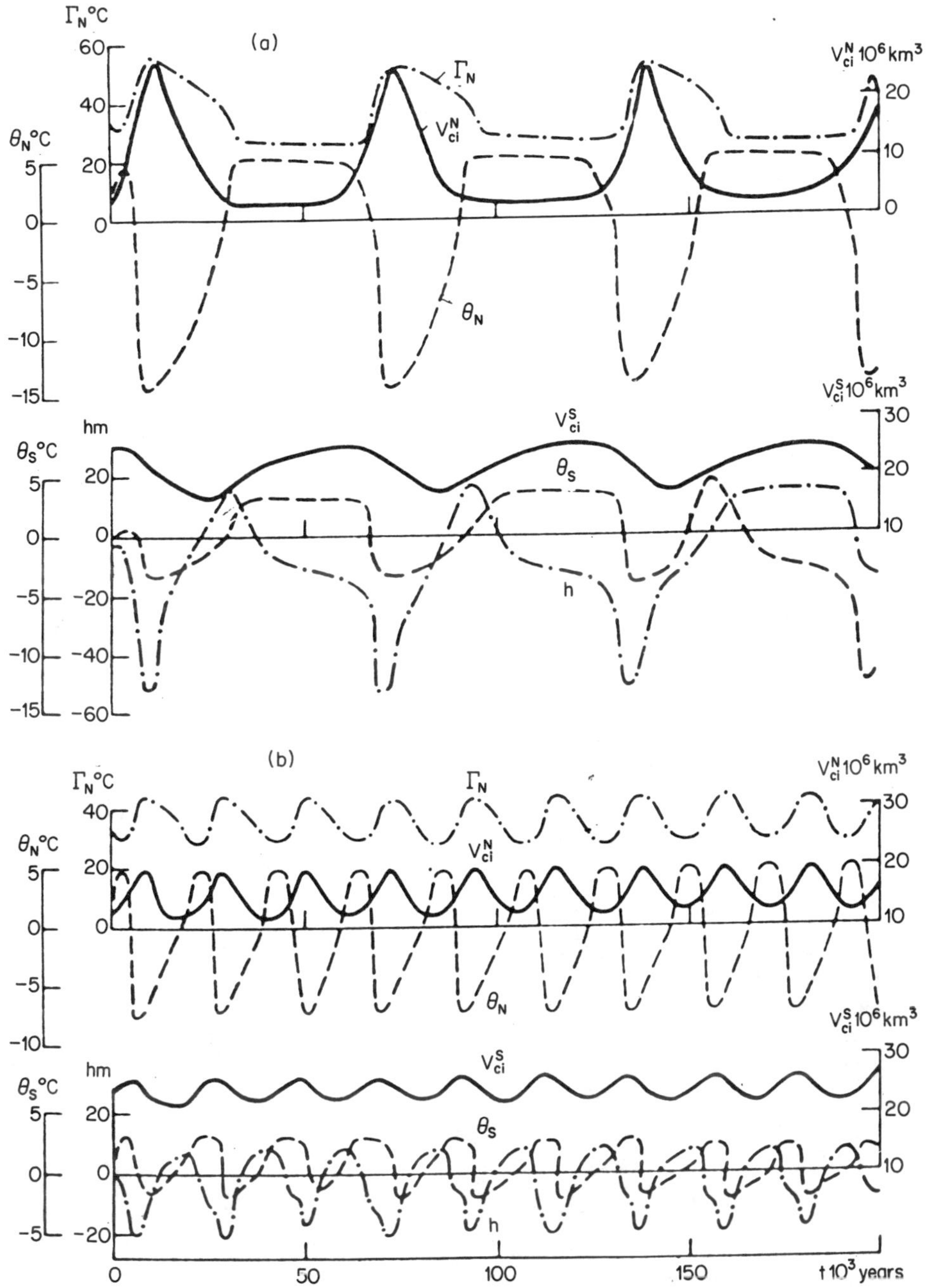

FIG. 16 Effect of the change of parameters on self-excited oscillations in the model of the planetary GOA system by V. Ya. and S. Ya. Sergin (1978). Self-excited oscillations computed by computer M-222.
(a) nominal values of parameters (including the time constant of the ice cover $T_{gl} = 10^4$ y; average depth of the ocean $H = 2$ km; average amount of precipitation in the glaciation zone $a_0 = 100$ cm/y.
(b) $T_{gl} = 6.10^3$ y; $H = 1.6$ km; $a_0 = 60$ cm/y.
For explanation of symbols see Fig. 15.

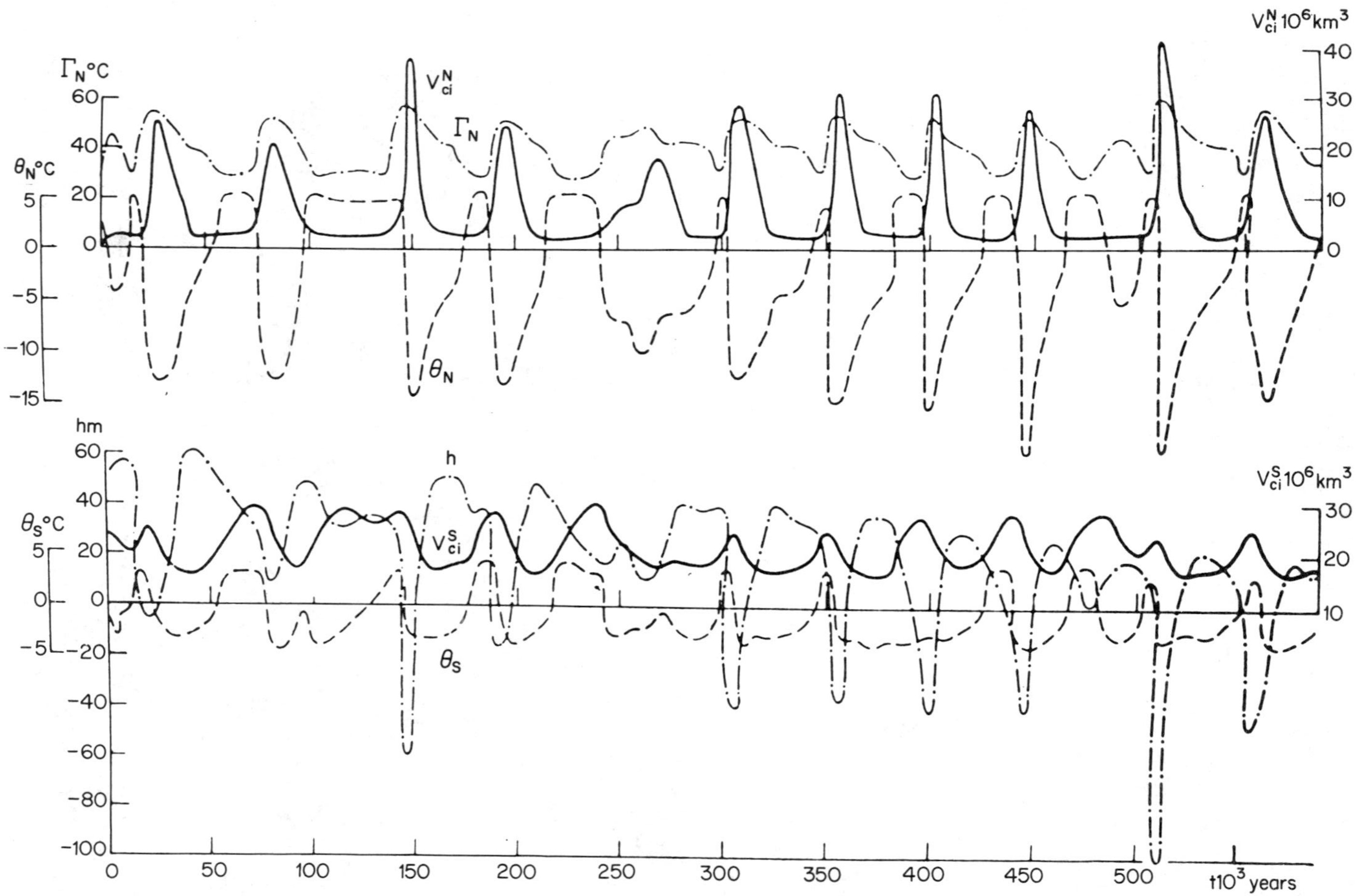

FIG. 17 Effect of external disturbances on the GOA model by V. Ya. and S. Ya. Sergin (1978). Oscillations in the GOA model at nominal values of parameters and external disturbances caused by changes in orbital parameters of the earth and by tectonic forces.

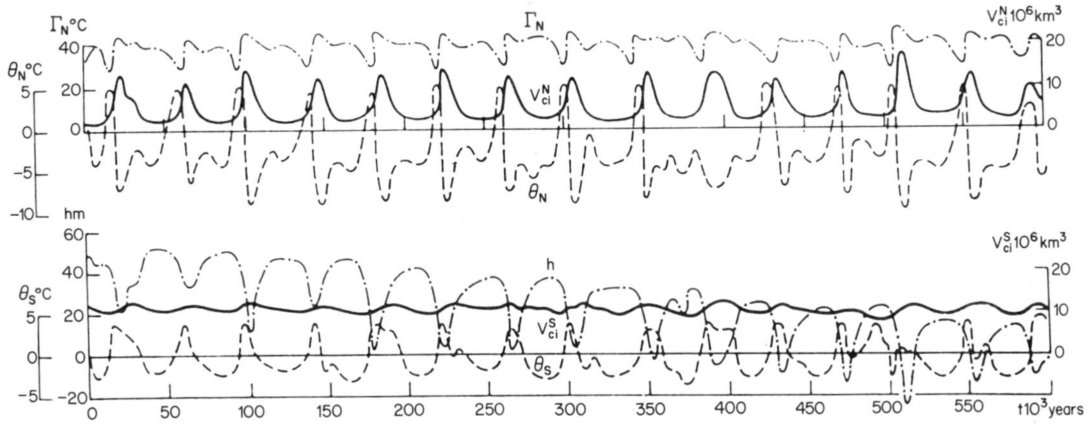

FIG. 18 Effect of external disturbances on the GOA model by V. Ya. and S. Ya. Sergin (1978). Oscillations in the GOA model at $T_{gl} = 6.10^3$ y; $H = 1.6$ km; $a_0 = 60$ cm/y and in the presence of external factors.

fuller account of the numerous climatic factors than the Gernet–Stokes theory. It also substantiates the basic conclusions of the latter.

For the study of the problem of correlation of geological events this model is of primary importance. It demonstrates that the self-excited oscillatory system, incited by individual peculiarities of a definite region, can influence the whole planet and become a global system; that concrete manifestations of such a global system in different regions can be expressed by different oscillations even shifted by phase; that reacting to external disturbances it generates new long-period oscillations and individual sharp peaks; and, finally, that small changes in external conditions can produce both the appearance and the deterioration of the self-excited oscillatory system. Naturally all these features can occur not only in the GOA system but also in other systems capable of developing in the self-excited oscillatory regime. This fact should be taken into account when working out global correlation of geological events.

CONCLUSION

The examples given in this paper demonstrate that there are ample opportunities on the surface and in the earth's interior for the appearance of self-excited oscillatory systems. These systems are of different natures, appear in different situations, and generate oscillations in such processes as heat transfer, transportation of sediments, tectonic movements and formation of climate. Every self-excited oscillatory system 'quantifies' the processes of material or energy transfer and makes them pulsate. In this process the frequency of generated oscillations is formed by the system itself and depends on its parameters.

The analysed models have different scales and purposes, that is, 'magmatic' models are interesting for understanding of large tectonic structure development, volcanism and metallogeny; 'sedimentary' models show the formation of various kinds of layering; 'deformation' models simulate the formation of different tectonic structures, especially of overthrusts (Lukyanov, 1972); 'climatic' models allow an insight into the global significance of local systems and of the modulating role of external factors. Reciprocal superposition of different kinds of oscillations forms a very complicated rhythmic pattern of the earth's evolution. One of the fundamental problems of global correlation of geological events is understanding this pattern and finding the role of individual self-excited oscillatory systems. The study of models is one means of finding new ways of solving the problem.

When correlating geological phenomena and looking for relations between them, the geologist is inevitably faced with the results of the activity of numerous self-excited oscillatory systems. He is often unaware of them because they are unknown to him. The systems cover regions of different size; they have oscillation periods determined by values of different orders; they appear in various geological processes. Every self-excited oscillatory system experiences the modulating effect of external factors, partly associated with the adjacent self-excited oscillatory systems and partly of extra-terrestrial origin. Reciprocal superposition of oscillations, generated by different self-excited oscillatory systems, forms a rather complicated spectrum not easily deciphered; moreover, it is dangerous to pick out similar events and correlate them without knowledge of their inner meaning. The oscillations merge into a united symphony played by the gigantic orchestra of the earth. When correlating various phenomena, the geologist should distinguish in that symphony 'the voices' of individual self-excited oscillatory systems. Otherwise he may correlate unrelated events and obtain misleading conclusions about a simultaneous and global occurrence which does not exist. And, on the other hand, he may miss genuine global phenomena, because their manifestation in different systems is different. The study of models allows one to 'hear' the self-excited oscillatory systems as they sound when played as solo, in a duet or trio, and to learn to recognize them. My effort will be justified if the reader makes an attempt to find in familiar geological objects the echoes of the self-excited oscillatory systems whose models were analysed in this paper.

REFERENCES

Akivis, T. M. and Devderiani, A. S. (1982). The mathematical model of sedimentation on continental margins of the Atlantic Ocean, in *Geology of the Seas and Oceans*, vol. 3, Moscow: pp. 10–11 (Russian).

Andronov, A. A., Vitt, A. A. and Khaikin, S. E. (1981). *The Theory of Oscillations*, Moscow: Nauka, p. 568 (Russian).

Bobrov, V. A. and Romanovsky, S. I. (1970). Factors controlling the formation of flysch, *Izvestia AN SSSR*, ser. geol., **7**, 108–14 (Russian).

Brooks, Ch. (1952). *Climates of the Past*, Moscow: Publ. House Foreign Lit., p. 357.

Chizhov, O. P. (1976). *Glaciation of the Northern Polar Region*, Moscow: Nauka, p. 240 (Russian).

Chizhov, O. P. (1981). Gernet's theory in the light of recent conceptions, in Gernet Ye. S. *Ice Lichens*, Moscow: Nauka, pp. 94–144 (Russian).

Devderiani, A. S. and Pitovranov S. Ye. (1982). Regularities in the formation of sedimentary cover of continental margins (from results of mathematical simulation), in *Avalanche Sedimentation in the Ocean*, Rostov: Rostov University, pp. 154–61 (Russian).

Dolgushin, L. D. and Osipova G. B. (1971). New data on the pulses of modern glaciers, in *Materials of Glaciological Research*, Chronicle, discussions, Moscow, issue 18, pp. 191–218 (Russian).

Donn, W. L. and Ewing, M. A. (1966). Theory of ice ages, III, *Science*, **152**, no. 3730, 1706–12.

Duff, P. McL. D., Hallam, A. and Walton, E. K. (1967). *Cyclic Sedimentation*, Amsterdam, London, New York: Elsevier.

Ewing, M. and Donn, W. L. (1956). Theory of ice ages, I, *Science*, **123**, no. 3207, 1061–6.

Ewing, M. and Donn, W. L. (1958). Theory of ice ages, II, *Science*, **127**, no. 3307, 1159–62.

Freindlich, G., (1939). *Thixotropy*, Leningrad, Moscow.

Gernet, Ye. S. (1930) *Ice Lichens* (a new theory of glaciation in popular presentation), Tokyo.

Gernet, Ye. S. (1981). Ice Lichens, Moscow: Nauka, p. 93.

Joly, J. (1930). *Surface History of the Earth*, Oxford.

Keller, B. M. (1949). The Palaeozoic flysch formation in the Zalair synclinorium of Southern Urals and similar formations, *Transactions IGN AN SSSR*, ser. geol., **104**, no. 33 (Russian).

Keondyan, V. P. (1980). The model of chemical-density convection in the mantle of the earth, *Izvestia AN SSSR, Fizika Zemli*, **8**, 3–21 (Russian).

Keondyan, V. P. and Monin, A. S. (1980). On concentration convection in the earth's mantle, *Doklady AN SSSR*, **253**, no. 1, 78–81 (Russian).

Khain, V. Ye. and Lomize, M. G. (1961). Transversal consedimentation faults on the boundary between the Central Caucasus and the distribution of Mesozoic and Cenozoic facies, *Izvestia AN SSSR*, ser. geol., no. 4 (Russian).

Khvorova, I. V. (1958). On the origin of flysch, *Bull. Mosk, obshch. isp. prirody* geol. section, **33**, issue 5, 3–21 (Russian).

Krenke, A. N. (1974). The existing concepts about rapid movements of glaciers, in *Materials of Glaciological Research*, Chronicle, discussions. Moscow, pp. 247–88 (Russian).

Kühnen, Ph. H. (1958). Turbidity currents a major factor in flysch deposition, *Eclog. geol. helv.*, **51**, no. 3, 91–127.

Leonov, M. G. (1972a). The Upper Eocene wild flysch of the Alpine belt, in *Reports of the Soviet Geologists at the XXIV Session of the International Geological Congress. Problem 6, 12. Stratigraphy, sedimentology and geology of the Quaternary period*, Moscow: Nauka, pp. 11–19 (Russian).

Leonov, M. G. (1972b). Flysch, a formation of submarine slope, *Lithol. and useful minerals*, no. 2, 44–54 (Russian).

Lukyanov, A. V. (1972). On the problem of dynamics and energetics of formation of tectonic covers, in *Internal Geodynamics*, Leningrad, issue 2, *Problems of Dynamics of Endogenous Processes*, pp. 22–4 (Russian).

Lukyanov, A. V. (1972). External and internal causes of the cycles in some geological phenomena, *J. Interdiscipl. Cycle Res.*, **3**, no. 3/4, 358.

Lukyanov, A. V. (1980a). Simulation of systems with cyclic development, in *Problems of Global Correlation of Geological Events*, Moscow: Nauka, pp. 28–32 (Russian).

Lukyanov, A. V. (1980b). Plastic deformations and tectonic flow of lithospheric rocks, in *Tectonic Layering of the Lithosphere*, Moscow: Nauka, pp. 105–46 (Russian).

Lukyanov, A. V., Zinkov, V. V. and Scherba, I. G. (1982). Self-excited oscillatory nature of rhythms in sedimentation, in *Geology of the Seas and Oceans*, vol. 3, Moscow: Nauka, pp. 33–5 (Russian).

Lukyanov, V. S. (1937). Calculations of technical parameters on Lukyanov hydraulic instruments, Moscow: Transzheldorizdat.

Lukyanov, V. S. (1959). The method of hydraulic analogies as the means of research and calculations, in *Proc. of the Scientific-Technical Meeting on the problems of engineering and exploitation in mining of deposits of useful minerals with complicated hydrogeological and engineering-geological conditions*, Moscow: Ugletekhizdat (Russian).

Lyubimova, Ye. A. (1968). *Thermics of the Earth and the Moon*, Moscow: Nauka, p. 279 (Russian).

Magnitskiy, V. A. (1964). Zonal melting as the mechanism of formation of the earth's crust, *Izvestia AN SSSR*, ser. geol., no. 11, 3–8 (Russian).

Milankovitch, M. (1941). *Kanon der Erdbestrahlung und seine Anwendung auf das Eiszeitenproblem*, Belgrad, 633p.

Pounder, E. R. (1965). *The Physics of Ice*, Oxford: Pergamon.

Romanovskiy, O. I. (1976). *Dynamics of Flysch Formation*, Leningrad: Nedra (Russian).

Savel'yev, A. A. (1977). *Layering of Sedimentary Rocks and its Transformation during Metamorphism*, Moscow: Nauka (Russian).

Savel'yev, B. A. (1963). *Structure, Composition and Properties of Ice Cover on Sea and Fresh-water Basins*, Moscow State University (Russian).

Sergin, S. Ya. and Sergin, V. Ya. (1966). 'The earth's surface-atmosphere' as the system of automatic regulation, *Doklady AN SSSR*, **171**, no. 4, 923–6 (Russian).

Sergin, V. Ya. and Sergin, S. Ya. (1978). Systematic analysis of the problem of great oscillations of climate and of the Earth's glaciation, Leningrad, Gidrometeoizdat, p. 279 (Russian).

Sharkov, Ye. V. (1980). *Petrology of Layered Intrusions*, Leningrad: Nauka, p. 183 (Russian).

Shatskiy, N. S. (1939). Orogenic phases and folding, in *Proc. of XVII Session of the International Geologia Congress, USSR, 1937*, Moscow, GONTI, vol. 2, pp. 329–30 (Russian).

Shatskiy, N. S. (1964). *Selected Works*, Moscow: Nauka, vol. 2 (Russian).

Shimazu, Y. (1959). A note on genesis and upward transfer of parental magma of igneous rocks. A thermodynamical aspect of the earth's interior, Part III, *J. Earth. Sci. Nagoya Univ.*, **7**, no. 1, 91–7.

Shimazu, Y. (1961). Physical theory of generation, upward transfer, differentiation, solidification, and explosion of magmas, *J. Earth. Sci. Nagoya Univ*., **9**, no. 2, 185–223.

Shumsky, P. A. (1973). Glaciers, in *The Large Soviet Encyclopedia*, 3rd edition, Moscow, vol. 14 (Russian).

Stille, H. (1964). *Selected Works*, Moscow: Mir (Russian).

Stille, H. (1924). *Grundfragen der vergleicheden Tektonik*, Berlin: Verl. von G. Borntraeger.

Stokes, W. L. (1955). Another look at the ice age, *Science*, **122**, no. 3174, 815–21.

Strakhov, N. M. *Basic principles of the theory of lithogenesis*, Moscow: AN SSSR, vol. 2 (Russian).

Tikhonov, A. N., Lubimova, Ye. A. and Vlasov, V.K. (1969). On evolution of melting zones in the thermal history of the earth, *Doklady AN SSSR*, **188**, no. 2, 27–31 (Russian).

Tikhonov, A. N., Lubimova, Ye. A. and Vlasov V.K. (1972). Radioactive energy and melting processes in the earth's interior, in *Energetics of Geological and Geophysical Processes*, Moscow: Nauka, pp. 27–37 (Russian).

Tsurikov, V. L. (1976). *Liquid Phase in Sea Ice*, Moscow: Nauka, p. 210 (Russian).

Vassoyevich, N. B. (1940). On the conditions of flysch formation, *Izvestia AN SSSR*, ser. geol., no. 4, 37–65 (Russian).

Vassoyevich, N. B. (1948). *Flysch and Methods of Study*, Moscow, Leningrad: Gostoptekhizdat (Russian).

Vassoyevich, N. B. (1951). *Conditions of Flysch Formation*, Moscow, Leningrad: Gostoptekhizdat (Russian).

Vassoyevich, N. B. (1960). About flysch, in *Materials of the Carpathians-Balkans Association*, no. 3, *stratigraphy*, Kiev: Academy of Sciences of Ukrainian SSR, pp. 26–49 (Russian).

Veviorovskaya, M. A., Kravchenko, I. P., and Rumiantsev, S. A. (1962). The method of hydraulic analogies of V. S. Lukyanov and the method of electrodynamical analogies of N. P. Pavlovsky as applied to filtration calculations, Moscow: Moscow State University (Russian).

Vinogradov, A. P. (1959). Meteorites and the earth's crust, *Izvestia AN SSSR*, ser. geol. no. 10, 5–27 (Russian).

Vojejkov, A. I. (1952). Climates of the globe, particularly in Russia (1884), *Selected Works*, vol. 3, Moscow: USSR Acad. of Sci., pp. 321–64 (Russian).

Wager, L. and Brown G. (1968). *Layered Igneous Rocks*, Edinburgh.

Whitman, W. G. (1926). The elimination of salt from sea-water ice, *Amer. J. Sci*., 5th ser. **11**.

Wilson, A. T. (1964). Origin of a ice ages: an ice shelf theory for Pleistocene glaciation, *Nature*, **201**, no. 4915, 147–9.

Index

Italic page numbers refer to Figures and Tables